THE BRITISH ARBORETUM:
TREES, SCIENCE AND CULTURE IN THE NINETEENTH CENTURY

SCIENCE AND CULTURE IN THE NINETEENTH CENTURY

Series Editor: *Bernard Lightman*

TITLES IN THIS SERIES

THE BRITISH ARBORETUM:
TREES, SCIENCE AND CULTURE IN THE
NINETEENTH CENTURY

BY

Paul A. Elliott, Charles Watkins and Stephen Daniels

LONDON
PICKERING & CHATTO
2011

Published by Pickering & Chatto (Publishers) Limited
21 Bloomsbury Way, London WC1A 2TH

2252 Ridge Road, Brookfield, Vermont 05036-9704, USA

www.pickeringchatto.com

BRITISH LIBRARY CATALOGUING IN PUBLICATION DATA

Elliott, Paul A.
The British arboretum: trees, science and culture in the nineteenth century. –
(Science and culture in the nineteenth century)
1. Arboretums – Great Britain – History – 19th century. 2. Trees – Collection
and preservation – Great Britain – History – 19th century.
I. Title II. Series III. Watkins, C. IV. Daniels, Stephen.
582.1'6'07341-dc22

ISBN-13: 9781848930971
e: 9781848930988

1006405309

This publication is printed on acid-free paper that conforms to the American
National Standard for the Permanence of Paper for Printed Library Materials.

Typeset by Pickering & Chatto (Publishers) Limited
Printed and bound in the United Kingdom by the MPG Books Group

CONTENTS

LIST OF FIGURES

PREFACE

This book was motivated by recent work in the history of science and cultural and historical geography and a belief that arboretums have had a profound impact on British cultural and scientific history. We embark upon an exploration of the historical and cultural geographies of nineteenth-century arboretums in order to try to understand the various principles and practices underpinning their design and the management and consumption of tree collections, particularly tensions between naturalistic and geometric aesthetics and botanical taxonomies. This book traces interconnections between horticulture, botany and forestry, the role of institutions and the relationships between arboretums and their various social and cultural contexts including the horticultural trade, scientific development of forestry and importance of international networks, exploration, trade and imperialism in tree collecting. Although the principal focus is British arboretums (with some reference to the Republic of Ireland), just as John Claudius Loudon's *Arboretum et Fruticetum Britannicum* (1838, hereafter *Arboretum Britannicum*) was something of an international collaboration, and like nineteenth-century planted arboretums, any study of British arboretums is inherently both national and international.

This book is the result of a major project on the cultural and historical geographies of the arboretum begun at the School of Geography, Nottingham University and funded between 2004 and 2007. We would like to thank the Arts and Humanities Research Council for financing the project, an international conference on arboretums and the publication of P. Elliott, C. Watkins and S. Daniels (eds), 'Cultural and Historical Geographies of the Arboretum', special issue of *Garden History* (2007), supplement 2 (RA1526/39). We would like to express our gratitude to Bernie Lightman for his patient editing and Pickering and Chatto for publishing this book, and particularly two anonymous referees who provided superbly detailed criticisms and comments on the typescript and Stephina Clarke for her help with copy-editing. We are also grateful to colleagues at the University of Nottingham and the School of Humanities at Derby University for help and encouragement, especially Susanne Seymour, Mike Heffernan, Stephen Legg, Alex Vasudevan, Ross Balzaretti, John Beckett and Harry Cocks at the former and Ian Whitehead, Ruth Larsen, Robert Hudson, Ian Barnes and Tom Neuhaus at the latter. Colleagues and friends in the School of Historical Studies at Leicester University past and present have also offered much help and support, especially Roey Sweet, Simon Gunn and Richard Rodger. Thanks also to James Hervey-Bathurst of Eastnor Castle.

The international conference on the history of arboretums took place at the Linnean Society of London, Burlington House, 6–8 September 2006 and we gained much inspiration from this. We would like to thank our fellow participants Max Bourke, Tom Schlereth, Beryl Hartley, Brent Elliott, Owain Jones, Nuala Johnson, Divya Tolia-Kelly, Simon Naylor, Chris Harris, David Whitehead, Sophie Piebenga and Simon Toomer and the staff at the Linnean Society for ensuring that it was successful. Most of the conference papers were published in the *Garden History* volume (2007) which complements this book and includes an additional essay on modern Irish arboretums by Finola O'Kane, and we are grateful to Barbara Simms, Cristiano Ratti and the staff of *Garden History* for their help in putting it together. The identification and naming of trees underpins the idea of an arboretum. In this book we have retained the original tree names, whether vernacular or botanical, in all quotations. Otherwise we have followed Alan Mitchell's *A Field Guide to the Trees of Britain and Northern Europe*.

Finally, it is worth noting that despite inevitable casualties, many arboretums analysed in this book survive in one form or another and we are dealing with living, breathing and constantly changing habitats and not a dead historical subject. Still studied and carefully tended by experts and enjoyed by hundreds of thousands of visitors, late-Georgian, Victorian and Edwardian tree collections flourish in botanical gardens such as those at Edinburgh, Kew, Sheffield, Glasgow and Glasnevin, Dublin and on great estate arboretums and pinetums like those at Chatsworth, Elvaston, Biddulph Grange, Eastnor and lush Bodnant in North Wales. In urban areas, trees and shrubs battle with the gravestones at Abney Park, Stoke Newington in densely-populated London championed by enthusiastic supporters of the Cemetery Trust and elegantly on the hillside of the Derwent Valley at Belper Cemetery, Derbyshire, visible for miles around. There are numerous opportunities for ecologically sensitive tree planting to challenge the weeds, buddleia and indestructible shrubbery in places that would have delighted Loudon, such as urban parks and gardens, roadside verges, disused cemeteries, country parks, railway cuttings and out-of-town shopping centres.

A core concern of the book is how it is possible to balance beautiful and scientifically significant systematic tree collections with public access for research, education and pleasure. Aside from some academic institutional or private arboretums, this is a perennial issue as the continuing success of the National Arboretum at Westonbirt, Gloucestershire and recent projects to restore John Claudius Loudon's Derby Arboretum and Samuel Curtis's Nottingham

Arboretum demonstrate. Modern arboretum designers, planters, curators and gardeners are wrestling with ongoing problems concerning the design, restoration and management of contemporary tree collections in the face of changing leisure patterns, urban deprivation, social inequality, pollution, industrial farming, neglect, lack of protection and shortsighted cutbacks. Yet we hope that in demonstrating the fundamental importance of arboretums to British society, culture and science, like the resilience and tenacity of living trees themselves and Loudon's work, we have demonstrated how crucial systematic tree collections remain in the contemporary modern world. John Ruskin argued that trees deserve 'boundless affection and admiration from us' serving as 'a nearly perfect test of our being in right temper of mind and way of life' and no one 'can be far wrong in either who loves the trees enough and every one is assuredly wrong in both, who does not love them'.[1] Freed from oppressive imperialistic connotations, established arboretums can be valued, adapted and nurtured whilst new 'living museums' like cosmopolitan Burnley Millenium Arboretum are created and bequeathed to future generations as living symbols of our modern multicultural interconnected world. Those who pursue this honourable endeavour will find inspiration from the great nineteenth-century arboretums and pages of Loudon's *Arboretum Britannicum*.[2]

Notes
1.   J. Ruskin, *Modern Painters*, 3rd edn, 5 vols (London, 1860) quoted in J. Ruskin, *Selections from the Writings of John Ruskin* (Edinburgh: W. P. Nimmo, 1907), p. 81.
2.   D. P. Tolia-Kelly, 'Organic Cosmopolitanism: Challenging Cultures of the Non-Native at the Burnley Millenium Arboretum', in P. Elliott, C. Watkins and S. Daniels (eds), 'Cultural and Historical Geographies of the Arboretum, *Garden History*, special issue (2007), pp. 172–84.

INTRODUCTION

Special places for the cultivation and display of a wide variety of both deciduous and coniferous trees, arboretums developed during the late eighteenth and
early nineteenth centuries. They were a combination of plantation, which usually consisted of a few varieties of trees, and botanical garden. Humphry Repton
adopted the idea for his landscape gardening commissions followed by John
Claudius Loudon. Repton included an arboretum in his red book for Woburn
in 1804 and in his design for Ashridge Park, Hertfordshire between 1813 and
1815 alongside a pomarium, rosarium and other eclectic features. Loudon mentioned arboretums in his *Treatise on Country Residences* (1806) but did not take
a special interest in them until later. In his *Hints on the Formation of Gardens*
(1812), for instance, Loudon carefully defined the differences between groves,
woods and plantations, without mentioning arboretums whilst in the third book
on arboriculture in the *Encyclopaedia of Gardening* (1824) they were not separately categorized and the term was hardly used.[1] The arboretum idea was always
intimately related to written textual manifestations from which it had arisen.
Loudon's *Arboretum Britannicum* (1838), with its wealth of drawings and information, was only the most famous of a series of works that were, in effect, virtual
page-bound arboretums, but which were based on detailed observations of trees
and collections in specific places.[2]

Largely through Loudon's encouragement, systematic tree collections and
arboretums became popular from the 1830s within larger private and public
gardens, estate parks and botanical gardens. Separately and inspired by Loudon, other gardeners and landscape gardeners also promoted versions of the
idea during the first decades of the century. George Sinclair, head gardener
at the Duke of Bedford's Woburn Abbey seat, under the auspices of the Society for the Diffusion of Useful Knowledge, enthusiastically recommended the
widespread introduction of arboretums. For Sinclair, the 'interest arising from
the adoption of foreign trees into domestic scenery' was 'not confined to their
picturesque effects', but reminded all 'of the climes whence they come' and
the 'scenes with which they were associated'. In exploring 'a well-selected arboretum', the 'eternal snows of the Himalaya', the 'savannahs of the Missouri ...

untrodden forests of Patagonia ... vallies of Lebanon, pass in review before us: we seem to wander in other climes, to converse with other nations'.[3] The popularity and fashionableness of tree collecting, encouraged by the important cultural status of trees in British myth, culture and society, estate economy and changing fashions in landscape gardening, especially the decline of formalism and advent of picturesque naturalism, made the acquisition of novel tree and shrub specimens, like works of art or antiquities, highly desirable for their own beauties or as a backdrop for parks, and illustrates the complex interrelationship of artistic and scientific approaches.[4] Uvedale Price, however, argued that 'it is not enough that trees should be naturalised to the climate, they must also be naturalised to the landscape, and mixed and incorporated with the natives'. He was an advocate of mixed plantations and thought that they should be planned carefully so that they added to the 'infinite richness and variety' of the landscape yet seemed 'part of the original design.' Indeed he was critical of those who ignored 'the spontaneous trees of the country' and 'excluded' them 'as too common'.[5]

Recent work in landscape history and cultural and historical geography has developed new approaches and emphases in studies of tree collections and arboretums, although the subject has been surprisingly neglected in analyses of nineteenth-century natural history, which have paid more attention to floral culture, field clubs and field observation, despite the cultural and economic importance of trees in British society. This is partly because traditional historical narratives have emphasized the disappearance and devastation of woodland in the face of enclosure and industrialization when more recent work has demonstrated that woodland began to increase again during the Georgian period fostered by agricultural improvement and encouragement from organizations such as the Society of Arts and Manufactures. Traditional institutional and garden history approaches have tended to focus upon the designers, owners and promoters of gardens, usually from the elite, aristocracy or government. Whilst this remains, of course, important, many individuals and agencies were usually involved in promoting, designing, using and shaping arboretums and, like urban parks, they cannot simply be regarded as outcomes of the will of landscape gardeners or promoters. The often contested nature of garden and arboretum designs, management and usage has tended to be obscured, just as controversies associated with many early Victorian parks were effaced by subsequent public pronouncements which presented them as the natural and inevitable outcomes of rational recreational sanitary and leisure needs and civic communal will. Prosopographical approaches help to illustrate the diversity of individuals and agencies associated with institutional arboretums where subscription lists, visitor records or other such data survive, demonstrating, for example, differences in gender, and religious and political characteristics.

The neglect of nineteenth-century arboretums and tree collections in the history of science is surprising given their relationship to museology, particularly those with systematic labelled displays, numbered plans and guides which served, with botanical gardens, as important inspirations for Victorian museums, rhetorically and spatially asserting their rational objective status through architecture, modes of classification, labelling and display. As we shall see, as laboratories, places for the production of scientific knowledge, arboretums are significant in terms of the sociology of scientific knowledges and the history of science. In growing specimens from around the globe, arboretums sometimes aimed to replicate foreign habitats in microcosm. Trees and shrubs within were usually clearly labelled and demarcated to prevent identity confusion and reduce interbreeding, whilst specimens within were usually placed apart from each other or separated by spaces or boundaries. Arboretums also tended to have their own kinds of characteristic spaces and might be divided by taxonomy, climate, zone or geography. In addition, like institutional laboratories, their nineteenth-century development was associated with particular tools, practices, supply and support networks and trained staff such as arboriculturists and gardeners.[6]

Whilst such features signify a kind of idealized and objectivized collection akin to printed arboretums in arboricultural treatises, like laboratories, the reality varied considerably according to different socio-cultures. Wrenched from original climatic and geological contexts, trees and shrubs may grow, of course, in very different ways from those originally observed in situ, thriving in unexpected ways, or succumbing to unanticipated diseases or predators. Unlike laboratories, museums or glass house collections, arboretums were vulnerable to climatic and seasonal conditions and, in the context of rapidly industrializing Victorian society, the grime of air pollution. Even within the relatively small confines of the British Isles, geological, climatic, meteorological and other factors militated against the establishment of systematic, representative collections in all areas. Palmetums, for instance, were usually confined to the warmer climates of the Channel Islands, the west and south west coasts and the south of England. George Nicholson, Curator of the Royal Botanic Gardens at Kew, during the 1880s, produced a guide to the different kinds of trees and shrubs that could be planted in various conditions including chalky, clay, sandy and peaty soils, marshy and boggy conditions and waterside. Across the vast expanses of the USA and Canada the variations were even more marked. Charles Sargent divided the North American continent from Arctic periphery to Mexican border into nine fundamentally different 'tree regions' defined according to the 'prevailing character of aborescent vegetation'. Tree collections in urban areas presented their own special problems and in his selection of trees and shrubs that were 'best calculated to withstand the smoke and chemical impurities of atmosphere' within manufacturing towns, Nicholson tried to distinguish between those best

adapted to withstand the industrial conditions of northern, midland and southern towns.[7]

One response to criticisms of botanical collections was to devise others
that placed greater emphasis upon formal features and claimed to marry competing taxonomies with aesthetics; another was to reject formal representation
completely. In 1812, Loudon published a design for a spiral botanical garden
'arranged so as to combine elegance and picturesque effect with botanical order
and accuracy' that was later, as we shall see, like his design for iron-framed cur-
vilinear glazing, adopted by the Loddiges nursery company in Hackney. The
garden was 'intended to comprise a complete collection of the vegetables growing in this country' and was arranged according to Carl Linnaeus's system with
the twenty-four orders planted in picturesque groups surrounding a central
hothouse to house the exotics (Figure I.1).[8] Various other forms of plan were
suggested to overcome the difficulties of interpolating systematic collections
into gardens including those founded, like some ancient and medieval gardens,
upon zonal or geographical representation.

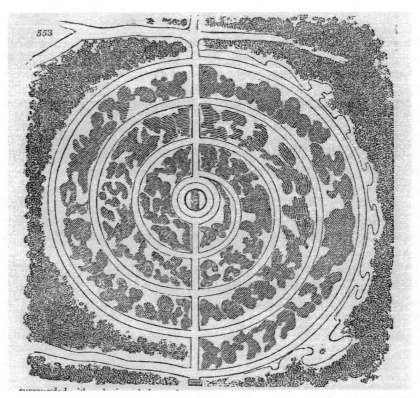

Figure I.1: Design for a spiral botanical garden from J. C. Loudon, *Encyclopaedia of Gardening* (1830), p. 801. Reproduced from the authors' personal copy

John Spencer's pinetum at Bowood grouped specimens according to country of origin and Charles Smith produced a remarkable design for a star-shaped arboretum based upon the display of families.[9]

Arboretums appeared in various different contexts such as the parks of landed estates, commercial nurseries, scientific and botanical societies, and cemeteries and we will examine how each encouraged differences in planting, design, management and consumption. These forms were, of course, not necessarily discrete and promoted by landscape gardeners including Joseph Paxton and Loudon there was considerable interaction and emulation between different versions. We will contend that the importance of trees in society, economy, culture and landscape gardening during the late eighteenth and nineteenth centuries fostered the development of arboretums as special places for the display of trees and shrubs, providing important foci for the dissemination of botanical and scientific knowledge. Tree acquisition gathered pace with aristocratic collectors prizing exotic specimens like works of art or antiquities for consumption, display, competition and learning. Landscape gardeners, such as Brown and Repton, placed new value upon trees and shrubs for their aesthetic qualities as much as horticultural economy. Floral, horticultural and agricultural societies and the Society of Arts also played an important role in promoting tree collecting for botanical study, agricultural improvement, cultures of natural history and the national economy. Post-Enlightenment botany was stimulated by the importation of novel specimens which encouraged further exploration and arboretum provision. As Chapter 2 demonstrates, taxonomic systems and display were required to accommodate novel British specimens and reinterpret familiar plants in the light of foreign discoveries and the precepts of natural theology and natural philosophy. We will examine the role of commercial nursery gardens in the development of the nineteenth-century British arboretum. Stimulated by networks in exotic plant importation, we contend that nurseries should be regarded as major botanical and scientific phenomena that helped to promote botanical education, for example, by encouraging the importation of novel specimens and the creation of new hybrids through crossing. Utilitarian and commercial imperatives underlay the design of nursery gardens and arboretums whilst seed and plant sales were promoted in numerous printed catalogues and gardening publications.

Natural history and gardening became popular middle-class pursuits fostered by ideas of self-improvement, rational recreation and the development of distinctive suburban villas with gardens modelled on larger aristocratic seats. These aspirations were encouraged by commercial nurseries and publications such as Loudon's *Gardener's Magazine* and Paxton's *Magazine of Botany*, which satisfied and stimulated the demand for botanical and gardening knowledge, supplying botanical information and details of the latest horticultural technology and practices. Local and national botanical, scientific and horticultural societies such as those at Glasgow, Dublin, Edinburgh, Cork, Hull, Liverpool, Manchester and Birmingham promoted botanical study by forming arboretums that were available to members and the public on various conditions. In fact, after the term arboretum had been employed by Linnaeus and his followers to denote published lists of national trees, it was through the adoption of the idea in British and Irish botanical gardens that the concept came to be applied to country estates.[10] An arboretum was projected for the Edinburgh Botanical Gardens during the 1780s under the arch-arboriculturist Professor John Hope and another formed in the Dublin Botanical Gardens at Glasnevin from around 1800. In both cases the term arboretum was employed contemporaneously rather than being retrospectively applied to tree collections years afterwards. Given that botanical society arboretums were intended for scientific institutions, arguments arose concerning the relative merits of aesthetics and botany driven by the need to retain subscription income which underscores the difficulties of distinguishing between the arts and sciences. Chapter 3 focuses on two influential examples of these, the garden of the Botanical Society at Glasnevin near Dublin established in 1798 and the gardens of the Royal Horticultural Society at Turnham Green, examining the design, management, consumption and significance of their arboretums.

Loudon's Benthamite enlightenment progressivism and passion for social and political reform encouraged him to sympathize with a younger group of natural philosophers and naturalists during the 1820s and 1830s who were, in general terms, challenging what has been defined as a broadly Anglican, Tory, aristocratic, amateur, scientific establishment. The challenge of scientific reformers in some respects paralleled attempts to promote education, factory reform, religious freedom and toleration, extend and standardize the franchise, and other measures attempted during the late 1820s and 1830s, including the Parliamentary Reform Act of 1832, Municipal Corporations Act (1834) and Poor Law Amendment Act (1835) of the Whig administration. One manifestation of this challenge was support from Whigs, reformers and many nonconformists for new scientific and educational establishments including University College, London, literary and philosophical societies, botanical gardens and mechanics' institutes. These were intended to provide new opportunities for the middle and labouring classes to obtain scientific education and participate in scientific culture as audience and practitioners, for pleasure and social improvement. Another aspect of the challenge posed by reformers, natural philosophers and naturalists with Whig and radical sympathies was the advocacy of novel and subversive scientific theories. Much inspiration came from government-led French science for instance through geology, Lamarckian evolutionism, more 'natural' taxonomies and comparative anatomy, which challenged the traditional British emphasis upon natural theology. Equally inspired by Continental science was the emphasis upon greater specialization and professionalism which sought to emulate the success of French and German institutions such as the universities with salaried professors. Some of the reformer's objectives were summarized by Charles Babbage's *Reflections on the Decline of Science in England* (1830), the famous manifesto projected and widely interpreted as an attack upon domination of aristocratic patronage and conservative values in the Royal Society.[11]

Loudon's arboriculture is the subject of Chapters 4–6, which examine the development and promotion of his arboretum concept primarily through the *Gardener's Magazine, Arboretum Britannicum* and design for the Derby Arboretum. Inspired by detailed knowledge of living trees and their situations and gardening practices as well as botanical ideas, the *Arboretum Britannicum* is unquestionably the most important and influential arboricultural work of the nineteenth century. Although a lavish and expensive book with volumes of plates, it was reissued in various editions and widely disseminated as the abridged *Encyclopaedia of Trees and Shrubs*, whilst the Derby Arboretum was regarded by Loudon as his most important commission because it provided an opportunity to demonstrate how systematic tree collections could be presented for public audiences to promote scientific education and rational recreation. We argue that Loudon's arboretum conception and the *Arboretum Britannicum* are significant because of the role they played in the development and dissemination of botany in gardening and horticulture, including taxonomy, physiology and anatomy. Although Loudon's ideas were avidly adopted by many country landowners and gardeners, they also impacted upon suburban and public gardens and parks, and dominated British arboriculture for decades. Botanical knowledge of trees and exotics were pivotal to Loudon's theories of landscape gardening as embodied in the gardenesque although, as we make clear, there were many tensions and inconsistencies inherent in his ideas and practices. Although Loudon was closely allied to reformers during the 1820s and 1830s and presented arboriculture in the *Arboretum Britannicum* as a secular form of wonder and science potentially for all social classes, it tended to be wealthy aristocrats and landowners who had the manpower and resources to realize his vision. Similarly, although Loudon argued that trees could not be understood without attention to global contexts in both civilized and 'wild' states of nature, the notion of global equalization that he promoted was clearly nurtured by empire.

The importance of Georgian aristocratic tree collecting, and the impact of this upon the sciences of horticulture and gardening, has already been emphasized. Chapter 7 examines the design, management and consumption of nineteenth-century estate arboretums and tree collections, especially the highly influential arboretums and pinetums at Chatsworth and Elvaston in Derbyshire and Westonbirt, Gloucestershire. We contend that, with agricultural schools, estate arboretums and tree collections played a major role in rural botanical education and facilitated the professionalization of agriculture, forestry and horticulture. The wealthiest nineteenth-century estates had the resources, staff and organization to nurture the careers of gardeners and arboriculturists, whilst in the second half of the century with government support from bodies such as the Department of Woods and Forests and the formation of the English Arboricultural Society in the 1880s, training in these fields began to be institutionalized on the landed estate model. Barron's experience enabled him to engage in private practice and design public and private parks and arboretums, whilst Paxton's work planning the major 2,000 species arboretum at Chatsworth helped to secure his position as the pre-eminent mid-Victorian landscape gardener. As Chatsworth, Elvaston and Westonbirt demonstrate, estate improvement and large arboretums provided a powerful symbol of continuing aristocratic cultural ambition and influence in the face of declining real political power during the nineteenth century. As well as satisfying the aristocratic desire for collecting and display, estates functioned as the main component of the rural economy, encouraging the development of arboriculture and the application of new techniques such as tree transplantation which were employed in the creation of arboretums elsewhere. Some of them also became celebrated examples of landscape gardening that were described in horticultural publications and visited by hundreds of amateur and professional gardeners, especially from the 1840s. Whilst some public urban arboretums declined during the nineteenth century, the expansion, institutionalization and professionalization of forestry encouraged the foundation of new types of arboretum associated with agricultural education, horticulture and commercial forestry.

Chapters 8 and 9 examine the development of public and semi-public arboretums beyond botanical and horticultural society gardens contending that they were the Victorian equivalent of museums, mechanics' institutes and other rational recreational associations which provided models for global public scientific institutions. They also explore the attention devoted to botanical education at public parks in London, Bath, Macclesfield and other towns from the 1840s and 1850s, through the provision of botanic gardens, arboretums and taxonomic planting schemes, a development that has hitherto received little attention. With their hundreds of labelled and carefully ordered specimens, public arboretums, arboretum cemeteries and some public parks were 'living museums' designed to promote an ordered and rationalized image of nature and vision of scientific progress which impacted upon local urban scientific cultures. This vision of order, of course, ignored the complex interplay of human and non-human agencies prevalent in public arboretums with flora and fauna and different individuals and social groups collectively appropriating arboretum spaces in different interlinked ways. It also masked considerable disagreements within the sciences and culture concerning taxonomies, evolution and access to scientific information which had an impact upon planting and management and erupted in particular contexts. Some of these factors help to explain the transformation of many public arboretums away from the kind of enlightenment vision promoted by Loudon towards urban leisure parks.

Despite the extensive work that has been done on garden history and nineteenth-century natural history and the fact that Loudon provides one of the most important links between the British enlightenment and Victorian natural history, arboretums and Loudon's arboricultural work have been neglected subjects. This book shows how, inspired by experiences of practical gardening, the formation of multiple public and private systematic tree collections and the *Arboretum Britannicum*, arboretums cemented the place of trees in British economic life, culture, society and the sciences, widening the public audience for natural history and reinvigorating links between the sciences, gardening and horticulture.

1 BRITISH TREE CULTURES IN THE NINETEENTH CENTURY

Introduction

Late seventeenth-century Britain was one of the least wooded countries in Europe. In broad terms woodland had fallen to less than 5 per cent of the land area compared to the 10 per cent estimated to have existed at the time of Domesday. Moreover, compared to many European counties, the number of trees that could be classed as native to Britain was very low with less than thirty broadleaved species and as few as five evergreens. The geography of tree species resulted from a complicated mixture of environmental conditions and past human activities. Although of small acreage, the woods found in different parts of the country varied considerably: intensively managed oak coppice grown for tannin in Cornwall and Devon; strips of coppiced alder along brooks and rivers in the Midlands; pollarded oaks and ashes and stripped elms found in many hedgerows; while significant areas of native Scots pine were only to be found in Scotland. Partly as a consequence of the small area of woodland, Britain was largely dependent on imported rather than home-produced timber and wood products. In addition, there was very little publicly owned forest apart from some small remnants of Crown forests, such as parts of the Forest of Dean and the New Forest. This was important, as it meant that throughout the nineteenth century the espousers of continental ideas of scientific forestry had the difficult job of attempting to persuade many hundreds of private landowners of the value of the new approach, rather than work with state forest officials as in parts of Germany.

Partly in response to the small area of woodland and the relative dearth of native tree species the eighteenth and nineteenth centuries saw a dramatic increase in interest in the management of trees and woods. Between 1750 and 1850 there was considerable anxiety concerning the decline in British woodland which had, of course, major implications for national security, but also for emerging industry and manufactures such as iron working which utilized charcoal. Landowners were encouraged to plant trees as part of the improvement of their estates by the Society of Arts which offered an 'honorary premium' of a gold or silver medal for those who had planted the greatest number of trees or the greatest area of ground during a given year. It was hoped that this would help to supply the navy, cover commons and waste ground, provide employment for the poor, a resource for industry and further 'the ornamenting of the nation'.[1] All aspects of trees and woods were the subject of debate and discussion, including whether selection should be founded upon aesthetic, scientific or economic grounds and whether existing woods should be managed profitably. Attempts were made to define and categorize trees scientifically or as native and exotic, and to address the question of how those outside woods in hedgerows, parks, gardens and fields should be managed. These questions were enmeshed with social and political considerations such as whether communities had access to trees or could use wood. This interest in trees was part of the enthusiasm for agricultural and rural improvement intimately connected with the rise of British imperial power, trade, industry and wealth. This enthusiasm was equally manifest in diverse representations of trees and forests in paintings, drawings, poetry and literature. Trees were valued as signifiers of property and wealth, of nature and beauty, of age and senescence and of British freedoms contrasted to continental tyranny and abstract systems. For Uvedale Price, trees were essential for the picturesque, improved landscape. Rising 'boldly into the air', in beauty they 'not only far excel everything of inanimate nature', but are 'complete and perfect' in themselves. Trees offered 'infinite variety' in their 'forms, tints ... light and shade', and the 'quality of intricacy', composed of 'millions of boughs, sprays and leaves, intermixed ... and crossing each other' in multiple directions. Through their many openings, the eye discovered 'new and infinite combinations', yet this 'labyrinth of intricacy' was no 'unpleasant confusion', but a 'grand whole ... of innumerable minute and distinct parts'.[2] Trees were a crucial element of the network of hedgerows, shelter belts, plantations and clumps that were employed to reconfigure the British landscape practically and visually through the late eighteenth and nineteenth centuries.

In this chapter we set the social and cultural context for the popularity of tree collections and arboretums. We consider the development of different types of forestry and the changing markets for different types of woodland produce uses. We then consider the fascination with newly introduced trees. The cultural values associated with trees are further examined through an analysis of changing attitudes to the ancient oaks of Sherwood Forest in the nineteenth century, the introduction and adoption of the Japanese larch towards the end of the century and the enthusiasm for evergreens.[3]

Ancient and Modern: Woods and Plantations

It was in the long-term interest of landowners to maintain control over the management of timber trees on their estates. The management of woodland and trees increasingly became the domain of landowners instead of farmers, although this varied geographically depending upon forms of landownership and arboricultural rights. In England there was a legal distinction between 'timber' trees, which were usually owned and managed by the landowner, and other trees that were managed by their tenants, and tree and woodland management became increasingly disassociated from other agricultural practices. Even hedgerow trees on farms were maintained and controlled by landlords, although not without frequent social protest and wood thefts.[4] Landed-estate forestry took two main forms in the nineteenth century: traditional woodland management, principally taking the form of coppice woodland or coppice with standards, and plantation forestry, which had developed its own strong traditions from the early eighteenth century onwards.

The trajectories of these two types of management diverged, encouraged by trading conditions and changing markets for wood products. The replacement of firewood and charcoal by coal for domestic and industrial energy production brought about massive reductions in two of the principal markets for coppice. Yet the rapid rise in population caused strong growth in markets for a wide variety of woodland products used in industry, agriculture and the home. Many of these markets were localized and there were considerable regional variations in the strength of the different markets.[5] Woodland managers recognized that many coppice woodlands were of ancient origin. Main noted in 1839 that both historical and geological evidence suggested that 'the greater part of the continent of Europe, as well as its islands, were at an early period almost entirely covered with wood' and that some tracts of forest had been preserved within the royal forests and private parks, while other tracts of natural forest 'are also in existence, occupying broken or marshy ground, or precipitous slopes inaccessible to the plough'. In 1843 J. West distinguished between ancient woods, which were mainly coppice or coppice with standards, and modern plantations. In general in the nineteenth century the use of exotics was almost entirely restricted to the making of new plantations: the replanting of ancient woodland with such trees was generally uneconomic and such replacement was not likely to take place while coppicing remained profitable. Standish and Noble argued in 1852 that

> It is often the object of proprietors to remove woods which are composed of the ordinary indigenous trees of the country, and to replace them with others of an exotic and more ornamental character. But the advantages of such existing woods are generally too great to allow their removal.[6]

William Ablett was able to emphasize the profitability of coppice as late as 1880: 'Where copse-wood is cultivated to any considerable extent, it is advantageous so to arrange matters to come on in perpetual rotation, which may be cleared and put to profitable use yearly.'[7] And John Nisbet noted in 1894 the regional importance of coppice, 'Oak coppice-woods are often by far the most remunerative form of silvicultural crop, when there is any favourable market near at hand for the disposal of the bark to tanneries'. Mixed coppices of ash, field maple, sycamore, sweet chestnut and hazel were 'also exceedingly remunerative throughout southern England, when near to favourable markets for saplings and poles, such as hop districts'. Alder coppice was similarly 'often a more remunerative form of crop than almost any other on land that is suited for the Alder, but which cannot be conveniently drained to serve higher purposes'.[8] However, the decline in coppice prices became marked towards the end of the nineteenth century and only nine years later, in 1905, Nisbet described a completely different and very bleak picture of the state of coppice woodland. He called such woods 'the national form of arboriculture' which 'have for the most part become practically transformed into game coverts ... Yet the copse-woods and coppices were at one time among the most profitable parts of large estates'.[9]

It was not until the turn of the century – by which time metal implements had almost totally replaced locally produced wooden ones – that the market for coppice finally collapsed in most areas. Many of the characters in Thomas Hardy's *The Woodlanders* (1887) gained their living from coppice trades. In a new introduction to the novel composed in 1912 Hardy wrote:

> in respect of the occupations of the characters, the adoption of iron utensils and implements in agriculture, and the discontinuance of thatched roofs for cottages, have almost extinguished the handicrafts classed formerly as 'copsework' and the type of men who engaged in them.

The final collapse of coppicing in the late nineteenth and early twentieth centuries was brought about by the further substitution of wood products by new metal and chemical products and changes in agriculture. This is not to say that the coppice industries did not manage to limp on well into the twentieth century in some localities. On the Eastnor estate in Herefordshire, for example, it was not until the early 1930s that the demand for coppice wood reached such a low point that auction lots did not attract any purchasers.[10]

Plantation forestry became a fashionable branch of British estate management in the eighteenth century. It was strongly associated with patriotism and improvement and continued to be a core aspect of rural estate management in the nineteenth century.[11] Its place was assured less by any inherent profitability, although this was always stressed by forestry publicists and professionals, than by a belief in the seemly nature of tree growing and the clear benefits for landscape, game and hunting. The planting practices developed by professional foresters on Scottish estates in the eighteenth and early nineteenth centuries were often used on English ones and by the mid-nineteenth century 'it was as much the correct thing for an estate to have a Scotch forester as it was for a nobleman's establishment to possess a French *chef* and 'the practice of early and heavy thinning, which prevailed in England for fifty or more years, was introduced by Scotch foresters'.[12] One of the most influential foresters was James Brown whose nursery at Craigmill House near Stirling specialized in 'the Coniferous kinds of trees only, all of which he rears from seeds brought from their native localities'. His text *The Forester* (1847) saw many editions and he strove to improve the quality of the timber produced from plantations, although A. C. Forbes, who wrote several forestry texts and lectured on forestry at the Durham College of Science, Newcastle upon Tyne, later considered that

> it is to Brown and his school that we owe the introduction of the mixed plantation – a system of planting that has led to some of the worst results that could possibly be attained ... Pure plantations, when such were planted, invariably consisted of larch ... The object in many cases was not so much the ultimate production of first-class timber, as the speedy growth of game cover or screens and belts for landscape effect.[13]

There was no established school of forestry in England and most foresters were practical men who were trained on the estates themselves. Often the long-term planning of forests was in the hands of land agents or the owners themselves. The situation was criticized by many commentators and A. C. Forbes thought that

> until quite recently (and to some extent even now) it was no uncommon thing to find all classes of men filling the position of estate forester. Any man with a general knowledge of estate work was considered qualified to manage the woods, more especially on those estates on which the area under wood consisted of coppice with standards. It required no great ability to manage a squad of half a dozen woodmen, to mark and measure the necessary number of trees for estate use or sale, and to see that hedges and fences were more or less in good condition ... [When] planting was carried on, it was, and still is, usual to get a nurseryman to do it by contract at so much per acre and leave the method of planting and choice of species to him.

He was especially critical of the many small estates where the 'commercial details and higher branches of the work' were in the hands of land agents, who would have only a broad training in forestry, and the 'practical woodcraft' was the responsibility of a foreman woodman, who was 'little more than a skilled workman at the best, with a rule-of-thumb acquaintance with the elements of planting, thinning, draining, and so forth'. The effects of employing such poorly qualified staff were, in his opinion, 'inevitably bad'.[14]

Many argued that one reason for the lack of a sustained and economic forestry was the lack of state forestry. Foresters and arboriculturists looked with envy at state forests, especially in Germany, where long-term experiments and schemes for improved forestry could be instituted. There was an orchestrated campaign to form what was termed a 'new forestry' based on scientific principles developed on the continent.[15] Modern scientific forestry was introduced to Britain via its Empire. The vestigial Crown forests were managed by a body known as the Office of Woods, but this was small-scale and did not employ trained foresters. The need for specialized personnel for the Empire was first recognized in India. The 'first attempts at forestry conservancy' began in Burma as early as 1826 and Mr Conolly, the Collector of Malabar, 'commenced planting teak on a large scale at Niambur' in the 1840s. In 1856, the German botanist Dietrich Brandis was appointed Superintendent of Forests in the Pegu Province of Burma and, six years later, he became Forestry Adviser to the Government of India. In order to gain well trained forestry officers from 1866 onwards, 'a number of selected Englishmen were sent annually for a term of two years and eight months to the Continent to study Forestry, half of them going to France, and the rest to Germany'. Brandis later urged that some form of British school of forestry education should be established, and in 1879 the Forestry School of Dehra Dun was opened.[16]

Following that first step, a forestry section was established at the school of military engineering at Cooper's Hill in Surrey. The first director was Wilhelm, later Sir William, Schlich, who eventually became director of the Oxford Forestry Institute at the University of Oxford. Schlich had a great influence on the orientation and development of forestry education, which he based mainly on textbooks translated from German, especially his *Manual of Forestry*. The end result was a first generation of trained foresters whose knowledge had been imported from Germany and France and who were essentially concerned with plantations rather than with coppice with standards. This circulation of ideas, formal and informal, and its implications for British forestry were recognized by Forbes in 1910: 'since about 1860, when Cleghorn and Brandis inaugurated the Indian Forest service, a small stream of continental trained youths has been going out to India, an equally small stream of retired Indian foresters ... has been returning from it'. According to Forbes, whatever the exact practical results of this inter-mixture of British and Anglo-Indian ideas there was 'little doubt that fresh ideas were instilled into British foresters and proprietors, and a wider knowledge of forestry as an industry instead of a hobby resulted'. Formal education was not, however, the only way in which Continental scientific modes of forestry were popularized in England: 'the constant visits made by the British landowning class to the Continent in search of pleasure, sport or health' although not directly connected with forestry 'can scarcely have failed to open the eyes of landowners to the possibilities of scientific forestry'.[17] These novel modes of scientific forestry had to compete against the powerful interests of estate owners in shooting and hunting which had for many years coexisted with traditional coppice and coppice with standard woods. The great wealth of many landowners, derived from urban property, trade and industry, allowed them to treat their estates as sites of conspicuous consumption rather than production. They were usually less concerned with the potential long-term profit from their woods and plantations than with the pleasures that could be gained from their beauty and their crucial importance for delivering exciting fox hunting and pheasant shooting. But whether for reasons of scientific forestry, sport or land-scape, the pressure for landowners to plant exotic trees in the British landscape became overwhelming.

Enthusiasm for Exotic Trees

From the seventeenth century onwards the range of trees available to be planted in gardens, parks and plantations increased rapidly and dramatically. P. J. Jarvis argues that there were four key stimuli to horticultural and silvicultural innovation in the period 1500 to 1900: First, scientific and technological advances, such as the dissemination of botanical knowledge through publication of classical and modern works on plants and trees, experimentation, the work of botanical gardens, improvements in green houses and the development of the Wardian case. Second, there were changes in attitude and taste and in particular the importance of fashions for particular tree species and styles of planting. Third was the development of an economic infrastructure, the successful establishment of important nurseries such as Gordon at Mile End and Kennedy and Lee at Hammersmith and the establishment of tree nurseries on private estates which assisted the rapid diffusion of newly introduced species. And finally there was the enormous increase in the number of introduced species. By 1550 it is estimated that there were thirty-six hardy and woody exotic species cultivated in England: 'by 1600, 103 species; by 1700, 239 species; by 1800, 733 species; and by 1900, 1911 species'.[18]

Writers and horticulturalists were thrilled with the opportunities for profit and pleasure provided by these new trees. Mark Catesby in his *Hortus Britanno-Americanus* (1763) extolled the advantages of American trees which could be

> usefully employed to inrich [*sic*] and adorn our woods by their valuable timber and delightful shade; or to embellish and perfume our gardens with the elegance of their appearance and the fragancy of their odours; in both which respects they greatly excel our home productions of the like kind.[19]

Loudon went so far as to say that 'no residence in the modern style can have a claim to be considered as laid out in good taste, in which all the trees and shrubs are not either foreign ones, or improved varieties of indigenous ones'. He summarizes thoroughly the dates of the introduction of foreign trees to the British Islands and the principal collections of trees in the second chapter of *Arboretum Britannicum*. Dr Henry Compton (1631/2–1713), Bishop of London 1675–1713, was 'the great introducer of foreign trees' in the seventeenth century and he 'may truly be said to have been the father of all that has been done since in this branch of rural improvement'. John Ray in his *Historia Plantarum* (1686) lists fifteen rare trees and shrubs, many from America in Compton's Fulham Palace garden, including the tree angelica and the tulip tree. The Bishop received many plants from John Banister – 'the first university-trained naturalist to send specimens, illustrations, and natural history data from North America to England'. In contrast to Compton, Loudon considered that John Evelyn was 'more anxious to promote the planting of valuable indigenous trees, than to introduce foreign ones'.[20]

Douglas Chambers argues that Virgil's *Georgics* 'first made widely popular in Dryden's 1697 translation, provided both a model for silviculture and an encouragement to the sort of botanical experimentation already taking place'.[21] Successive editions of John Evelyn's *Sylva* in 1664, 1670, 1679 and 1706 extolled the introduction of new trees. Evelyn's recommendations for the planting of trees at Euston for Lord Arlington in 1671 included firs, elms, limes and ash which Chambers convincingly argues is 'more limited than would have been the case in by the second quarter of the eighteenth century'. He goes on to note that Evelyn 'had a voracious interest in new species, chiefly trees. Throughout his library any reference to the introduction of new species is marked or annotated'.[22] The growing knowledge of introduced trees is also shown by comparing Stephen Switzer's essays in *The Practical Husbandman* (1733) to his *Ichonographia Rustica* of 1718. In 1718 recommended trees included oak, ash, beech, chestnut, hornbeam, Scotch pine, silver spruce, elm, lime and poplar. By 1733, Switzer was insisting

> any one that would strive to bring the raising and planting of Forest Trees to their utmost Perfection he ought not to be content with treating barely on those plants that grow at Home, but ought by all means to endeavour at such Introduction of foreign Trees and Plants from Climates of equal Temperature, or (if possible) from Climates which are cooler than ours.[23]

Many enthusiastic botanists such as Samuel Reynardson (who lived at Cedar House, Hillingdon from 1678 until his death in 1721) and Dr Robert Uvedale (1642–1722) of Enfield had large collections of exotic trees, but their collections were dispersed after their deaths. Reynardson 'kept them for the most part confined to pots and tubs, preserving them in green-houses in winter, never attempting to naturalize them to our climate'. They were not laid out in gardens and grounds in the form of planted arboretums.[24]

The very rapid growth in the number of introduced species led to the practical need to identify, classify and label trees so that nurserymen, gardeners and owners could be relatively secure about which trees they bought and sold, discussed and displayed. Pulses of new trees arrived initially from Europe and Asia Minor, then in the seventeenth and eighteenth centuries for eastern North America, and finally a great surge from western North America, China, India, and lastly Japan. Initially the classification and display of trees took place in a complex paper landscape of trade catalogues, botanical treatises and manuscript notebooks; descriptions taking the form of dried leaves and seeds, competing botanical nomenclatures and drawings of flowers, seeds, leaves and eventually whole trees. Loudon thought that in the seventeenth century the 'taste for foreign plants was confined to a few, and these not the richest persons in the community; but generally medical men, clergymen, persons holding small situations under government, or tradesmen'. In the following century, however, 'the taste for planting foreign trees extended itself among the wealthy landed proprietors' influenced by the Dowager Princess of Wales at Kew and by several aristocrats.[25] One of the most prominent landowners with a fascination for growing and displaying introduced trees in the eighteenth century was the Earl of Islay (later third Duke of Argyll) at Whitton on Hounslow Heath in Middlesex. He was best known 'as the personification of unionist Scotland in the first half century after union' but also was a keen classical scholar. He 'had one of the largest private libraries in Western Europe' and was seen by some as one of the fathers of the Scottish Enlightenment.[26] One of Linnaeus's pupils, Pehr Kalm, noted that the Duke was particularly interested in '*Dendrologie*' and visited Whitton in May 1747: 'there was a collection of all kinds of trees, which grow in different parts of the world, and can stand the climate of England out in the open air, summer and winter'. He pointed out that the Duke had 'planted very many of these trees with his own hand', that 'there was here a very large number of Cedars of Lebanon' and 'Of North American Pines, Firs, Cypresses, Thuyas' there was 'an abundance which throve very well'.[27]

Other key eighteenth-century arboreal enthusiasts include the ninth Lord Petre at his estate at Thorndon in Essex and as an advisor to the Duke of Norfolk at Worksop Manor, Nottinghamshire; the second Duke of Richmond at Goodwood, Sussex, Lord Bathurst at Cirencester Park, Gloucestershire and the ninth Earl of Lincoln at Oatlands, Surrey. The latter was described by the Duke of Richmond in a letter of 1747 as 'quite mad after planting'.[28] An early form of plant labelling is described by Dr Richard Pococke who visited Lord Lincoln's Oatlands on 29 April 1757. Visitors went down 'a winding walk' through shrubberies to a nursery 'laid out like an elegant parterre'. Near this was 'lately made another enclosure for all sorts of exotic plants that will thrive abroad, with boards plac'd over them on which their names are cut'.[29] Some of the plantings were on a very extensive scale. A letter of 1 September 1741 from the English botanist and gardener Peter Collinson (1694–1768) to the American botanist and explorer John Bartram (1699–1777) notes,

> The trees and shrubs raised from thy first seeds, are grown to great maturity. Last year Lord PETRE planted out about ten thousand Americans, which being at the same time mixed with about twenty thousand Europeans, and some Asians, make a very beautiful appearance – great art and skill being shown in consulting every one's particular growth, and the well blending the variety of greens.[30]

The nurseries at Thorndon were 'the most extensive private nurseries in the country'; the plants were catalogued by Philip Miller in about 1736 and Lord Petre also had a herbarium of species introduced from America. Some time after the death of Lord Petre in 1742, at the early age of twenty-nine, the contents of his nurseries were sold to fellow tree enthusiasts the Duke of Richmond, Lord Lincoln and Sir Hugh Smithson, later first Duke of Northumberland.[31] From the late seventeenth century onwards well-known collections of trees on private estates were celebrated and much visited. However, with the massive increase in tree species in the nineteenth century it became necessary for rapidly expanding collections to be formalized and displayed scientifically and artistically in planned, labelled and mapped arboretums.

John Claudius Loudon was the most important exponent of the value of introduced trees in the nineteenth century. Loudon's knowledge and experience of arboriculture underwent considerable development during the course of his career, but it is clear that he received much inspiration from his Scottish enlightenment background. Although he took some time to adopt and promote the concept of an arboretum, arboriculture was central to Loudon's conception of landscape gardening and horticultural improvement right from the start in published works and early commissions such as the plan for the gardens at Scone. This is evident in his 'Observations on Laying out the Public Squares of London' (1803) and *Observations on the Formation and Management of Useful and Ornamental Plantations* (1804). In the former, he argued that the sombre appearance of London squares planted with too many evergreens could be corrected by mixing in deciduous varieties such as Oriental and Occidental planes, sycamores, almonds and other ornamental varieties that would survive the metropolitan smoke.[32] Loudon considered that he was applying Scottish methods of agricultural improvement to England and emphasized his nationality in publications promoting his methods and landscape gardening commissions. He amassed the very large sum of £15,000 through such commissions and the income gained from the improvements he undertook on his estate at Great Tew in Oxfordshire which funded his first major European tour in 1813 and 1814 through Sweden, Germany and Russia. This and Loudon's various other Continental tours, including one undertaken to France and Italy in 1819, invariably included visits to the principal public, private and botanical gardens, and meetings with gardeners and landowners and eminent botanists such as Augustin Pyramus de Candolle at Geneva. They supplied numerous valuable examples of European agriculture, horticulture and gardening stimulating lengthy descriptions in his journal and fostered a network of botanical and gardening contacts. This is evident from his election to various scientific societies including the Imperial Society of Moscow, the Natural History Society at Berlin and the Royal Economical Society at Potsdam.[33]

The importance of trees and shrubs for Loudon and the experience of creating and managing plantations for the Earl of Mansfield at Scone provided the inspiration for the *Observations*. Large parts were devoted to arboriculture on the basis that trees were 'the most striking objects that adorn the face of inanimate nature'. He imagined how barren and joyless would be the European surface if 'totally divested of wood', and how 'bleak, savage and uninteresting' would be features such as 'hills and valleys, rivers and lakes, rocks and cataracts'. From the start of Loudon's career, it was primarily trees and shrubs that made natural and artificial landscapes enjoyable, meaningful and civilized. 'Cold and barren' mountains covered with wood and water shaded by trees became 'rich, noble and full of variety', offering 'continually changing scenes' to the weary eye of the traveller, new 'pleasing combinations' and full engagement of the mind. An individual tree was equally pleasing and fascinating through the 'intricate formation and disposition of its boughs, spray, and leaves, its varied form, beautiful tints, and diversity of light and shade'. This made it 'far surpass every other object' and produced a 'general effect that was 'simple and grand' despite the 'multiplicity of separate parts'. Wood was the 'greatest ornament on the face of our globe'.[34]

Economic Value of Trees

Trees were not only aesthetically pleasing and intellectually inspiring, but through the multiplicity of their agricultural, manufacturing, commercial, building and naval applications they were according to Loudon, 'the most essential requisite for the accommodation of civilised society'. The pleasure 'attending the formation and management of plantations' was a 'considerable recommendation to every virtuous mind' and young trees could be regarded as akin to offspring with 'nothing' more satisfying 'than to see them grow and prosper under our care and attention', examine their progress and mark their peculiarities. As they 'advance to perfection', their 'ultimate beauty' is foreseen and a 'most agreeable train' of 'innocent and rational' sensations are excited in the mind, so that they 'might justly rank with the most exquisite of human gratifications'. At the start of his career in the *Observations*, as his wife Jane recognized when she included lengthy extracts in her memoir, Loudon was already arguing passionately for the planting, managing and improvement of trees as one of the major hallmarks of human civilization.[35]

Loudon developed some of these ideas in his *Hints on the Formation of Gardens and Pleasure Grounds* (1812) which included a series of plans for laying out gardens and pleasure grounds for large villas and an analysis of plantation management.[36] Although Loudon had a profound interest in trees individually and collectively in landscape gardening and in the presentation of systematic collections, it was not until the later 1820s that he promoted arboretums as a special form. One of the most important aspects of the *Arboretum Britannicum* (1838) was the attention devoted to the cultural and economic uses of trees and tree products which represented and strove to satisfy the insatiable demand for facts characterizing early-Victorian natural history. The utilitarian economic arboricultural applications were one reason why Loudon wanted to introduce new species and varieties from across the globe. Just as cereal crops and edible roots supplied food, so trees were scarcely less essential for providing timber 'without which' there would be no 'houses and furniture of civilised life, nor the machines of commerce and refinement'. 'Herbaceous productions alone' provided enough for humans to 'live and be clothed in a savage and even in a pastoral state', but without wood, they could not 'advance further', farm, or construct houses or ships. Trees and shrubs supplied many foodstuffs and drinks from tropical to temperate climates, including 'exquisite fruits', 'all the more delicate luxuries of the table' and the 'noblest of human drinks'. In sum, there was 'hardly an art or a manufacture, in which timber, or some other ligneous product' was not, in some way, 'employed to produce it'.[37] With regards to commercial and industrial arboriculture, Loudon examined not so much how plantations and individual trees were used in arts and manufactories but types of timber and other tree parts and the modifications most effectively obtained through breeding and cultivation.[38]

Loudon synthesized much evidence of the different uses of trees and shrubs in relation to the arts of construction, machinery, the chemical arts of dyeing and colouring, in 'domestic and rural economy' and medicine in the *Arboretum Britannicum*. The latter had long provided an incentive for the introduction and study of foreign trees and shrubs and Loudon supplied information concerning the applications of these within medicine including drug preparation and other forms of treatment. Trees were essential for civil, military and marine architecture, engineering, carpentry, rope-making, weaving, joinery, cabinet-making, carving, modelling, the brewing industries, making locks, mathematical instruments and many other objects including machinery in manufactories, carriages, and a multitude of other objects. Some indication of the role of wood in agriculture was provided – although this was detailed more fully in the *Encyclopaedia of Agriculture* – including the creation of agricultural implements, gates, fences and the harvesting of wood products for human and animal food and drink. In chemical manufactories, tree products were employed in tanning, dyeing, colouring and other processes including the expression of oils, the extraction of sugar and the fermentation of some alcoholic drinks. The domestic use of timber also received attention including everything from wood for fuel and basket making to toys, walking sticks, fishing rods, chests, desks and coffins to wooden pieces for games and sports. Finally, Loudon regarded the cultural and aesthetic uses and associations of trees and shrubs as another form of utility, emphasizing their use in religion, gardening, architecture and the many 'poetical, mythological, and legendary associations' they had. Long-established British trees and shrubs such as the rose, elm, yew, lime or oak had multiple economic uses and major cultural significance. The 232 page section on the different kinds of oak in the *Arboretum Britannicum*, for instance, is long enough to be a book in its own right, with numerous literary and botanical quotations, references and 307 illustrations embedded in the text. The exclusion of much cultural and historical information from the *Encyclopaedia of Trees and Shrubs* and concentration on botanical information meant that chapters were much more even, although much of the cultural significance of well established trees and shrubs was missed. The *Quercus* section, for example, occupied just fifty-eight pages.[39]

To supplement and illustrate this utilitarian information, Loudon included statistical data of two kinds defined as geographical and commercial. The former summarized information contained in the forms returned from around the country in preparation for the book, providing details of the comparative hardiness of trees and shrubs and suitability for particular soil types and conditions prevailing around the British Isles. Loudon's tree regions consisted of the metropolitan environs with a radius of ten miles, the south of London, English counties north of London and Wales, Scotland and Ireland, which were sometimes noticed separately but more often treated with the northern English portion. Beyond the British Isles, there were separate notices of trees and shrubs in other countries and continents either already imported or which might be suitable for a temperate climate, including some from India, South America, Australia and Polynesia. The commercial statistical information noted the prevalence and the cost of species in nurseries in London, New York and at Messrs. Baumann at Bollwyller on the Rhine, chosen for its central European situation. It also tried to summarize the general importance of particular species and ligneous products in foreign and domestic commerce.[40]

Ancient Forests and Historical Culture

Although some of the remnants of the medieval royal forests such as the New Forest and the Forest of Dean remained actively managed by the Office of Woods throughout the nineteenth century, most, including some of the most famous, such as Sherwood Forest, were disafforested and sold off to private land-owners. It was well known that the medieval Royal Forests were areas where the monarchy retained special hunting rights.[41] Some forests, such as Exmoor, had very few trees; most were made up of tracts of land which could contain villages, heaths, arable land, pasture and woodland. A survey of Sherwood Forest made by Richard Bankes in 1609 shows that by far the greater part of the forest was settled agricultural land and heath, and that it included the whole town of Nottingham.[42] There was no direct connection between the idea of forest and the concept of woodland: medieval forests were administrative units more akin to a modern national park than a plantation of trees. With the decline in Crown interest, especially from the eighteenth century onwards, the term forest became increasingly associated with those wooded areas which survived in areas that remained or were once Royal Forests in the legal sense.

In the Victorian period the ancient trees found in many forests became increasingly celebrated by tourists and authors. Many popular works with titles such as *English Forests and Forest Trees, Historical, Legendary and Descriptive* (1853) and *English Woodlands and Their Story* (1910) were published.[43] Ancient trees were celebrated as silent witnesses of historical events and used as a didactic device in children's books, such as Emily Taylor's *Chronicles of an Old English Oak; or, Sketches of English Life and History, as Reported by those who Listened to them* of 1860. The popularization of the Christmas tree in Britain encouraged by Germanic royal example is another example of the symbolic significance of trees at the heart of Victorian domestic life. Trees mirrored human existence; Strutt (1826) restated a commonplace when he noted that

> among all the varied production with which Nature adorned the surface of the earth none awakens our sympathies, or interests our imagination, so powerfully as those venerable trees ... silent witnesses of the successive generations of man, to whose destiny they bear so touching a resemblance, alike in their budding, their prime and their decay.[44]

Some areas which had many old trees, especially if they were of a grotesque appearance and had legendary connections, such as Sherwood Forest, became important tourist attractions. In two instances large areas of old woodland were protected by Acts of Parliament because of their landscape or recreational importance. The New Forest Act of 1877 gave protection to the 'ancient and ornamental' woods in that forest, while the terms under which Epping Forest was saved from clearance provided that the woods should remain in a 'natural state'. Queen Victoria, who officially opened the Forest in 1882, noted in her journal that it had 'been given to the poor of the East End, as a sort of recreation ground'.[45]

We can examine the ancient oaks at Sherwood to explore the different cultural and social meanings that ancient trees and forests gained throughout the nineteenth century. The Sherwood oaks were championed by the resolute antiquarian Major Hayman Rooke, who sketched, dissected and publicized them during the late eighteenth century. Rooke supported the aristocratic Whiggish landed interest and identified even the old and decrepit Sherwood oaks with the greatness of the British navy and elided their Druidic ancestry with the new plantations being made by the local aristocracy. He saw with great 'pleasure' the efforts being made

> to adorn this ancient Forest in a manner truly patriotic and worthy of imitation; the many respectable Persons, whose Mansions and Parks border on the Forest, *have* made, and continue to make, large Plantations in honour of the splendid Victories gained by our gallant Admirals.[46]

This rise in plantation forestry was predicated by the collapse of traditional common grazing rights in conjunction with the increasing domination of private landed estates. Rooke's antiquarian understandings of the ancient oaks were well established by the first years of the nineteenth century, but they were soon to be usurped by picturesque and romantic sensibilities.

The picturesque became an increasingly powerful and influential way of understanding the landscape from the late eighteenth century onwards. William Gilpin and Uvedale Price presented and codified a vocabulary with which polite society could enjoy the landscape. Ancient trees were especially valued and while John Evelyn and Rooke celebrated old trees for their dimensions, age, and classical and patriotic associations, Gilpin and Price used Flemish and Italian landscape painting to emphasize their aesthetic value. In his *Remarks* on *Forest Scenery* Gilpin 'laments' the 'capricious nature' of picturesque ideas which often ran 'counter to *utility*'. He argues, 'What is more beautiful for instance, on a rugged foreground, than an old tree with a *hollow trunk?* or with a *dead arm,* a drooping *bough,* or a *dying branch?*' He extols the 'blasted tree' as having a

> fine effect both in natural, and in artificial landscape. In some scenes it is almost essential. When the dreary heath is spread before the eye, and ideas of wildness and desolation are required, what more suitable accompaniment can be imagined than the blasted oak, ragged, scathed, and leafless; shooting it's [*sic*] peeled, white branches athwart the gathering blackness of some rising storm?[47]

In his *Essay on the Picturesque* Uvedale Price compares the 'tameness of the poor pinioned trees (whatever their age) of a gentleman's plantation drawn up strait

and even together' with old pollarded trees which 'stretch out their limbs' in
'every wild and irregular direction'. He is delighted by their 'large knots and
protuberances' which 'add to the ruggedness of their twisted trunks' and hollow trees whose mosses, and 'decayed substance, afford such variety of tints, of
brilliant and mellow lights, with deep and peculiar shades, as the finest timber
tree, however beautiful in other respects, with all its health and vigour cannot exhibit'.[48] The picturesque soon became the dominant arboreal aesthetic
throughout Britain and the ancient Sherwood oaks, however hollow and rotten,
found themselves in the forefront of fashion.

Old trees could also be valued for their historical connections. Several individual ancient oaks, such as the Boscobel Oak in which the future Charles II
hid after the battle of Worcester in 1651 or the Birnam Oak near Dunkeld,
thought to be a remnant of Macbeth's Birnam Wood, became nationally famous.
At Sherwood the large number of ancient oaks began to attract literary tourists
in search of Robin Hood, the hero of medieval ballads. Gilpin related that 'this
forest was also the retreat ... of the illustrious Robin Hood ... who ... making the
woody scenes of it his asylum, laid the whole country under contribution', whilst
in 1810 F. C. Laird emphasized the importance of

> that famous, but legendary character, ROBIN HOOD whom tradition records as
> having made [Sherwood] his principal haunt, and of whose popular and interesting
> story but little is known to any degree of certainty, though his exploits have been
> celebrated in ballad in every succeeding age.[49]

It was in Walter Scott's novel *Ivanhoe* (1820), that the association between
Robin Hood and Sherwood was fully celebrated.[50] This was a runaway success
in England, the United States and on the Continent, and saw many editions.
An inveterate tree planter on his own estates, Scott was enormously influential
in bringing about a revival of interest in the Middle Ages and a famous episode
in *Ivanhoe* tells of Robin Hood and his meeting with King Richard I in Sherwood.[51] Soon, tourists began to flock to Sherwood in order to explore the Forest
so vividly described by Scott. The impact of the novel on visitors to Sherwood
can be seen through the eyes of Washington Irving, the American author who
rode from Lord Byron's former home Newstead Abbey to the Forest:

> among the venerable and classic shades of Sherwood. Here I was delighted to find
> myself in a genuine wild wood, of primitive and natural growth, so rarely to be met
> with in this thickly peopled and highly cultivated country. It reminded me of the
> aboriginal forests of my native land. I rode through natural alleys and greenwood
> glades ... What most interested me, however, was to behold around the mighty trunks
> of veteran oaks, the patriarchs of Sherwood Forest. They were shattered, hollow
> and mossgrown, it is true, and their "leafy honours" were nearly departed; but, like
> mouldering towers they were noble and picturesque in their decay, and gave evidence,
> even in their ruins, of their ancient grandeur.

He relishes the literary associations:

> As I gazed about me upon these *vestiges* of once 'merry Sherwood' the picturings of
> my boyish fancy began to rise in my mind, and Robin Hood and his men to stand
> before me ... The horn of Robin Hood again seemed to sound through the forest. I
> saw his sylvan chivalry, half huntsmen, half free-booters, trooping across the distant
> glades, or feasting and revelling beneath the trees.[52]

Tourist guides celebrating the medieval cultural associations, such as James Carter's *A Visit to Sherwood Forest* (1850), began to appear. The great landowners were quick to appropriate this democratic symbol. The Duke of Portland built a new lodge at Clipstone in the forest in 1844. The main room over the arch was 'dedicated by its noble founder to the cause of education, for the benefit of the villagers of Clipstone'; 'The prospects from this room are most beautiful, including Birkland with its thousand aged oaks, the venerable church of Edwinstowe, and a wide expanse of forest scenery'. The popular author January Searle noted that, 'on the north side, there are statues of King Richard the lion-hearted, Allan o'Dale and Friar Tuck; on the south side there are similar sculptures of Robin Hood, Little John and Maid Marian'. Another guide describes three of the sculptures as 'the ancient frequenters of the neighbourhood: one its presiding deity, Robin Hood; the other Little John; and, bearing them pleasing company, as was her wont formerly, Maid Marian'.[53] A few years later the great new mansion built in 1864–75 at Thoresby for Earl Manvers, designed by Salvin, incorporated a vast library fireplace celebrating Sherwood Forest. Its iconography confirmed the historical connection between Robin Hood and the ancient oaks for the mid-Victorian mind. The huge chimney piece

> consists of an elaborately carved representation, in Birkland oak, of a scene in Sherwood Forest, in which are introduced the venerable 'Major oak' with his knotted and
> gnarled branches, a foreground of botanical specimens, and a herd of deer – all chiselled with much similitude to Nature. This monument of patience and ability was cut
> by Mr Robinson, of Newcastle; the wood being from an oak which once flourished
> in the forest in which the leading feature in the subject forms so proud an ornament.
> Statuettes of Robin Hood and Little John support each side of the piece.[54]

The combination of medieval legend and trees old enough to have witnessed the scenes depicted by Walter Scott was enormously potent. By the mid-nineteenth century the ancient oaks of Sherwood had become firmly fixed in the popular imagination as medieval icons. Within a few years individual trees, such as 'Robin Hood's Larder', were imaginatively named and gained credence through being printed on Ordnance Survey maps. By the end of the nineteenth century, the eighteenth-century naval and Druidic cultural associations of the Sherwood oaks had been largely obscured by an accumulation of picturesque, romantic and literary associations.

Coniferous Novelty: The Japanese Larch

As areas of the world were opened up to trade in the eighteenth and nineteenth centuries, the range of new species of tree that could be tested for growth in Britain rapidly expanded. British landowners and foresters began to expect and demand novel trees that could be established on their estates and supplement the trees that they already grew. We can examine the cultural appropriation of a new species through the case of the Japanese larch.

In the late nineteenth century the fashion for Japanese plants and gardens spread throughout Europe and the United States, part of a larger cultural moment included the vogue for design, arts and crafts known as 'Japonisme'.[55] Japan was increasingly identified in Britain as a global partner, with a comparable imperial history and interest in horticulture. Japanese gardens were displayed in many international exhibitions and there was extensive coverage in the popular press, including the many illustrated garden magazines. This led to a flourishing export market in mature plants, bulbs and stone ornaments. In Japan large export nurseries were established, such as L. Boehmer's and the Yokohama Nursery Company. British based nurseries stocked an increasing range of Japanese plants and ornaments, as did the leading arts and crafts design store Liberty's in London, which advertised Japanese stone lanterns.[56]

Before 1858 Japan had been largely secluded from the world from the early seventeenth century, except through links with the Netherlands and China via the artificial island of Dejima at Nagasaki. Most plant introductions and knowledge in Europe about Japanese plants arose through Dutch trading connections. The botanical writing and descriptions of Engelbert Kaempfer (1712, 1727), Carl Peter Thunberg (1784) and Philip Franz von Siebold (1850) played a major role in raising expectations about the new plants that could be found once trade took place more regularly.[57] *Flora Japonica* (1843), written by Philip Franz von Siebold and Joseph Gerhard Zuccarini, drew on the experience of von Siebold who lived in Japan between 1823 to 1829 and sent specimens of Japanese plants to Professor Zuccarini at Munich University, who classified and named them.[58] The European collectors derived much of their knowledge by collecting plants and botanizing with local experts, consulting texts, illustrations and herbaria written, drawn and collected by them.[59] Early Victorians therefore had some tantalizing knowledge about the diverse flora of Japan and also had considerable experience from earlier introductions of the likely potential of Japanese plants in British gardens and parks. It was, however, only after 1858, when Japan opened three treaty ports to the West that British collectors and horticultural traders were able to experience the Japanese landscape first hand and fully exploit the commercial potential of its trees and shrubs.

Western names were applied to plants that already had established Japanese botanical and horticultural names, although these latter were also used by western nurseries to enforce the exoticism of newly introduced Japanese plants. The Japanese name for *Larix leptolepis* is frequently given as *karamatsu*,[60] although in the important Japanese plant encyclopedia *Kai* (1763) this larch is given three different names: '*Kin sen shou*', '*Fuji matsu*', '*Nikko matsu*'. The first name refers to the golden colour of the autumn foliage while '*Fuji*' and '*Nikko*' are both areas where the tree grows naturally, and '*matsu*' means pine. The tree is described as having a thick, scaly bark similar to the Japanese White Pine. It is noted that after frost the needles fall, and the golden autumn colour of the needles is particularly praised.[61] The tree had a limited distribution, mainly at altitudes of between 1,000 and 1,400 m, especially on dry, volcanic soils. The main natural stands were found in central Honshu, especially in Yamanashi Prefecture, including the slopes of Mount Fuji. Lindquist notes that larch 'has long been planted in Japan. For several centuries there have been plantations of it in northern Honshu' and in southern and central Hokkaido after colonization many more plantations were made.[62] The Japanese larch, though known to Kaempfer and Thunberg in the eighteenth century and mentioned by Lambert, was first described by Lindley in 1833. The tree was included in von Siebold and Zuccarini's *Flora Japonica* and illustrated with a spray of twigs and needles, and cones with details of the needles and seeds, but it was not actually grown in Britain until the 1860s.[63]

Arriving in Japan in 1860, John Gould Veitch (1839–70) was one of the leading British plant collectors. He worked for his family company at the Royal and Exotic Nurseries based in Exeter and Chelsea.[64] In 1860 *The Gardeners' Chronicle* reported that

> Of all parts of the earth in which vegetation is vigorous and little known, Japan stands pre-eminent. With a climate like that of England, and a half Siberian or Himalayan and half Chinese Flora, it offers the greatest inducement to Europeans to investigate its productions.[65]

In September 1860 Veitch discovered *Abies leptolepis* with three other conifers and his climbing companion the diplomat Rutherford Alcock reported that it was

> at an elevation of 8,000 to 8,500 feet on Mount Fusiyama. It is remarkable as being the tree which grows at the greatest elevation on this mountain. Its greatest height is 40 feet; but on ascending the mountain dwindles down to a bush of 3 feet. (Japanese name is Fusi matsu.)[66]

Initially there was come confusion as to whether the Japanese larch discovered by Veitch was the same species as the *Larix leptolepis* described by earlier botanists. *The Gardener's Chronicle* of 12 January 1861 noted that the cones described in von Siebold and Zuccarini's work were 'four times larger than those sent home

by Mr. Veitch' and that 'there is some doubt whether his plant is not distinct'.
Henry Elwes and Augustine Henry stated that 'A stunted form, growing on the
higher parts of Fuji-yama, was collected by John Gould Veitch, and was considered to be a new species by A. Murray; and is recognised as a variety by Sargent'.[67]
They provided eight different botanical names[68] for the Japanese larch, eventually plumping for *Larix leptolepis*, Endlicher (1847):

> *Pinus leptolepis* was the name preferred by Endlicher; but he quotes *Larix leptolepis*,
> Hort., as a synonym; and as this is the first publication of *Larix leptolepis*, Endlicher
> is responsible for the name, and it is credited to him; and being the first published
> name under the correct genus is adopted by us. Moreover, it is the name by which this
> species is universally known; and the adoption of Sargent's name, *Larix Kaempferi*,
> would cause great confusion, as this has been used for *Pseudo-larix Kaempferi*, the
> golden larch of China.[69]

The difficulties over naming the newly imported tree did not hinder its appreciation by foresters and landowners. Elwes and Henry noted that although few
trees were established from the seeds collected by Veitch in 1861, trees grown
from other seeds 'grew so well generally that it is now being planted almost everywhere, and some of the older trees have produced good seed for ten years or
more'. Elwes was keen to assess the growing conditions and uses of the tree in
Japan and its suitability as a tree for forestry plantations. He saw the trees growing
in volcanic soils in Japan in 1904 and thought they 'were very similar in habitat
to the larch in the Alps, and had not an excessive development of branches'. He
noted that the timber was used for 'ship- and boat-building' and 'railway sleep-
ers and telegraph poles'. The plantations in Japan were also closely connected to
the demands of modern development. Elwes saw many young plantations which
'were very similar to larch plantations in England in growth and habit. I also
saw it planted experimentally in Hokkaido, along the lines of railway, where it
seemed to grow as well in this rich black soil as in its native mountains'.[70]

The tree became very popular in Britain and was 'looked upon by many foresters as likely to replace the common larch'. Elwes and Henry thought that 'no
conifer of recent introduction has attracted so much attention among foresters
as the Japanese larch, which, during the last ten years, has been sown very largely
by nurserymen'. Elwes himself successfully sowed seeds collected from trees from
three different British estates, Dunkeld, Perthshire; Hildenley, Yorkshire and
Tortworth, Gloucestershire in 1890, and after six years they had grown to four to
eight ft in height. In his view the Japanese larch had three main advantages: First,
its establishment as a plantation at 1,250 ft in Scotland where it grew 'very vigor-
ously in mixture with Douglas fir' showed it to be hardy. Second, it appeared to
be immune from the canker *Peziza willkommii* which affected European larch.
Henry examined in 1904 'six plantations of Japanese larch of ages from five to

sixteen years, and in none could detect any sign of canker'. Third, it was a vigorous tree suitable for economic plantations as it grew in its first twenty years
quicker than European larch, although it appeared to have 'a great tendency to
form spreading branches'.[71] By the mid-twentieth century Japanese larch had
become 'one of the most important exotics planted in Britain' with about fourteen million plants used annually, a number only exceeded by Sitka spruce and
Scots pine.[72]

The case study of the Japanese larch demonstrates that there could be several stages in the reception and transculturation of a new tree as it was named,
collected, transported, acclimatized and naturalized. The stage when new trees
were conceptualized as exotic can itself be recognized as one form of encultur-
ation. A second stage is when the tree became 'culturally assimilated'. At this
stage the trees are not physically changed or modified, but they have been grown
long enough to demonstrate that they are well adapted to live in the open air,
and propagate well enough to become common plants in Britain. A key factor
allowing a plant to become culturally assimilated is its hardiness. Hardy plants
were useful for ornamental and commercial planting in existing gardens, parks
and woods; they became common in the British landscape and were no longer
seen as exotic. In some cases there was a third stage of physical hybridization.
The crossing of exotic and native species of plants was one way to produce new
varieties of plants; sometimes, however, hybridization occurred naturally. *Larix
leptolepis* became culturally assimilated through its economic timber value and
eventually became physically hybridized with the European larch at Dunkeld
in Scotland to produce *Larix eurolepis*, which itself became an important commercial tree species.

Evergreen Aesthetics

Eighteenth- and nineteenth-century tree enthusiasts were particularly susceptible to the pleasures of evergreens and conifers. The native flora was bereft of
significant evergreen trees other than the broadleaved holly (*Ilex aquifolium*)
and the coniferous yew (*Taxus baccata*) and Scots pine (*Pinus sylvestris*). The
box (*Buxus sempervirens*) and juniper (*Juniperus communis*) were also native but
were usually grown as shrubs rather than substantial trees. This paucity of evergreens meant that Victorian gardeners prized any introduced trees that could
provide varied foliage during the long winter months. One of the greatest enthusiasts for evergreens was the nurseryman and horticulturalist William Barron
(1805–91) who worked for the earls of Harrington at Elvaston castle, Derbyshire in the 1830s and 1840s before setting up his own nursery business. Barron's
British Winter Garden (1852) promoted the use of evergreens in public and private spaces helping to drive the new fashion in British, European and American

gardens. He was attracted to conifers for their cultural associations and natural characteristics, which he was able to exploit for economic and moral purposes. In turn, Barron and his company helped to foster the mid-Victorian fashion for evergreen planting, promoting them for their economic value and as special and ornamental specimens. Although he propagated and popularized a huge variety of exotic conifers, the evergreen that became most closely associated with Barron was the yew and the cultural associations and ornamental value of the yew, as well as the relative ease with which it could be moved, attracted him to favour it for many of his landscape gardening and transplanting commissions.[73]

Barron was particularly critical of the types of deciduous trees that dominated many parks and plantations. He complained at how common it was to see 'close to our mansions, such commonplace things as elms, ashes, sycamores, poplars, or any other rubbish that the nearest provincial nursery may happen to be overstocked with: all stuck in to produce either immediate or lasting effect!' Deciduous trees provided a 'continued litter of decayed leaves', 'an unwholesome effluvia' during the winter and 'an assemblage of leafless stems' with no shelter or protection 'from bleak winds for seven months in the year'. He claimed that as fond as he was 'of the study of trees, 'the construction and consequent peculiarities of the human mind' was 'a subject of deeper interest still'. Barron argued that individuals were differently constituted with varied tastes and enjoyments and different capacities to form correct estimates of 'form, size, proportion, texture, and colour of natural objects' depending upon capacities and cultivation of different faculties. The 'divine author' had constituted people to be capable of high enjoyments which depended upon knowledge and fulfilment of natural laws and this was proportional to the degree of organization or structural development of natural objects perceived. Hence, coniferous trees and shrubs excited admiration by providing an infinite variety of form, size, colour, texture and outlines 'from the formal araucaria and fastigiate Junipers, to the wild grandeur of the pine, and even to the delicate, graceful, and flowing habits of the Cryptomeria Japonica, Funebral Cyprus, Deodar Cedar and Hemlock Spruce'. Gigantic 'Lambert and Bentham pines', *Sequoia sempervirens* and Douglas Fir towered 'their lofty heads a hundred feet above the pride of British forests' providing pleasure from their unique dimensions and also produced massive amounts of timber demonstrating the economic importance of coniferous trees. Placing different forms or colours of individual specimens or groups against different backgrounds could be particularly satisfying, such as the combination of large and small yews behind golden yews, Irish yews, variegated white cedars and different junipers. Evergreens were superior on grounds of health, practicality and neatness, providing enjoyment for an entire year whilst gardening, park and cemetery commissions provided an opportunity for

Barron to further the polemical objectives of the *British Winter Garden* and 'arrest the attention of the public' towards evergreens.[74]

Clipped yews as hedges or garden ornaments had been fashionable during the seventeenth century, but this declined rapidly during the eighteenth century with the exception of a few places such as Oxford and Wroxton. There was, however, resistance to the idea that yews were miserable and solemn trees, which must have encouraged Barron. William Gilpin professed himself 'contrary ... to general opinion', an admirer of the 'picturesque perfections' and 'form and foliage' of the yew although he condemned the 'indignities' of clipping under which it suffered. In a 'state of nature', the yew was one of the 'most beautiful evergreens we have' possibly even superior to the Cedar of Lebanon as it had been grown in England. It grew well on most soils, yet was seldom visible in its natural state. Gilpin admitted, however, that 'though we should be able to establish the beauty of the yew with respect to form and foliage' it was difficult to combat the view that 'its colour, unfortunately, gives offence'. Its 'dingy funereal hue' it was thought made it 'only fit for a churchyard' but this attachment to colour was 'an indication of false taste' through which rose the 'numerous absurdities of gaudy decoration' whilst dislike towards particular colours 'shows a squeamishness, which should as little be encouraged'. The key was that colours should 'act in concert' so that soft and pleasant colours were combined with reds, blue, yellow, light green and dingy green; the 'virtue of each consists solely' in its 'agreement' with its 'neighbours'.[75]

The importance of yews for bow making and their sacred symbolic value, being customarily planted in churchyards, ensured that many celebrated examples existed by the eighteenth century. As the Manchester naturalist John Bowman put it in the *Magazine of Natural History*, 'it seems most natural and simple to believe that' from the 'perennial verdure' longevity, and the durability of yew wood, how it had been 'at once an emblem and a specimen of immortality' used by 'our Pagan ancestors' to adorn graves and other sacred purposes. The practice was continued into the Christian era and yews became associated with ghosts and fairies. In early Christian society churches became associated with old and striking yew trees and it was believed that yews were vestiges of ancient druidical groves and were 'already sacred, venerable for size and indispensable for religious rites'. The decorated evergreen Christmas tree at the heart of mid-winter celebrations, of course, came to have similar significance and Christian associations in the Victorian household.[76] The longevity and sacred cultural associations of yews were celebrated by Loudon in a lengthy analysis in the *Arboretum Britannicum* and in the poetry of Robert Blair, Walter Scott and William Wordsworth. In Blair's enormously popular Gothic poem *The Grave* (1743) the yew is characterized as a 'Cheerless unsocial plant! that loves to dwell / 'Midst skulls and coffins, epitaphs, and worms'. An edition of 1785 notes that various reasons have been assigned to the custom of planting yew trees in church yards but that one of the most probable is 'this tree being an *evergreen*, may in some respects be esteemed no unfit emblem of the immortality of the soul, as it never dies'. Scott described 'A dismal grove of sable yew, / With whose sad tints were mingled seen / The blighted fir's sepulchral green'. Whilst the yew retained its sorrowful associations for Wordsworth, he also celebrated its longevity and appearance. The yew tree of Lorton Vale was 'of vast circumference and gloom profound ... a living thing, produced too slowly ever to decay' whilst four other yews of Borrowdale were to be celebrated for their 'solemn and capacious grove' of 'huge trunks' and 'intertwisted fibres serpentine' where 'ghostly shapes' might meet 'as in a natural temple' to worship.[77] Loudon too emphasized the beauty and longevity of the yew, arguing that the neglect of evergreens in landscape gardening and horticulture was unjustified. He provided accounts of the most celebrated British yews with illustrations including those at Fountains Abbey, the Buckland Yew in Kent, the two Tytherley Yews in Wiltshire, the Tisbury Yew in Dorset and the Iffley Yew near Oxford. In the right context the yew was valuable for instance as an avenue tree; it was 'suitable for approaches to cemeteries, mausoleums, or tombs; and as a single tree, for scattering in churchyards and burial-grounds'.[78]

Whilst appealing to Gothic and Romantic sensibilities, such associations prevented the yew from enjoying general popularity in Georgian landscape gardening. There were also practical reasons why yews were retained in gardens and churchyards and were rarely planted in parks and plantations: the foliage was known, when cut, to be poisonous for cattle and there was only a limited market for the timber. However, the rapid importation of new coniferous species and the speed that most of these grew helped to encourage change. Many new varieties were first detailed in Aylmer Bourke Lambert's *Description of the Genus Pinus* (1803) which, along with the new conifers on estates such as Painshill and Woburn, encouraged a change in attitudes and practices towards evergreens. For example, by the 1780s Linnaeus thought that there was 'a greater variety of the fir' to be found at Painshill 'than in any other part of the world', whilst Lambert thought Charles Hamilton's estate, which helped to inspire his book, to be 'the most remarkable gardens for the cultivation of pines in this country' and superior to anything in Europe.[79]

Barron's work was motivated by deeply held moral and religious beliefs. He remained hostile to evolutionary ideas and became a major figure in the temperance movement and vice-president of the national Temperance League, seeing his arboricultural, landscaping and nursery businesses as a means of furthering particular moral, religious and temperance objectives. Temperance associations were encouraged to undertake visits and to hold galas in parks and gardens, including Elvaston.[80] The emphasis upon evergreens, and the concept of the winter garden utilizing evergreens in novel ways, was intended to provide all year round opportunities for rational recreation for urban populations. Encouraged by nurserymen and landscape gardeners such as Barron, evergreens emerged from the background and the role assigned to them by many picturesque improvers tying together ornamental and usually deciduous specimens, to become the primary focus. Like the colourful decorated Christmas tree at midwinter, Barron's propagation and breeding of more colourful varieties of coniferae removed one of the major objections to the widespread ornamental use of evergreens: that they did not have sufficient variety to delight the eye. Like the swelling Victorian cities, evergreens defied time and the passage of the seasons, like the growing industrial economy they were not idle for large parts of the year but always showing their productivity and beauty. As we shall see, this sense of the spiritual importance of trees in the Victorian soul mediating between human beings and divine nature was most powerfully asserted in the work of John Ruskin.

Conclusion

British arboreal landscapes were transformed in the eighteenth and nineteenth centuries. The amount of woodland increased owing to a spirit of planting galvanized by the cultural, social and economic interests of the increasingly important landed estates. The larger these estates became, often enriched by injections of wealth from trade and industrial enterprises, the greater the potential for extensive new plantations of trees. These plantations developed alongside the ancient woodlands with their long history of coppice management. The income from the traditional coppice woods remained of considerable importance until the end of the nineteenth century, and indeed, was to some extent revivified by the demand for coppice products from new industries. Ancient trees, such as churchyard yews and historic oaks were celebrated by poets and artists and their links with historical figures became almost a commonplace of literary allusion, whilst influential writers such as Ruskin saw trees as symbols of an enduring spirituality in the face of modernity, mediating between nature and humanity. At the same time the rapidity and scale of the introduction of new trees produced a hothouse atmosphere of enthusiasm for new species, varieties and forms of tree. The multitude of exotic trees caused great excitement amongst gardeners and foresters keen to enjoy the new forms of growth, foliage and bark. Evergreen trees were especially welcomed as they could at last provide a much greater variety of greenery in the winter and spring gardens. New variegated conifers and broadleaves eponymously symbolized the fashion for diverse, novel trees. However, the dynamic and vibrant market for new trees led to confusion over classification and authenticity. How could a putative new species be validated? How should trees be classified and named? What system of nomenclature should be used? How could one be certain that a new tree was hardy or likely to produce timber of high quality? One of the most important ways in which these questions was answered was through the collection and ordered display of the new trees in arboretums in private gardens, public parks and scientific institutions.

2 TREES AND TAXONOMY

Introduction

Botany became one of the most fashionable and popular British Georgian pursuits, particularly after the translation and publication of the works of Carl Linnaeus. The Linnaean system helped to make botany accessible for men, women and children alike by providing a key through which plants in flower could be identified easily without requiring a profound knowledge of botanical characteristics. There were, however, other important aspects to British botany inspired by the economic, patriotic, cultural and fashionable emphasis upon horticultural and agricultural improvement. During the later Georgian period, and inspired by French botany and British arboriculture, increasing importance was attached to the interconnectedness of taxonomy, vegetable physiology and anatomy. The popularity of botany is exemplified by the number of natural historical works that were published and disseminated, the development of private gardens and commercial nurseries, creation of herbariums, practice of flower painting and formation of botanical societies. Introductions to the Linnaean system, intended for both sexes, sold widely and there were various attempts to translate key Linnaean works into English, whilst Erasmus Darwin's *Loves of the Plants* (1789), a poetical representation of the Linnaean world personified, proved especially popular during the 1790s. Although various botanical systems competed internationally in the period between 1760 and 1820, Linnaean taxonomy was dominant, particularly in Britain. The greatest challenge came from those who proffered what they claimed to be more 'natural' systems, especially French botanists such as Antoine-Laurent de Jussieu, although their work was partly inspired by suggestions offered by Linnaeus. The proliferation of botanical systems initially encouraged British botanists to adhere to the Linnaean framework ever more emphatically. Thomas Martyn, reader and later professor of botany at the University of Cambridge admitted to despairing at the widespread 'system madness' represented by the sixty different taxonomies listed in

Michel Adanson's *Familles des Plantes* (1763) which he adjudged, however, to be not 'worth one farthing'.[1]

Popularity of Botany and Natural History

During the first decades of the nineteenth century natural history became an even more popular pastime encouraged by writers, such as Loudon, and numerous publications catering to this expanded audience. Gardening, horticulture and floriculture had been important in Georgian society and polite culture, but the perceived utilitarian and practical importance of these subjects grew fostered by natural theology, Romanticism and the drive for rational recreation and personal improvement. Whilst these cultures remained in many ways domestically centred through gardens, glasshouses and improving literature and urban-centred scientific institutions provided foci and leadership, there was also a proliferation of sites where natural history was experienced and encountered by all social classes and ages, including the fields and countryside, rivers and lakes, cliff faces, and the sea and seashore. Specimens were eagerly collected as objects for glass cases, herbariums and cabinets and facts were laboriously recounted and detailed in numerous popular best-selling natural history texts by the mid-nineteenth century. These pursuits and encounters were also driven in a negative sense by rapid urbanization, industrialization and population expansion which fostered a nostalgia for nature and the countryside that was, somewhat later, to hasten the development of public urban parks, suburban gardens and allotments. On the other hand, industrialization provided wealth and prosperity for the middle classes and increasing opportunities for national and international travel and mass-produced, widely-disseminated and cheaper literature, facilitating the collecting and preservation of plants and trees in glasshouses, glass cases and museums.[2]

There were tensions between more popular forms of experiencing natural history and the tendency towards specialization, professionalization and institutionalization in the nineteenth-century sciences which saw a bifurcation between natural history and the biological sciences. These developments were, in some respects, hastened by the challenge to the Anglican, aristocratic and amateur scientific elite from a new group of professional naturalists and natural philosophers with reforming sympathies, particularly from the 1820s and 1830s. Although these divisions should not be exaggerated they were manifest by the professionalization of botany, an increasing division between biology and natural history and a greater emphasis upon scientific utility rather than natural theology through which it was hoped that botany and biology could attain an equivalent status to the physical sciences. One manifestation of this was disagreements concerning the nature of botany and taxonomy. Proponents of different taxonomies claimed that they were striving for objective forms of classification based on the nomenclature of the objective Linnaean binomial system. In actual fact, of course, there were considerable disagreements concerning the nature of botany and different forms of classification, particularly concerning the respective merits of the Linnaean and different systems that claimed to be more naturally-founded and less 'artificial'. In many ways, stimulated by the influx of exotic specimens from the colonies, classification represented a claim for imperial dominion over flora and fauna. Although botany and taxonomy did not have the status that they had enjoyed in the later Georgian period, they still had great cultural and political significance. Despite the claims for objectivity, naming practices and designations were often intended to celebrate national sciences and reinforced social and cultural divisions within British society as well as internationally (such as British and French colonial rivalry for instance). Inconsistencies remained and as Ritvo has contended, 'nomenclature remained inconsistent and multiple, through the nineteenth century', the conquest of nature proving 'as protracted, troublesome and ambiguous in its results as the political enterprises that it paralleled'. Ambiguities resulted from the difficulties of identifying and distinguishing between species and varieties internationally using preserved specimens and literature owing to problems of language, the proliferation of synonyms as well as other factors. The Jussieuian or natural taxonomy that many professional British botanists came to favour by the 1840s was not a unified system but promoted in a variety of forms by various individuals using different divisional categories and terminology. Until at least the 1850s, various types of numerical system such as the quinarian and septennial systems, for instance, competed with interpretations of Jussieuian taxonomy modified in different ways.[3]

After examining some of the differences between Linnaean and Jussieuian taxonomies, this chapter argues that it was British arboriculture and particularly the practical experience of managing systematic tree collections that encouraged the adoption of the Jussieuian 'natural' system of botany, particularly as defined by the Swiss naturalist Augustine Pyramus de Candolle (1778–1841). Although recent analysis of the introduction and development of the natural systems in the British context has emphasized the importance of textbooks, it was in botanical gardens, tree collections and nurseries that problems surrounding the application and development of Linnaean and natural systems were first faced.[4] As we shall see, because of their cultural significance and the stimulating challenge they presented to British botany through global importation and utilization in gardening and horticulture, trees became crucial for reforming botanists and their friends, including John Lindley, Joseph Hooker and Loudon. They also presented a much greater opportunity to demonstrate the efficacy of the natural system and interdependency of physiology, structure and classification than the bewildering, unmanageable and often ephemeral variety of plants and crypto-gams that faced early nineteenth-century botanists. Cryptogams such as ferns, mosses, fungi and algae presented special difficulties for Linnaean botany, but there was no consensus concerning how they should be classified in the period between 1800 and 1850, or how they related to other plants. Widely visible, known and loved, and the apex of the plant world, the status of trees mirrored notions of order and hierarchy. Trees satisfied the aesthetic demands and expectations of late-Georgian landscape gardening, providing shelter from sun, rain and wind, reinforcing and obscuring boundaries and serving as objects for taxonomic, physiological and anatomical vegetable studies. The chapter emphasizes the dramatic impact of the discovery and importation of novel trees and shrubs upon British botany and arboriculture, which is evident in the design, classification and management of tree collections in early nineteenth-century botanical society gardens. Whether in books or gardens, systems of labelling, classification and positioning tried to present an ordered image of external nature, which was easier in the former, but this masked considerable taxonomic disagreements that were particularly acute when placing novel specimens or closely-related varieties.[5]

Linnaean and Natural Systems

The Linnaean system became accepted as the standard form of botanical classification in Britain by the late eighteenth century. It was an attempt to rationalize analysis of the natural world by classifying living organisms in terms of a binomial system of nomenclature according to a hierarchy of six principal groups: kingdom, class, order, genus, species and variety. Following John Ray, Tournefort and other botanists, Linnaeus used the reproductive organs of plants, namely the flowers, fruit and parts of fructification as the basis for his 'artificial' system of classification, although he continued to believe that progress would be made towards establishing a more 'natural' system. The Linnaeans believed that reproductive organs were more philosophically and theologically constant and fundamental than other characteristics, being central to the beauty of plants and bearing the imprint of the creator.[6] The system of classification detailed by Linnaeus in the *Systema Naturae* (1735) was based upon flower characteristics, particularly the number and arrangement of stamens and carpels, classes being mainly determined by the number of stamens and orders by carpel numbers. Subsequently, in the *Species Plantarum* (1753), Linnaeus adopted the binomial system of classification which helped make botany simple and practical, fostering its popularity. Linnaean botany was championed in Britain by Sir Joseph Banks, President of the Royal Society and James Edward Smith, who purchased the Linnaean herbarium and helped to found the Linnaean Society of London.[7]

Linnaeus distinguished between the artificial method of classification and a more natural system in his *Fragments Towards a Natural Method* (1738). According to Smith, Linnaeus had recognized that proceeding on this path placed him within a 'labyrinth' because the 'natural orders and families of plants, so far from being connected in a regular series', came together in such a complex fashion 'as to bewilder instead of directing us'. According to Smith, the genera of the Linnaean system were 'really founded in nature' and in any case, Linnaeus had suggested that although a more natural system might be accomplished it was 'scarcely ever to be completely discovered'.[8] Linnaean botany was never, of course, fully discarded, but the versions of the natural system developed by Bernard de Jussieu (1699–1777) for the Royal Garden at Trianon and his nephew Antoine Laurent de Jussieu's *Genera Plantarum* (1789) impacted upon British botany by the early nineteenth century. The Jussieuian system adopted acotyledons, monocotyledons and polycotyledons from Linnaeus, but replaced the latter with Ray's dicotyledons. Jussieuian botanists and arboriculturists, particularly in the French-speaking world, searched for underlying common natural types but in practice combined these with artificial taxa. Jussieu's work was extended by other botanists, particularly with respect to the nomenclature of the characters of fruits and seeds.[9] As Watson emphasized the 'elegant introductory works' of Charles Francois Mirbel, Nicaise Auguste Desvaux, Louis Claude Richard and de Candolle were 'far superior to any thing we possess in our language' and were 'in every person's hand who made a study of this part of botany'. The works of de Candolle had 'become the classic standard' and supplanted those of Linnaeus and Willdenow, and Watson strove to present a detailed elucidation of the French master's arboricultural terminology in his *Dendrologia Britannica* (1825). Similarly, Mirbel's *Elements de physiologie et de botanique* (1815) was widely employed by Watson. The inspiration from French arboriculture is also evident from the number of French botanical works translated at this time. These included Lindley's rendering of Richards' *Observations on Fruits and Seeds* (London, 1819), which Watson contended was 'by far the most difficult part of botany' yet the most poorly served in the English language.

The most influential version of the natural system prevailing was that of the Swiss natural philosopher and botanist de Candolle. Professor of botany at Montpellier from 1808, de Candolle composed works on lichens (1797), the medical application of plants, and produced a much enlarged edition of Lamarck's *Flore francaise* (1805). Resident between 1798 and 1808 in Paris and between 1808 and 1816 in Montpellier, he was employed by the French government to undertake a six-year botanical and agricultural survey of the nation between 1806 and 1812, before returning to Geneva to assume the position of professor of botany. His version of the natural system was detailed in the *Theorie elementaire de la botanique* (1813), the *Regni Vegetabilis Systema Naturale*

(1818–21) and the many volumes of the *Prodromus Systematis Naturalis Regni Vegetabilis* from 1824. Parts of the former were translated into English and published in 1821. Drawing upon his experience of other parts of natural history and natural philosophy including chemistry and mineralogy, de Candolle strove to develop and expound the principles underpinning the Jussieuian natural system. He coined the term taxonomy to describe the development of natural botanical categories formed according to morphological rather than *a priori* or physiological principles. Although not always consistent, he placed emphasis upon divisions between taxa regarding them as discrete entities that had existed through time rather than interlocking or connected divisions. Hence, when discussing relationships he used the analogy of political divisions upon a map of a country to represent discrete taxa, differences in closeness of affinity being akin to distance from population centres, with extremities being unpopulated.[10] According to Lindley, by the 1840s de Candolle's methods had 'now almost superseded all others' partly through their 'easiness and simplicity', but mostly because they were followed in the *Prodromus* or 'celebrated description of species'. De Candolle stated that he had proceeded by placing 'Dicotyledons first, because they have the greatest numbers of distinct and separate organs. Then, as I find families where some of these organs become consolidated, and consequently seem to disappear, I refer them to a lower rank'. The series was devised as 'least removed from a natural sequence' and 'convenient and easy to study' although did not attach 'the least importance to it'. This was because 'the true science of general natural history consists in the study of the symmetry peculiar to each family, and of the relation which these families bear to each other'. The rest was 'merely a scaffolding, better or worse suited to accomplish that end'.[11]

Candolle's extensive work on the geographical distribution of plants and taxonomy encouraged him to develop laws of natural classification with what Sachs considered a 'clearness and depth such as no one before him had displayed' within the constraints of his continued belief in the constancy of species from creation. Although not fully and consistently applied throughout his work, for de Candolle, morphology, the basis of his natural system of classification, depended upon symmetry of form in plants, or examination of the 'relative position and numbers of the organs, disregarding physico-physiological properties as of no account' morphologically. Only the entire system of the organization of plants or the symmetry of organs defined plant classes which could be masked by difficulties in the morphological comparison of organs, or establishment of homology, owing to abortion, degeneration or adherence. De Candolle made attempts to rationalize the natural system, introducing larger subdivisions, for instance, into the class of dicotyledons, uniting families essentially related. Belief in the constancy of species encouraged him to agree with Linnaeus that the vegetable kingdom could be compared to cartographical divisions with quarters of the globe representing classes and kingdoms representing families which also facilitated representation of the natural system in systematic planted collections. As we shall see, it was de Candolle's version of the Jussieuian system that inspired Lindley and Loudon to try to rationalize and simplify the delimitation of arboricultural species.[12]

The end of the revolutionary wars helped to renew international botanical contacts between British and Continental botanists that had never been severed entirely. One key figure was Robert Brown (1773–1858) who travelled to Australia with a naval scientific expedition between 1801 and 1805, served as librarian to Joseph Banks and the Linnean Society and was appointed curator of botany at the British Museum. Although he never fully articulated his general botanical principles, Brown employed studies of plant anatomy, morphology and biogeography in the wake of Alexander von Humboldt to promote the natural system, encouraging various younger botanists to pursue the subject.[13] One of Brown's pupils William Jackson Hooker, later Regius professor of botany at Glasgow University, stayed with Dawson Turner and his family in Paris in 1814 and travelled to Switzerland and Italy and the south of France, mixing with leading botanists and exchanging ideas. The impact of this is evident in his study of mosses (1818) and especially his *Flora Scotica* the second part of which was one of the earliest British botanical textbooks to examine flowering plants with cryptogams and other orders arranged according to the natural system.[14] Similarly, Robert Sweet organized his *Hortus Suburbanus Londinensis* (1818) according to the 'Linnean system' but included references to 'natural orders', which it was claimed did not disturb 'the sequence and facility of the Linnean method', but were useful in revealing for the practical 'cultivator' the 'vegetable relations' which provided the 'greatest probability of success' in attempting unions.[15] Encouraged by Irish, Scottish and English provincial botanical societies and nurserymen, academic and institutional botany was converted to the natural system during the 1820s. John Edward Gray, botany lecturer at the Borough school of Medicine in London, taught the natural system at various metropolitan institutions and published a *Natural Arrangement of British Plants* in 1821. At Cambridge, Thomas Martyn, professor of botany and a convinced Linnaean, was succeeded in 1825 by Rev. John Stevens Henslow who was more sympathetic to newer botanical ideas.[16]

Writing during the 1820s, Loudon, John Lindley and their collaborators on Loudon's *Encyclopaedia of Plants* still maintained that no comprehensive English language introduction to the natural system existed that was suitable for botanical novices. The only options were French-language works such as de Candolle's *Théorie elementaire de la botanique*, which explained 'the principles upon which the orders of plants are constituted' or Jussieu's *Genera Plantarum* which detailed 'their characters as determined in 1789'. Although Charles Darwin studied de Candolle's natural system at Edinburgh University during the 1820s, Loudon considered that the *Theorie elementaire* had become 'too obsolete to be very useful to the tyro'. The only English-language work available that could be 'consulted ... with advantage' was Hooker's *Flora Scotica*, which provided the characters of the natural orders of Scottish plants as determined by Lindley.[17] In his influential translation of Richard's *Analyse du fruit* (1819), Lindley maintained that although the works of Linnaeus had provided materials for many of his followers, the unthinking 'veneration' for him had retarded the progress of botany. Followers of the Swedish botanist had regarded his works as being 'almost sacred' and it had been 'held almost profane either to suggest amendments' to the system or to 'expose the defects in his works'. Some had gone to 'extraordinary lengths' to 'conceal the important improvements' undertaken by 'his great opponent' Jussieu but the latter's system had an 'important advantage' over that of Linnaeus by drawing attention to 'every modification of the organs of reproduction' and opening the mind to new 'fields of investigation' enlarging the 'sphere of comprehension'.[18]

Along with Smith, Lindley and Hooker's works, it was Loudon's *Encyclopaedia of Plants* (1829), *Gardener's Magazine*, and *Magazine of Natural History* that disseminated Jussieuian methods most fully to British audiences during the 1820s and 1830s, the first and second volumes of the latter, for instance, contained detailed expositions of the natural system. Loudon explained how he had devised the plan of the *Encyclopaedia of Plants* whilst Lindley 'determined the genera and the number of species' arranged under them, 'prepared the specific characters, derivations and accentuations', wrote or examined the notes, and corrected the whole. The drawings were made by Sowerby, assisted by David Don and Messrs Loddiges of the large commercial nursery company of Hackney. Loudon and his collaborators utilized the collections of the Linnean Society and Messrs. Loddiges, and Loudon claimed 'without the herbarium of [Aylmer Bourke] Lambert and hothouses of Messrs Loddiges' the *Encyclopaedia of Plants* 'could not have been produced'.[19]

Although Loudon argued that those requiring an elementary botanical education could most effectively utilize the Linnaean arrangement in order to determine the name of an unknown plant in flower which remained 'the most perfect artificial system', he advised learning the natural system as soon as possible. The efficacy of this was obvious when identifications had to be made using other parts of plants such as leaves because only the natural system provided information concerning the general physiology, structure and other aspects. Crucially, even if the name could not be found, it was still better to use the natural system as even small parts would reveal 'something of [the] nature' of the entire plant. The Linnaean system facilitated the discovery of names whilst the natural system taught the 'natures of plants' and 'how to recognise them as belonging to certain classes of groups'. Loudon's *Encylopaedia of Plants* provided sections based upon each with a system of cross-referencing and indexing to prevent repetition.[20] According to Lindley there would always be 'more difficulty in acquiring knowledge of the natural system of botany than of the Linnaean'. The latter only skimmed the surface of things, leaving students of botany 'in the fancied possession' of easily obtained information 'of little value'. In contrast, the natural system required and encouraged a 'minute investigation of every part and every property known to exist in plants' and when understood conveyed to the mind a great deal of highly useful information. Therefore, 'whatever the difficulties may be of' becoming acquainted with the natural system, it was 'inseparable from botany' which could not usefully be 'studied without encountering them'.[21]

According to Loudon and Lindley, the purpose of the natural system was to discover the species or variety of a plant through observation of the forms and different conditions of the leaves, stems, and other parts of the bodies of plants as well as with their flowers.[22] Difficulties with the natural system were recognized stemming from the fact that characters of many orders remained 'imperfectly known' and depended upon physiological and structural analysis 'not to be determined without much labour and ... considerable ... practical skill' with microscopes and dissecting knives, not least of which were in relation to crytogams. However, the system encouraged the 'habit of viewing all natural bodies with reference to the relations' they bore to others and not as 'insulated individuals merely possessing certain peculiarities' through which they could be referred to a station in an artificial system. It revealed the 'hand of nature' rather than the mind of Linnaeus and provided useful descriptions rather than a description of particular classes in one system including the structure and physiology and characteristics of petals, flowers, seeds, ligneous matter, leaves, roots and so on, often revealing whether the products of plants had medicinal, economic or other applications.[23]

Global and British Tree Encounters

Through its union of botanical systematics with physiology and anatomy, and the inspiration provided by agricultural improvement, horticulture and trees in landscape gardening, British arboriculture demonstrated clearly how the most useful aspects of artificial and natural systems could be combined. Arboriculture, particularly the study and experience of tree places such as landscape gardens, estate plantations, botanical gardens and arboretums, hastened the crisis of incommensurability between natural and artificial systems. By the time Loudon published his *Arboretum Britannicum* in 1838, he felt no need to organize any of it according to the Linnaean system, the Candollean natural system being adopted throughout without any special justification.[24]

It was largely arboriculture in Britain that destroyed the uneasy truce that had subsisted between practical pragmatic botany and theoretical systematics, and it was arboriculture that made it difficult for botanists to ignore wider theoretical questions of natural affinity. There are a number of reasons why arboriculture encouraged the adoption of the natural system in Britain. These include the importation and study of new species and varieties from around the globe, arboricultural studies inspired by horticultural and agricultural improvement, the importance of trees in British culture and landscape gardening, the tendency to regard systematics, physiology and anatomical studies to be inextricably intertwined, the relationship between arboriculture and natural theology, and the problems of managing botanical garden tree collections between 1800 and 1850.

The huge influx of new species from around the globe between 1750 and 1850 was exciting for British arboriculture. If anything, the British turned to empire with renewed vigour after the loss of the American colonies and encouraged by French attempts to exclude them from European affairs between 1793 and 1815. Looking back from the 1840s, Lindley emphasized the profound impact of the thousands of specimens identified and collected by botanical travellers since the eighteenth century. These had demonstrated that 'the few thousand ill-examined plants which inhabit Europe gave a most imperfect idea of the vegetation of the globe' and rendered older methods of classification 'useless', although he accepted that 'in its day', the Linnaean system had 'effected a large amount of good'.[25] For Watson, by the early 1820s botany and arboriculture were in a completely 'different state' from previously when 'little else but European plants were known, and these very imperfectly' and had become a more 'philosophical' subject. He considered that perhaps 50,000 plants had been 'scientifically described' but the number was 'fast increasing by the communications of bold and scientific travellers penetrating North and South America and exploring many other regions'. Comprehension of this required a tremendous 'extent of intellect' in order to 'characterise, class and arrange the half of these' and devise an 'attendant extensive nomenclature', so that botany retained the status of a philosophic study. The amount of work required was overwhelming even though 'an immense extent of country must be traversed to furnish 3–4,000 distinct species'. Just as botanists and planters struggled to get to grips with this formidable influx and facilitated by patronage from wealthy individuals, nurseries and institutions, so natural history had become 'an indispensable part of a traveller's education', demarcating enlightened philosophical explorers from dilettantes. The 'scientific adventurer' had to be 'pre-educated' in different taxonomies and it was 'the special business' of botanical publications, botanical gardens and other scientific institutions to 'furnish the preliminary necessaries to such enlightened and adventurous travellers'.[26]

The special qualities of trees and most shrubs, which also have ligneous branches but, of course, do not tend to develop one large central trunk, ensured that they had a major impact upon British taxonomy between 1790 and 1840. Although Linnaeus had encouraged some observational work on trees amongst his followers, British botanists such as Lindley claimed that the Linnaeans had known 'little of vegetable physiology' because they had applied system rather than experimental understanding. Studies of vegetable physiology and structure fostered primarily by arboricultural observation and experiment, which were 'the foundation of all sound principles of classification', had produced a new understanding of the philosophy of botany.[27] The variation exhibited by trees, even within single species, founded upon a complex interrelation of environmental, climatological and cultivation factors, coupled with the shock of reality caused by the discovery and importation of exotics, threatened the apparent harmony displayed by abstract Linnaean perfectionism. The implications for western botany of global plant encounters in the past few centuries and how these were shaped by Linnaean botany still need to be fully elucidated. Efforts continue to be made to remove vestigial anthropocentric elements from western-centred botany and come to terms with the so-called 'latitudinal species-diversity gradient'. These are reflected in continued ignorance of tropical species compared to knowledge of those in Europe, North America and the rest of the northern hemisphere with their comparatively smaller levels of biodiversity.[28]

The difficulties presented by Linnaean arboriculture are apparent when we explore how trees and shrubs were positioned within the twenty-four classes and 113 orders, excluding the twenty-fourth class, Cryptogamia, distinguished by hidden, uncertain or non-existent sexual organs. Taking Loudon and Lindley's presentation of the Linnaean system in the first edition of the *Encyclopaedia of Plants* (1829), it is apparent that whilst there are some significant groupings of trees and shrubs in more natural orders, many are scattered across the classes and mixed with smaller plants with little in common in terms of structure, form, physiology, habitat and so forth. For example, Diandria, the second class, was distinguished by plants bearing two stamens on the flower included some hardy herbaceous bushes such as common privet (*Ligustrum vulgare*) and common lilac (*Syringia vulgaris*) and many smaller plants. The fourth class, Tetrandria, distinguished by its four stamens, included ornamental shrubs such as common holly (*Ilex aquifolium*) and foreign trees including sandalwood (*Santalum album*) and the iron tree (*Siderodendrum triflorum*) and was a 'miscellaneous assemblage of species, with few characters in common'. The very extensive fifth class, Pentandria, cut across natural orders and included exotic trees such as the coffee tree (*Coffea arabica*), the giant and enduring teak (*Tectona grandis*) and large spreading mango tree (*Mangifera indica*) of Asia. The small seventh class, Heptandria, distinguished by its seven stamens, was also a mixture of trees, shrubs and smaller plants including diandra, a trailing plant with bright yellow flowers, the prickly pisonia (*Pisonia aculeata*), a thorny West-Indian tree, and the elegant horse chestnut (*Aesculus hippocastanum*), staple of nineteenth-century parks and gardens. Other classes were equally mixed. The twelfth class, Icosandria, distinguished by many stamens which were perigynous or inserted into the calyx, included many common garden and domestic plants as varied as the genus Cactus, Rosa (roses), Rubus (brambles) and Eucalyptus with ornamental fruit trees such as the common almond (*Prunus dulcis*), common plum (*Prunus domestica*), common cherry (*Prunus cerasus*) and common pear (*Pyrus communis*). The thirteenth class, Polyandria, with many stamens inserted under the ovary (hermaphrodite flowers) included several extensive natural orders such as Ranunculaceae, Magnoliacaea, clematis, peonies, poppies, delphiniums and the tulip tree. The seventeenth class, Diadelphia, distinguished by stamens united in two separate parcels inconsistently mixed artificial and natural characters. It included many important agricultural plants such as common peas (*Pisum sativum*), kidney beans (*Phaseolus vulgaris*) and chick peas (*Cicer arietinum*) alongside numerous ornamental trees and shrubs such as the Indian red sandalwood (*Pterocarpus santalinus*), the common laburnum (*Cytisus laburnum*), and the fast-growing locust tree (*Robinia pseudoacacia*). The nineteenth class, Syngenesia, distinguished by five stamens and anthers united by their edges, was an extensive and well defined Linnaean division but still placed popular ornamental garden plants such as dahlias (*Dahlia*), chrysanthemums (*Chrysanthemum coronarium*), marigolds (*Calendula*) and sunflowers (*Helianthus*) with trees and bushes. Many Syngenesian plants grew as herbaceous or low bushes in the northern hemisphere and towering trees in the tropics illustrating the problem of establishing universal distinctions between the two (and an advantage of the Linnaean system).[29]

The Monoecian, Dioecian and Polygamian classes contained the largest concentrations of hardy British trees and shrubs, but still intermixed with 'a variety of plants of all kinds, natures and affinities'. The former, distinguished by the presence of male and female organs in distinct flowers upon the same plant embraced 'nearly all the most important timber trees' of temperate countries, such as the oak (*Quercus*), pine (*Pinus*), Fir (*Abies*), birch (*Betula*), hornbeam (*Carpinus*), beech (*Fagus*), plane (*Platanus*), box (*Buxus*), mulberry (*Morus*) and cyprus (*Cupressus*). However, it also included duck weed (*Lemna*), annual plants that float upon stagnant water, bur reed (*Sparganium ramosum*), plants with long ribbon-like leaves that live in ditches or ponds and carex (*Carex*), another family of plants that favoured wet swampy ground, bogs or fens, coarse grasses, nettles (*Urtica*) and maize (*Zea*), none of which had any obvious similarities with large timber or ornamental trees except for one part of their floral structure. Similarly the Dioecian class, distinguished by male and female flowers upon different plants, included trees and shrubs such as junipers (*Juniperus*), yew (*Taxus*), weeping willow (*Salix babylonica*), Norfolk Island pine (*Araucaria excelsa*), and the poplar (*Populus*) alongside mistletoe (*Viscum album*), Candleberry myrtle (*Myrica*), hemp (*Cannabis*) and montinia (*Montinia caryophyllacea*) 'a little worthless weed-like Cape plant'. Finally Polygamia, the twenty-third class, distinguished by flowers male, female or hermaphrodite upon the same or different plants, was equally heterogeneous placing many grasses, the mimosa or sensitive plant (*Mimosa sensitiva*), maple (*Acer*), ash (*Fraxinus*), acacia (*Acacia)* and the common fig (*Ficus carica*) with ornamental sub-tropical plants such as the dwarf fan palm (*Chamaerops humilis*).[30]

It is true that such dissimilarities between plants within the Linnaean classes – and to a lesser degree orders – were taken as a demonstration of scientific merit and superior botanical discernment. In many respects in terms of structure, growth, nutrition and reproduction, trees do, of course, resemble smaller flowering plants upon which much previous botanical study had been founded. Such fundamental differences help to explain why pre-Linnaean classification systems tended to group them together.[31] However, as Loudon's spatial representation of the Linnaean system suggests (see Figure 2.1), it was difficult to reconcile its artificial botanical groupings with landscape gardening aesthetics, especially with regard to the scattering of trees and shrubs within the higher taxa. Uvedale Price warned against those who thought that they could achieve variety by exhibiting 'in one body all the hard names of the Linnaean System'. Whilst 'in a botanical light' this might be 'extremely curious and entertaining', such a 'collection of hardy exotics' was 'part of the improver's pallet' rather than a picture. This resulted in 'a sameness of a different kind, but not less truly a sameness than would arise from there being no diversity at all'. More natural arrangements could be more easily presented as picturesque clumps of trees and shrubs (see Figure 2.2).[32]

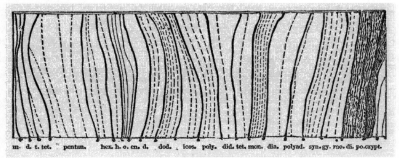

Figure 2.1: Spatial manifestation of Linnaean arrangement from J. C. Loudon, *Encyclopaedia of Gardening* (1830), p. 825. Reproduced from the authors' personal copy.

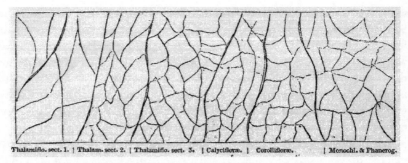

Thalamiflo. sect. 1. | Thalam. sect. 2. | Thalamiflo. sect. 3. | Calyciflоræ. | Corolliflоræ. | Mcnochl. & Phanerog.

Figure 2.2: Spatial manifestation of Jussieuian arrangement from J. C. Loudon, *Encyclopaedia of Gardening* **(1830), p. 825. Reproduced from the authors' personal copy**

Trees can be divided into two common natural types, broadleaf varieties being in the dictotyledenous group of hardwood plants and coniferous trees being mostly softwood and evergreen with narrow, needle like and leathery leaves as part of the gymnosperm flowering plants. However, there are many other ways in which trees have been regarded as forming a distinctive group. Their large ligneous stems, of course, distinguish them from smaller and far more ephemeral flowering plants, and they usually have a much longer life cycle. In marked contrast with herbs, grasses and most other flowering plants, trees grow relatively slowly when young and produce highly varied quantities and qualities of seed at fluctuating intervals rather than annually. As we have seen, these characteristics were taken to define dicotyledenous plants in the Jussieuian system.[33] Tree reproduction normally occurs naturally by seed but can also take place by vegetative reproduction from a portion of the living tree, coppice shoots from stumps, and in some cases, such as English elms and the locust tree, sucker shoots from roots can also be important. Grafting, of course, necessitates human intervention. The sexual reproduction of trees is similar to that of other flowering plants, although there are more differences amongst conifers which are gymnosperms or 'naked seed plants' where the ovule is exposed on the scale of the female flower at the time of fertilization instead of being concealed within a carpel, as in the case of angiosperms. As the cone grows and becomes green and swollen, then hard and woody, the developing seed become concealed, not coming to view until the cone ripens and the seed falls out.[34] Conifer flowers are male and female separate structures, and usually (but not always) found on the same tree. Pollination always occurs by wind and never insects, and the flowers are therefore not intended to attract insects. Some cones ripen in a single season, others, including pines, require a couple of years to do so.[35]

Conifers, in some important respects, posed greater difficulties for Linnaean classification because of their frequently insignificant flowers. In contrast, most broadleaved trees were insect pollinated and structured accordingly in a manner appreciable to those with a Linnaean botanical education with more striking flowers than many of their coniferous counterparts, both sexes being represented by single hermaphrodite flowers. However, there are variations in floral structure that might confuse amateur botanists as other broadleaf tree flowers were wind-pollinated and bore catkins, that is greenish clusters of single-sexed flowers. The male flowers usually incorporated green bracts or scales and anthers, female versions having similar scales and a pistil. Most catkin-bearing trees bore, on smaller branches, male and female flowers together, although some trees only bore these flowers separately. Broadleaf trees also develop distinctive fruits from female flowers such as apples, cherries and acorns, seeds being sometimes dispersed by wind (as in sycamore), but more usually scattered by animals and birds. These variations between and within coniferous and deciduous varieties mitigated against identification by flower parts alone, encouraging broader character descriptions, fostering doubts about the arboricultural efficacy of Linnaean botany and encouraging arboretum planting.[36] The 'discovery' and importation of novel trees and shrubs to Europe exacerbated the difficulties for Linnaean arboriculture because most of these did not bear flowers during their first few years of growth as seedlings and saplings, whilst others did not flower until they were over twenty years old. Even when trees became mature enough to flower, the season in which they appeared tended to be very short. Given that most introductions of new varieties occurred through the growth of seeds, cuttings, seedlings or saplings because of the obvious difficulties presented by transportation, and that only very experienced botanists could identify trees from seedlings, these youthful trees and shrubs were hardly ever mature enough for Linnaean identification. The centrality of flowers to the Linnaean system therefore meant that despite the distinctive nature of many tree flowers, observation of other characteristics were often necessary for arboricultural categorization. As early nineteenth-century British botanists such as Watson noted, Linnaean efforts to identify arboricultural families such as *Amentacea* and *Coniferae* were considerably hampered by the fact that 'many species, enumerated in our catalogues' did 'not flower, and of course not fruit in our island'. Those that intended to study many exotic hardy trees were therefore required to 'have resource to the floras and itineraries' of their indigenous countries or to 'perfect specimens', or had to 'wait until the extant works have been consolidated into one uniform publication'. Watson provided 'tables of floration' in his *Dendrologia Britannica* (1825) intended to 'enable the cultivator to regulate his expectations as to what exotic trees will really flower'.[37]

North American genera discovered by western botanists and explorers between 1750 and 1850 presented acute difficulties to Linnaean botany because of the small and inconspicuous nature and periodicity of tree flowers. Excluding modern Mexico and the tropical regions, there are an estimated fifteen genera of

North American conifers containing ninety-five native species. Monocotyledons are represented by only a few forms, including the odd palm, palmatos and some yuccas. Dicotyledons, the other group of angiosperms however, are represented by some hundred genera. They are spread across and extremely varied continental forest with regions of tropical, coastland, the Rocky Mountains, Pacific coast and desert. There are also the vast northern and eastern forest regions sweeping up through Canada towards the Arctic Circle. Botanists in North America became frustrated with the Linnaean system and adopted replacement or supplementary means of identification and classification. The American edition of Smith's *Grammar of Botany* (1822), for instance, included a 'reduction of all the genera contained in the catalogue of North American plants to the natural families' of Jussieu by Henry Muhlenberg originally prepared by Abbe Correa for botany lectures in Philadelphia in 1815.[38] A modern guide to the identification of North American trees emphasizes that whilst reproductive organs provide the 'most accurate means of identifying most trees', they are 'little used in the field because the period of bloom is so short'. Instead, a large number of other characteristics are recommended as more useful, although these vary considerably according to age and environment. These include height, diameter, the nature and character of fruit, seeds, the qualities and appearance of bark and twigs and the character of the wood itself. The appearance of leaves using their characteristic patterns was probably the most useful means of identifying both deciduous and evergreen trees, but of course, because they were absent for five months from the former there were limitations. A further set of 'silvical' characteristics can also be decisive including the tolerance of trees to particular conditions, sites of growth, associated flora and fauna, enemies, roots, life zones and the altitudinal distribution of species, although similar habitats can, of course, produce parallel features in different species.[39]

Trees display a very wide range of size and manner of growth even within single species and amongst the progeny of a single plant and there are often major differences between wild, semi-wild and domestically cultivated examples. This is apparent when we consider the breadth and character of tree life cycles and how differently most seedlings develop compared with smaller plants, particularly the fact that the juvenile leaves which follow the cotyledons, or hypocotyls take a distinctive form from successor adult leaves. Although the juvenile stage might only last for a short period, identification of tree seedlings can therefore be very difficult, requiring attention to all characteristics. It might therefore have to be left until they mature, which was not aided by the fact that most gardening and arboricultural works failed to provide descriptions or illustrations of seedlings.[40] Furthermore, unlike most of their smaller flowering cousins, trees usually take a considerable amount of time to flower and bear seed, although this varies greatly between species. Willows, for instance, take five years from seedlings to fruition stage whilst other trees such as ash and Scots Pine might often only begin seed dispersal after fifteen or twenty years and other kinds, such as beech, take much longer to do so. In practice, most trees flower and fruit after about fifteen years in temperate and average conditions, but location and manner of cultivation, such as exposed or shady locations, produce wide variations, and it might take many more decades and planting experiments to determine optimum conditions. Landowners, gardeners and others who planted private or institutional arboretums were therefore very conscious of the fact that they were creating something only fully appreciable to future generations. They could also not be sure how foreign trees would react to British conditions. The large size of trees and the length of their life cycles compared to other flowering plants meant that they might be more affected by adverse conditions with greater consequences. Even having reached the stage where they fruit and flower, the process of tree growth is not linear and uniform following sometimes apparently idiosyncratic cycles, patterns and periods. Trees experience alternately bountiful and barren years of abundant and scarce seed caused by a range of climatic, geological and environmental factors which cannot be predicted by even the most meticulous observers.[41]

To a greater degree than other plants, the growth of individual trees reflects heredity and multiple factors of environment, such as quality of soil, competition, light, rainfall and geology, which was increasingly recognized between 1750 and 1850.[42] However, although there have been considerable fluctuations in previous centuries, the fact that the British climate was essentially temperate and mild with annual seasonal changes encouraged the planting of arboretums and the study of trees. The mild climate of the south and west, such as Cornwall, Devon and the Channel Islands, which hardly experiences any hard frosts, allowed semi-tropical trees and shrubs to thrive, which is reflected in the number of tree collections successfully cultivated there. On the other hand, the climate of northern and mountainous zones such as the Scottish Highlands and Northumberland provided ideal conditions for many cold-loving trees such as North American coniferae, and Erasmus Darwin suggested that they could accommodate vast plantations which might be harvested. Overall, climatic variations often found across continents were exhibited in microcosm within the British Isles resulting in seasonal tree growth and wood formation. Where climatic conditions made the planting of tropical and semi-tropical tree and shrub collections outside impossible, British industrialization and particularly the development of iron and glass structures, closely tied to agricultural and horticultural improvement, provided the answer. It was not to house the latest examples of industrial machinery that the mighty central circular transept roof of the Great Exhibition in Crystal Palace was constructed in 1850 but to accommodate large veteran Hyde Park trees.[43]

Tree Physiology, Anatomy and Taxonomy

Inspired by internal traditions of practical arboriculture, scientific experiment and French botany, British late-Georgian natural history also emphasized the union and interdependency between taxonomy, field botany, living plant physiology and anatomy, which tended to differentiate it from Linnaean and Jussieuian botany. Beyond the sexual organs, Linnaeus did not concern himself much with physiology and anatomy in his studies of plant systematics, regarding these subjects as separate and unconnected and this tendency was still reflected, to a lesser degree in Jussieu's work. Tellingly, it was only in the posthumously published second edition of his *Genera Plantarum* (1837) that Jussieu incorporated an account of plant physiology and anatomy, the first edition (1789) having paid little attention to it.[44] Studies of plant physiology and anatomy, which were pursued with the greatest avidity in France and Britain, were easier to undertake using trees. For leading British botanists such as James Edward Smith, who emphasized the scientific, economic and natural theological importance of natural history, anatomy and physiology were integral to botany, and taxonomy could not be studied in isolation of them or an appreciation of plants as unified 'organised beings' with important similarities to animals. Motivated by British conceptions of experimental natural philosophy and the importance of trees for improvements in agriculture and industry, an influential group of British investigators including Nehemiah Grew, John Evelyn, Stephen Hales, Erasmus Darwin and Thomas Andrew Knight explored plant anatomy and physiology by studying and experimenting upon living and dead trees. Encouraged by French philosophers, they constantly drew analogies between animals and plants, Darwin, Smith and Knight emphasizing the inadequacies of Linnaean arboriculture. Indeed Knight initially paid little attention to botanical textbooks he was so determined to learn from observations and experiments on living trees, especially the movement of sap, to uncover the laws that governed their growth. This practical experimental and observational approach, drawing upon the experience of French and British agriculture and horticulture was contrasted with Linnaean methods.[45]

The role of British practical arboriculture in promoting the adoption of natural systems is equally evident in the first systematic English-language works on the subject, such as Watson's *Dendrologia Britannica* (1825), which we will examine further. Although motivated primarily by identification and classificational requirements, Watson paid some attention to morphology and growth, encouraged by close and deliberate attention to individual living specimens. Watson was also inspired by contemporary French botany and German *Naturphilosophie*, which, partly inspired by Goethe's works on plant growth, emphasized morphology and structure rather than post-Linnaean single, external characteristics. The vitality of trees as living, changing entities is also manifest in Watson's emphasis on the force of 'impulsion' or growth, and he praised Carl Ludwig Willdenow's *Berlinische Baumzucht* (2nd edn, 1811) for its analysis of the 'development of all the parts' of trees and shrubs 'from nature' in addition to its analysis of floration. All ought to be: 'above all ... struck with the force of nature in the production of objects, in many cases so extremely bulking, from such minute origins, replete with such various and often contrary properties'. This was true even where trees grew close to each other imbibing 'as far as we can perceive, the same substances, fluids, and gasses from the earth and atmosphere', yet some providing 'nutritive and agreeable food' whilst others yielded 'most deadly poison'.[46] Equally redolent of *naturphilosophie*, Watson asserted that it was apparent that

> each being, animal or vegetable, has a distinct and peculiar power (property) to separate, combine and modify the few original elements (perhaps only one! Newton thought three) so as to constitute, when aided by that wonderful vital principle (life) whose essence is not in our power to comprehend, their various forms, figures, properties and products.[47]

Watson also described how trees radically changed and developed through the different stages of their life cycles, most starkly represented by the fact that 'these often immense masses' had been the 'offspring most frequently of very small seeds, whose embryos are frequently invisible to the naked eye'. What 'apparent relation' was there between 'the embryo of the minute seed of the birch (betula) or poplar (populus) or even of elms (ulmus) oaks (quercus)' and the 'various progenies that nature elicits from those minute bodies, by her unknown modifications of their pristine, material elements'. For Watson, a 'good work on the production of forest trees in the various sites and soils of our island' which paid attention to their 'impulsion (growth) [and] morbidity at stated epochs', was 'yet a desideratum'. Partly motivated by economic considerations, he was anxious to communicate 'information on cubic density, tenacity, durability and various uses of such timber trees, as can be grown in Britain'. Like his close friend Adrian Haworth, whose work on *Lepidoptera Britannica* (1803–28) seems to have provided the model for the *Dendrologia Britannica*, Watson emphasized the importance of ascertaining identification from living examples. He contended that whilst 'directions respecting propagation and cultivation' might be had by 'perusing the popular treatises on gardening' such works were 'generally very deficient in botanical identification'. He therefore devoted his attention 'to the latter arduous but necessary task', presenting entirely original descriptions and observations, which, as well as the drawings, were always made from identical specimens 'preventing that confusion too evident in many otherwise valuable productions', many of which were compilations.[48]

Conclusion

Whilst long-established varieties of indigenous trees presented few problems even for arboricultural amateurs, the importation of multiple foreign varieties required more specialist knowledge. Identification was of fundamental importance as even some closely-related varieties could look markedly different when mature. The Linnaean system, based on the successful identification of flowers, was of limited use for tree identification, providing virtually no information concerning the qualities, attributes and characteristics of trees and shrubs necessary for landscape gardeners and their patrons to visualize how incorporation could be successfully accomplished. It was necessary to know, for instance, the different shapes and colours that trees assumed through their growth cycle, annual periodicity, the type of soil and other conditions under which they flourished best, the size they grew to, the nature of roots and bark, the ease with which they could be transplanted and other matters. The adoption and development of 'natural systems' was encouraged by the importance of trees and shrubs in British culture, society and economy between 1750 and 1850 and the perception of the economic value of trees and their products which facilitated study of tree anatomy, physiology, experiment and observation. Although the Linnaean system remained useful for identification purposes, fuller knowledge of characteristics was much more useful in landscape gardening, helping to explain why landscape gardeners such as Loudon took a keen interest in Jussieu's work. The union of systematics with plant physiology and anatomy also facilitated the adoption and development of the natural system, and inspired by French botanists, British botanists became the most assiduous promoters of the latter. It is also likely that general British suspicion towards abstract Continental philosophical systems and emphasis upon experiment, empiricism and individual experience also contributed towards the adoption of the natural system just as it encouraged practical studies of arboricultural structure and physiology.

Finally, the use and importance of arboriculture in British landscape gardening also played a major part in the introduction and development of the natural system. Romanticism fostered an emphasis upon trees as whole living organisms, whilst Buffon, Goethe and the *naturphilosophen* facilitated a new aesthetic appreciation of trees, although this conflicted at times with the rationalism and experiment of studies on arboricultural physiology and anatomy. Just as woodland was being lost due to the demands of agriculture and industry and the results of enclosure so trees assumed a greater part in landscape gardening for aesthetic and practical purposes, particularly as the ideas of Uvedale Price and later Loudon gained currency represented by the picturesque and the gardenesque. The picturesque emphasized the importance of the effects of plantations as a totality, the impact of trees as individual specimens with their own unique seasonally changing features and life cycles, and competition and interaction between trees. Knarled, knotty and twisted trunks and freakish features set in 'naturalistic' – but artificial if necessary – craggy eminences were celebrated in Gothic and Romantic literature. Hence it was landscape and botanical artists who frequently came to have a greater appreciation and conception of trees as organic beings than many botanists during the first half of the nineteenth century. Trees were also, of course, important symbols of English and British identity despite the lack of tree coverage compared to most other European countries. Trees provided meaning, purpose and direction and a degree of controlled uncertainty in garden designs as well as screening, shelter from sun and rain and demarcating spaces within and between parks and gardens. Specimen trees provided botanical and aesthetic interest and contrast with their growth cycles and patterns, seasonal changes, colours, flowers, bark and other features. Indigenous and long-established varieties such as the oak and the elm remained of fundamental importance, but the multiple exotics transformed and challenged landscape gardening aesthetics and practices. Exotic trees were celebrated centrepieces in botanical gardens, arboretums and private tree collections and served as specimens for scientific study and education.

3 BRITISH ARBORICULTURE *c.* 1800–35

Loudon's *Arboretum Britannicum* has had a profound impact on the history of British arboriculture ever since its publication, unavoidably obscuring arboricultural works published prior to this and shaping subsequent interpretations of nineteenth-century tree collections. To fully contextualize Loudon's *magnum opus* and to appreciate its significance in the history of natural history we will examine some preceding arboretums and arboricultural works. This chapter argues that botanical society gardens were the first type of public and semi-public urban institutions with significant tree and shrub collections prior to the 1820s and that the challenges of applying Linnaean botany to arboriculture were first confronted leading to experiments with natural arrangements. Newly imported trees and shrubs were eagerly sought and presented but challenged taxonomic ideas and gardening practices, underscoring tensions between practical botany, education, popular natural history and publications. From the late eighteenth century, British botanical gardens began to present natural alongside Linnaean arrangements which impacted upon arboricultural publications such as Aylmer Bourke Lambert's *Genus Pinus* (1802–24) and Peter William Watson's *Dendrologia Britannica* (1825).

Older purposes of physic gardens were not entirely forgotten and, aided by active patronage from medical men, the potential to exploit plants for medicinal uses remained important, rhetorically if not often in practice. However, British botanical gardens and arboretums helped to develop and satisfy a new audience for botanical and arboricultural education. Private and semi-private institutions relied upon income from members, patrons and visitors whose expectations of the gardens had to be satisfied. They also had to meet the needs of natural philosophers, botanists, landowners, landscape gardeners and others with more formal stakes in arboriculture. Hence, public expectation required that aesthetics and science be married with regard to the design, layout, planting and uses of gardens. This is why they fostered the movement towards more naturalistic taxonomies before English-language botanical textbooks. Flowering plants and herbs had typified many earlier physic gardens but provided insufficient variety for nineteenth-century botanical audiences and so, to preserve income levels if nothing else, more trees and shrubs were introduced. Botanical gardens and arboretums were expected to be places for utilitarian investigations of plant systematics, anatomy and physiology, and trees and shrubs were the principal subject for these experiments and observations. With their living tree and shrub collections as well as dried herbariums (which remained crucial), botanical garden arboretums were also regarded as places where landscape gardening and gardening plants could be selected and studied.[1]

The Establishment of British Botanical Society Gardens

From the seventeenth and eighteenth centuries, many British botanical gardens included specially designated areas for trees and shrubs. The earliest examples at Oxford, Edinburgh and Cambridge were associated with universities, whilst other botanical gardens such as Glasnevin, Dublin were partly motivated by nationalism. The appearance of botanical societies and the need to raise income for gardens and other ventures such as lectures necessitated networks of patronage and support. The Horticultural Society of London, for instance, brought together gardeners, landowners, nurserymen, landscape gardeners, medical men and others as well as landed patrons with amateur and professional interests in the collections.[2] Differences between university and semi-public urban botanical gardens help to explain the relative importance of trees and shrubs in their collections. Subscribers of the latter tended to expect more pleasing displays, sometimes resulting in conflict between curators and other staff. Labelling and too rigid an adherence to systems sometimes placed curators and gardeners in conflict with subscribers and gardens that managed to combine both were especially praised by Loudon and other commentators (Figure 3.1). University botanical gardens were usually directed by a professor of botany and managed by head gardeners or curators, whilst the organization of semi-public gardens was placed under the direction of committees at the behest of paying subscribers. Curators in both institutions had to manage and augment collections. University botanical garden staff were expected to keep herbariums of dried specimens in order to supply academics for botanical and medical experiments and lectures. Subscription botanical gardens raised money by selling printed catalogues, holding special events, charging for admission, and selling and exchanging seeds, cuttings and plants with private individuals and commercial nurseries. They therefore tended to be more commercial ventures than university equivalents.[3] Some British botanists complained that the need to satisfy patrons and subscribers impacted detrimentally upon the collections by encouraging the selection of beautiful and striking rather than botanically significant varieties, thereby tolerating amateurism. In this respect, despite the fact that Britain had the advantage of colonial trade, exploration and importation, Continental botanical gardens were regarded as superior by some British botanists, such as Peter Watson. As the principal European gardens in Paris, Vienna, Berlin were 'under the direction of professors, who are in some measure accountable to the public for the accuracy of the names' so although 'they may be deficient in many of the Tropic, Cape and Australasian species', yet the scientific value was often greater. Watson argued that it was all too easy to gratify the senses 'at much lighter expense from ornamental borders, containing about a dozen sorts of the most striking, gaudy species; but to the naturalist, the most insignificant are of equal value as forming a link in the great chain of nature'.[4]

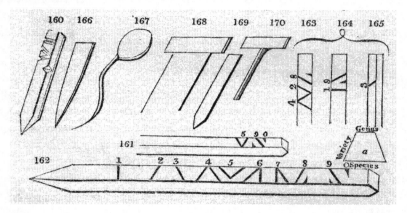

Figure 3.1: Different forms of ground tally for demarcating divisions between genera, species and varieties, J. C. Loudon, *Encyclopaedia of Gardening* (1830), p. 280. Reproduced from the authors' personal copy

Commercial nurseries also played an important role in the introduction of new trees and shrubs but could not usually do so with the authority of botanical gardens. As Loudon emphasized, the role of defining the nomenclature and history of plants 'to all eager enquirers' of scientific education helped to 'induce a taste for botany' by 'pointing out striking peculiarities of plants to superficial observers', utilizing things best that were likely to attract their attention through novelty whilst recognizing that sexual subjects and 'matters bordering on the marvellous are the most generally attractive to volatile or vacant minds'.[5] Loudon maintained that by the 1820s, botanical gardens were lynchpins of a new botanical educational network and suggested that printed catalogues should be produced and exchanged by British and international institutions so that 'by comparison of riches', global exchange could be made 'for mutual advantage'.[6] Botanical and nursery garden tree collections most impressed Loudon and served as the model for his public arboretums.

Tree Collections in Botanical Gardens

Before the advent of semi-public botanical gardens it was difficult for the British public to gain detailed scientific knowledge of trees and shrubs. Until the publication of Loudon's *Arboretum Britannicum* there was no accessible comprehensive work on the subject in the English language, treatises such as Lambert's *Genus Pinus* and Watson's *Dendrologia* being too expensive for most botanists or individuals with botanical interests to afford. Tree collections in botanical, horticultural, nursery and private gardens remained the only means by which botanists and most of the public could gain access to systematic arboricultural knowledge. Twenty years after the first edition of Lambert's *Genus Pinus*, Watson emphasized how few British arboricultural works were available. Some of these were general botanical journals such as James Edward Smith and James Sowerby's *English Botany* (1790) in thirty-six volumes or Henry Charles Andrews's *The Botanist's Repository for New and Rare Plants* (1797) in ten volumes or C. Loddiges & Sons *Botanical Cabinet* (1817) which together featured thousands of coloured illustrations of plants, including many trees and shrubs, although flowering plants was their primary focus. *The Botanist's Repository* included 664 coloured plates 'generally from hot climates' with 'a few showy, hardy shrubs ... amongst the number' at a cost of £30. *English Botany* was a 'very extensive, useful and indispensable work to the indigenous botanist' which provided thousands of coloured plates illustrating native British plants including about 100 trees and shrubs at a cost of £5 6s. 6d. Watson emphasized that although these works were 'an indispensable part of the apparatus of the student', because arboricultural analysis was scattered across them, students of the 'dendrologic department of botany (hardy trees and shrubs)' faced a difficult task in utilizing these 'voluminous and expensive works'.[7]

First established as a small physic garden near Holyrood Palace in 1670, the Edinburgh Botanic Garden illustrates the importance of botanical gardens in the development of British arboriculture during the eighteenth and nineteenth centuries. Many trees and shrubs were planted at the new five acre Leith Walk site from 1763 under the Regius keeper and professor of botany, John Hope (1725–86), who had studied botany in Paris under Bernard de Jussieu and combined interests in plant physiology with taxonomy. Under Hope and John Williamson, the principal gardener, experiments on the ascent and circulation of sap in different meteorological conditions similar to those previously conducted by Stephen Hales and later undertaken by Erasmus Darwin and Knight were conducted upon trees in the garden, multiple incisions being made and closely observed. Also under Hope a plan was prepared by Principal Gardener Malcolm McCoig in the early 1780s to redesignate what was known as the 'Autumnal Area' of trees as a 'Fruticetum Arboretum', which may be the first time that a planted tree collection designated as an arboretum was contemplated, however it never seems to have been fully executed. McCoig had interests in systematic botany and was preparing a *Flora Edinburgensis* of plants growing wild within fourteen miles of the city with Linnaean and English names when he died in 1789.[8]

The movement of the Botanical Gardens to a much larger fourteen and a half acre site on the Broompark estate at Inverleith in 1820 provided an opportunity to nurture a larger collection of trees and shrubs. Taking almost two years, large well-established trees were transplanted through the streets of Edinburgh and replanted on the new site including a 43-ft cut-leaved alder (*Alnus glutinosa var.*) and a 40-ft weeping birch (*Betula pendula*). Under Robert Graham (1786–1845) and William McNab (1780–1848) during the 1820s and 1830s, the new Botanical Gardens became a centre of Scottish field botany and international plant gathering and the importation of new species, which had to be accommodated in a completely new site at Inverleith with extensive hot houses placed under Barron's direction. McNab, who had been gardener at the Royal Botanic Gardens, Kew (1801–10), had collected an extensive herbarium and was regarded as 'the greatest practical gardener in Europe', whilst his son, James McNab (1810–78), who succeeded him as curator at Edinburgh, journeyed to North America where he collected numerous new specimens. The McNabs developed special expertise on conifers, William publishing *Hints on the Planting and General Treatment of Hardy Evergreens in the Climate of Scotland* (1830) whilst James designed and planted a pinetum, transplanting trees from other parts of the garden to form distinct groups. The impact of botany and arboriculture at Edinburgh in this period is evident in the career of William Barron who managed the Inverleith hothouses before obtaining his appointment at Elvaston.[9]

By the 1820s, semi-public botanical gardens at Glasgow, Dublin, Cork, Hull, Liverpool, Manchester and Birmingham invariably incorporated major tree collections whilst Cambridge renewed its arboretum.[10] The design of these arboretums was strongly determined by the demands of botanical education and plant experiment and the need to attract visitors and subscribers. Beyond the requirements of botanical education, taxonomy and economic utility, it was believed that botanical gardens would lead the mind towards higher pursuits, exalting the liberal arts, refining the character and, following Paley's authority, the contemplation of the divine as the 'supreme and intelligent author' of creation.[11] Changing arrangements within collections was stimulated by the demands of botany and scientific education and the need to present enlightenment taxonomies. However, the breadth of professional and amateur interests apparent in both London and provincial botanical societies, and the need to attract subscribers and paying visitors, meant that there was pressure to apply the principles and practices of landscape gardening to create visually interesting and attractive places for the fashionable to stroll.

Some of the earliest examples of semi-public botanical gardens were in Dublin and provided important statements of Irish cultural identity just when the Irish parliament was being dissolved against the wishes of some nationalists and Anglo-Irish loyalists such as the speaker of the parliament. The garden of the Botanical Society at Glasnevin was established in 1798 and largely designed by the physician and professor of botany, Walter Wade (Figure 3.2). It included a 'Hortus Linnaeensis' (a), a 'Hortus Jussieuensis' (b) comprising all the orders of hardy specimens in Britain, 'Hortus Hibernicus' (c) of native Irish plants, 'Hortus Exoticus' (p) and the professor's house and lecture room with library and hortus siccus (q). Indeed, this may be the first time that the term 'arboretum' is used to designate a specific place where trees are grown rather than being a retrospective application. The arboretum took the form of a linear strip along the south west side of the gardens, a similar width to and adjoining the 'plantation skreen'; within this border, the arboretum was intermixed with a fruticetum and takes up much of the south western half of the garden.[12] Begun in 1807 and designed by Mackay, the botanic garden at Trinity College in Dublin featured separate arrangements for trees, shrubs and herbaceous plants arranged according to the Linnaean system and medicinal plants arranged by the natural system. Each specimen was labelled with systematic and English names following the classes and orders of both the Linnaean and natural systems. In 1833, the significance of the tree collection was enhanced when two more acres were laid out which were 'principally occupied by ornamental trees on a grass lawn with surrounding borders for showy herbaceous plants and trees and shrubs which require the protection of a wall'.[13]

Two botanical society gardens established in the major port towns of Liverpool (1802) and Hull (1812) in the wake of the Glasnevin foundation set the pattern for tree collections in such institutions between 1800 and 1850. Liverpool had grown from a relatively small early eighteenth-century town into the largest British port as the docks grew, whilst Hull was the fourth largest port by the early nineteenth century. Although economic prosperity did not immediately translate into the growth and enrichment of urban public culture in Hull as it had at Liverpool where literary and philosophical culture flourished, the foundation of the botanic gardens in the former can be associated with the appearance of a subscription library, zoological gardens and a Hull Literary and Philosophical Society in the early nineteenth century. Although inspired by local naturalists and patrons of the sciences with connections to merchant elites including William Roscoe, William Spence (*c.* 1782–1860), Adrian Hardy Haworth (1768–1833) and Peter William Watson, the closely-related subscription botanical gardens in Liverpool and Hull were important contributions to

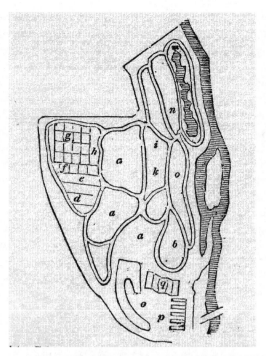

Figure 3.2: Plan of the Dublin Society botanic garden at Glasnevin
from J. C. Loudon, *Encyclopaedia of Gardening* (1830), p. 1097.
Reproduced from the authors' personal copy

middle-class public urban culture, helping to affirm aspirations towards polite knowledge amongst the middling sort. Plant collections were also intended to serve important utilitarian purposes and the commercial classes discerned many potential benefits for maritime botanical education and the international exchange and economic exploitation of specimens.[14]

Despite their relatively small size, the Liverpool and Hull botanic gardens both had significant tree and shrub collections. The five-acre Liverpool garden was designed by the wealthy merchant, botanist and scholar Roscoe and laid out by the curator John Shepherd, who also designed the five-acre Hull gardens so that there were considerable similarities between the two (Figure 3.3). Gravel walks and different compartments including stove (2), rock plants (3), bog plants (4), conservatory (6), aquarium (7) and herbaceous plants (8) were provided in both gardens with libraries, herbariums and lodges (1) whilst William Don, nephew of the late curator of the Cambridge Botanical garden, was appointed as curator at Hull. Roscoe's opening address, which was published with a catalogue of plants and a description of the Liverpool institution, was especially influential, expounding the benefits of botanical gardens for practical botany, natural theology, scientific education, polite culture and commercial utility.[15] He emphasized that public institutions were especially able to plant and preserve trees, providing the continuity, skills and knowledge necessary for their study and exploitation and part of the Liverpool garden featured a collection of labelled 'trees, shrubs, and hardy herbaceous plants' representative of those from around the world in keeping with Liverpool's international maritime links and sufficiently hardy to survive, augment and beautify the garden. Significantly, these were arranged under their own 'general and popular division' which cut across Linnaean classes and were represented as 'A' in the plan that it contained, a simple throwback to pre-Linnaean as well as natural taxonomies. Observers could use the labels to identify individual specimens and then consult the catalogue to appreciate their classification, either through the Linnaean divisions or English language index of common names that was necessarily appended.[16] Similarly, part of the Hull collection was also not arranged according to the Linnaean plan but followed natural schemes, initially that of Christian Hendrick Persoon (1761–1836) before being changed to a Jussieuean plan and then an arrangement by Don the curator.[17]

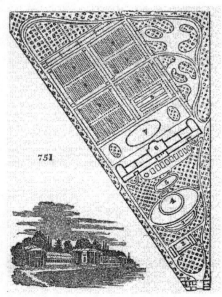

Figure 3.3: Plan of the Liverpool Botanic Garden showing hothouses
from J. C. Loudon, *Encyclopaedia of Gardening* (1830), p. 1081.
Reproduced from the authors' personal copy

'Rare and valuable' trees were placed in the Hull garden, affording shade and shelter to plants and humans alike, being visible for miles around and providing opportunities for local gentry to observe before planting in country seats. The garden included areas for bog, alpine and greenhouse plants, a pond for the growth of aquatics and a 12-ft high mound on the south west corner of the garden which allowed, though surrounded by trees 'an extensive view of the Humber, the Lincolnshire coast and the Wolds'. The Hull garden also helped to inspire Watson's *Dendrologia Britannica* (1824), one of the most important arboricultural treatises published in the period and an inspiration for Loudon's *Arboretum Britannicum*. The interrelationship between the tree collection in the gardens and the virtual arboretum in the book is also underscored by the fact that the latter provided a detailed account of the history and collections in the garden.[18] By 1814 as the report of the governing committee read at the annual general meeting emphasized, 'seventy species of trees and shrubs' had been purchased in London including many 'very rare and valuable' which 'on account of their growth it was highly desirable no time should be lost in procuring'. Very little remained to be done, it was reported, to render this 'department of the garden' which was 'unquestionably the most important', as 'nearly complete as possible'.[19]

Larger collections of trees and shrubs were established with more ambitious landscaping at the Glasgow and Sheffield botanical gardens. The Glasgow Botanic Garden was established in 1817 near Sauchiehall Street by the Royal Botanic Institution of Glasgow, and maintained very close links with Glasgow University, especially through William Hooker who was professor of botany before his removal to London. David Douglas (1798–1843) who trained at the gardens became, of course, one of the most prolific plant hunters and imported many American varieties to Britain including the Sitka Spruce, the Douglas Fir, the Red Alder and the Sugar Pine.[20] The gardens of the Royal Botanic Institution of Glasgow were the joint property of the University and a group of subscribers including local nurserymen who devised the plan of the garden. This combined aesthetic, scientific, commercial and medicinal concerns and incorporated various collections surrounded by a plantation belt with a section for medical plants for sale and others configured according to the Linnaean and Jussieuian schemes.[21]

The impact of the gardenesque upon botanical garden tree collections before Loudon's Derby Arboretum is perhaps most strongly evident at the Sheffield Botanical Gardens, designed by the Scottish landscape gardener Robert Marnock with contributions from the local architect Benjamin Brooke Taylor, and advice from Joseph Paxton (who helped adjudge Marnock's design winner of a competition) and Loudon. Occupying a hillside, the Sheffield gardens featured a large glass house, plantations around part of the perimeter and numerous labelled specimen trees and shrubs upon sweeping lawns between the gravel paths. Many of these were placed upon mounds to reveal the pattern and development of root structure, a practice, as we shall see, recommended by fellow Scotsmen Loudon and utilized by William Barron in his various commissions (Figure 3.4). It was found, however, that the soil of the gardens was 'not so well suited for conifera as that in the gardens of Birmingham', furthermore, by the 1850s, though 'at a distance of five miles from the body of the town', the smoke had a detrimental effect upon 'such clean air-loving plants'. However, some araucarias and some '*Pinus Austria*' and two or three '*Pinus excelsa*' looked 'healthy, promising plants' by then and 'all deciduous trees, including roses, thrive as well here as anywhere else'. An attempt was made to introduce a zoo into the botanical gardens and elegant large glass houses dominated views as noisy animals and birds disturbed onlookers, a pattern quite common at the time. The twelve-acre gardens of the Bristol, Clifton and West of England Zoological Society, for instance, also included, besides the animals, an arboretum representative of 'all the native British and hardy naturalised species of trees and flowering shrubs' which was 'the foundation of a botanical garden', satisfying the demands of 'shelter, ornament and science'.[22] Like most botanical gardens however, apart from on gala days, admission to the Sheffield gardens was limited to shareholders, their families and a guest, which limited their impact upon the general population, although the *Floricultural Magazine* begun by Marnock in 1836 incorporated details of trees and shrubs, conveying the gardenesque message.[23]

Figure 3.4: Tree planted on mound showing roots in Sheffield Botanical Gardens .
© Paul A. Elliott (2009)

Trees in the Horticultural Society Gardens, London

The gardens of the Horticultural Society at Turnham Green in London were regarded as 'the first in England' and most influential being visited by thousands of members, subscribers and their guests annually and serving as an important centre for training gardeners and labourers (Figure 3.5). The arboretum was laid out by William Atkinson who had assisted Sir Walter Scott in landscaping and tree planting his estate at Abbotsford in the Scottish Borders and contrasts with many botanical society tree and shrub collections because of its lack of systematic arrangement. Although established by a national scientific association with the nurseryman George Loddiges sitting on the planning committee, the design of the Horticultural Society gardens were inspired more by aesthetics than taxonomy. The buildings served as the meeting rooms of the society and occupied the position of a country house on an estate, such as those where Atkinson had gained his experience, with the view from the windows during meetings being carefully contrived in the design. The society was dependent upon the subscriptions of members and donations from patrons and therefore had to accommodate their wishes. Trees were arranged as 'clumps irregularly disposed upon turf' with further ornamental plants on the grass, whilst through the centre ran a canal supplied by a well in which aquatic plants were grown. The arrangement was admitted to be 'not systematic' although species of each genus were arranged 'as much as possible in the immediate vicinity of each other' to afford 'comparative examination'. The whole of the eastern sides consisted of separate groups of oak and elm plantations in front of a long strip that also included oaks and elms whilst paths surrounded but did not traverse the arboretum.[24]

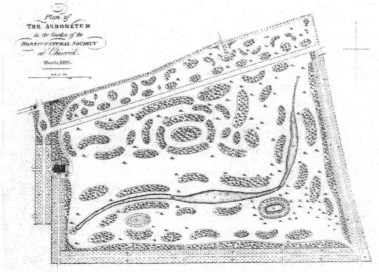

Figure 3.5: Plan of the arboretum in the Horticultural Society gardens at Chiswick, Report of the Garden Committee (1826). Reproduced by permission of the Lindley Library of the Royal Horticultural Society, London

The emphasis upon aesthetics rather than botanical principles in the arboretum of the pre-eminent English horticultural society was criticized by Loudon who noted that although the arboretum was the first of its kind in England it was 'to be regretted' that the space devoted to it 'was originally much too small'. For Loudon the Horticultural Society garden was 'most defective' in its general arrangement' and offered a 'want of grandeur and unity of effect as a whole, and of connection and convenience in the parts'. One 'obvious error that must strike every one that has had no part in making it, is the forming of the arboretum in a large rectilinear clump; and another is scattering the hot-houses and other buildings here and there over the garden'. In Loudon's view there should have been three main parts: a 'centre for all the buildings of every description'; a 'circumference displaying the arboretum, fruticetum, and ornamental flowers'; and the 'intermediate space laid out as culinary, dessert, floricultural, experimental, naturalisation and nursery gardens'. As a result of this, according to Loudon, 'the different kinds have not had an equal chance of displaying themselves, or of attaining that magnitude and character which they ought to have to answer the ends of an arboretum'. Loudon further recommended that the 'grand entrance' ought to have 'presented three carriage roads: one to the centre, to which visitors could drive and inspect the hot houses of all the departments' and survey the open gardens belonging to them and 'two others proceeding to the right and left and forming a circumferential one along which visitors might drive round the whole arboretum or shrubbery'.[25]

Lambert's *Genus Pinus* (1803) and Watson's *Dendrologia Britannica* (1824)

As we have seen, it was usually in botanical and horticultural society gardens rather than private collections between 1790 and 1840 that problems relating Linnaean and natural taxonomies to arboriculture were confronted. These tensions are also evident, to a lesser degree, in botanical and arboricultural books most of which claimed to follow the Linnaean system whilst pragmatically incorporating material that underscored the inadequacies of sexual classification such as Lambert's *Genus Pinus* (1803–24) and Watson's *Dendrologia Britannica* (1824). Encouraged by landscape art and the subject matter both works featured lavishly produced colour plates illustrating various tree parts with accompanying descriptions. As we shall see, the production of both these works was severely challenged by the impact of global encounters with novel trees and shrubs and institutional living and dried collections upon botany, arboriculture and taxonomy. This helps to explain the complexity, inconsistency and confusion surrounding the various editions of Lambert's book, the difficulties faced by Watson in producing and selling copies of his *Dendrologia*, and the prohibitive production expenses and poor sales that never remotely covered publication costs. Faced with the botanical, economic and aesthetic requirements of British arboriculture, both books included large quantities of additional information concerning tree characteristics that were unnecessary for Linnaean taxonomy.

Both Lambert and Watson were members of national and international botanical networks, the former usually being regarded as the most senior figure in English botany after James Edward Smith. Watson was more of an outsider with provincial and metropolitan connections with nurserymen and botanists, and it was the publication of the *Dendrologia Britannica* that resulted in his election as fellow of the Linnean Society in 1824. A member of the landed gentry who inherited estates in Jamaica and in Ireland through his mother, Lambert was sufficiently wealthy to survive independently and devote his life to botany whilst his colonial connections provided sources for the supply of specimens for his herbarium. At Oxford he befriended the botanists Daniel Lysons and John Sibthorp, later becoming acquainted with Joseph Banks and James Edward Smith. Friendship with the latter resulted in an invitation to become a fellow of the newly established Linnean Society and in 1796 Lambert became one of the four vice-presidents of the society. Other fellowships and membership of learned institutions in Britain and internationally followed, notably election as a fellow of the Royal Society in 1810, helping him to maintain his international scientific connections and attain the status of leading natural historian whilst publishing little. His first important botanical work was *A Description of the Genus Cinchona* (1797) written with the guidance of Joseph Banks, with illustrations by Ferdinand Bauer from specimens in the Banksian Herbarium. He also contributed to botanical periodicals such as the *Botanist's Guide through England and Wales* (1805) by Dawson Turner and L. W. Dillwyn, and Sowerby and James's *English Botany*.[26] By 1832 Lambert's list of international honours and institutional appointments included FRS, AS, GS, HS, MRAS and HMRIA, vice-president of the Linnean Society, member of the Imperial Academy Naturae-Curiosum, of the Royal Academy of Sciences, Madrid, and of the Royal Botanical Society of Ratisbon.

Lambert's status as a botanist was also reinforced by the increasing value and scientific importance of his dried herbarium and living botanical collections and the acquisition of one of the largest botanical libraries in Britain. The importance that Lambert attached to his botanical and institutional status is evident from the frontispiece engraving of the *Genus Pinus* which shows him as vice-president of the Linnean Society with a coat of arms and seated with the plate of 'Pinus Lambertiana' from the book on a table. The status of the library, natural history and living collections increasingly attracted British and international natural philosophers to his London residence and Lambert was generous in allowing visitors to utilize his unrivalled scientific materials. These were placed under the management of David Don (1799–1841), librarian at the Linnean Society, later professor of botany at King's College in London from 1836, and author of works such as the *Flora of Nepal* (1825). Lambert's status in the Linnean Society and international botany meant that like Smith, he was extremely reluctant to reject the precepts of Linnaean taxonomy which impacted upon his arboricultural work. It also explains why the attribution of authorship of the *Genus Pinus* is not straightforward; Lambert being more a compiler or editor than author in the normal sense. Although Lambert's own living and dried herbaria furnished much of the material and he wrote some of the descriptions, especially those for the first volume of the first edition (1803), many of the rest were composed by Don. Don also provided a full account of the sources for Lambert's herbarium. Lambert's role in the production of the book was more akin to that of a patron commissioning a work to celebrate their estate plantations than that of a writer, and he characteristically commissioned the plates without producing them himself.[27]

Watson's *Dendrologia*, on the other hand, was intended to be financially self-supporting and initially appeared in parts with 112 coloured drawings from living plants growing in the London vicinity completed in 1822. Publication began in January 1823 issued in royal octavo numbers on the first of every month, each containing eight coloured plates with analytic descriptions priced at four shillings and sixpence. The plates appeared in the Linnaean order which placed 'those majestic objects forest trees' towards the end of the series. The descriptions were intended to 'present as simple an appearance as the nature of the subject will admit', synonyms being included in an index at the end. The completed bound edition published in 1825 included a critical preface 'tracing up the subject', which reviewed arboricultural literature since the time of Evelyn, provided a general account of botanical principles illustrated with diagrams and presented a 'carpolegic concordance of the terms and definitions used by Gartner, Mirbel, Richard, Decandolle, Desvaux and others, as applied to fruits and seeds'. The second volume concluded with English and Latin indexes, a list of abbreviations, tables of 'properties and artificial uses of some of the plants' featured and a 'Linnean index with notices on the soil, exposure (site) and propagation of the trees and shrubs'. Watson intended to expand the work incorporating forty-eight new drawings prepared during summer 1823 provided that the sale of the parts would cover the cost of their production and claimed that by issuing 'departmental publications' dealing with particular parts of botany in parts, the cost would be reduced and sales increased, however this was not borne out by his experience.[28]

Lambert and Watson both emphasized the importance of utilizing observations of living trees and shrubs rather than merely relying upon books and dried specimens, although each made considerable use of the latter. According to Watson, dendrology (arboriculture), was a 'striking and most useful branch of botanical science' that had not hitherto been served by a single accessible book to satisfy the lovers of botany, planting, and decorative and rural scenery which allowed comparisons to be made between the delineated subjects and living trees.[29] With some exceptions, the *Dendrologia* claimed to provide drawings of 'most of the exotic trees that flower in Britain' with descriptions that followed 'strict Linnaean principles' to which the author claimed to have 'steadily adhered in all his botanic works', taking plant parts 'in the same order ... preserving the same sequential order of the parts' in contrast to the 'often vague, slovenly and reiterated way of many botanic writers'. Watson retained faith in the power and authority of single 'type' specimens in Linnaean manner rather than accepting the variety and individuality of trees and shrubs as they grew in situ, suggesting a reluctance to face the challenges posed by novel exotic trees.[30] In this, however, he followed the system pursued in works such as C. L. Willdenow's *Berlinische Baumzucht* (1811) which provided a 'plain description of each plant ... from the living subject, as far as it had developed itself in the Berlin Botanic Garden'.[31] In some ways, despite his experience developing the Hull Botanic Garden, Watson's foray from the comfort and security of the study, the dried herbarium, Aristotelian logic and Linnaean perfection remained tentative. Although he got his shoes dirty in the nurseries, botanical gardens and private collections around London, his single type specimens remained wedded to the page. The descriptions, observations and drawings were always made from the same identical specimen, supposedly preventing 'that confusion too evident in many otherwise valuable productions edited on the principle of computation'.

Lambert and Watson sought to distinguish their more 'scientific' approach to arboriculture from their predecessors primarily upon the basis of the systematic Linnaean organization, scientific descriptions and the quality and detail of their plates. Watson contrasted the science of the *Dendrologia* with the 'defective manner' and 'paucity of ... materials' of Evelyn's *Sylva* and the ponderousness and unoriginality of William Hanbury which ignored the work of Linnaeus. By 'the present state of botanic science', the 'numerous superstitious legends from early writers' detracted from their 'intrinsic merit, whilst those who had attempted to bring Evelyn's *Sylva* to date such as Alexander Hunter were criticized for 'shackling' themselves with his text and 'not compiling 'an entirely new systematic work' drawing upon the 'excellent materials of Linnaeus'.[32] Watson professed astonishment that given the 'present advanced state of botany', none in Britain since the time of Evelyn with the necessary botanical skills had taken up the 'dendrologic department of the science' and the trees and shrubs 'composing our forests, woods, and plantations; ornamenting our parks and pleasure grounds' whose qualities were hailed with the passage of every season. The 'most majestic inequalities of the globe, the most sublime mountain scenery, when naked and destitute of vegetation, and particularly of those noble objects trees, presented but 'arid sensations'.[33] Lambert claimed that 'confusion', 'difficulty and obscurity' surrounded the subject of the genus *Pinus* which he attributed to the fact that most botanists, including Linnaeus and his followers, had founded their analysis upon descriptions derived chiefly from 'dried and mutilated specimens'. No British works comprehensively treated the subject and even the *Hortus Kewensis*, although it distinguished the species 'better ... than in any other work', did not 'enumerate all that are now known' or discriminate between characters correctly. To correct this, Lambert strove to produce a 'new form of arrangement', consulting relevant botanical publications and visited 'every plantation within many miles of the metropolitan' in order to ascertain 'the most accurate specific distinctions' and to 'collect every fact relative to the culture and uses of every individual species'.[34]

Whilst professing loyalty to the Linnaean framework, Lambert and Watson recognized its limitations and emphasized the cultural, aesthetic and economic value of trees and shrubs. According to Watson the beauty of trees arose from their 'multifarious varieties of structure in stem, branches, foliage and fruit'. They provided protection 'particularly in tropic countries, from the ardent sun' and had 'manifold uses' in the 'numerous purposes of the arts' from their 'solid parts and chemical properties'. Trees supplied 'food, medicine, fuels, clothing, durable materials for the construction of our stationary habitations' in addition to providing the construction material for ships upon which, as Evelyn had emphasized, English prosperity depended.[35] The different requirements of British audiences for arboricultural works such as those by Lambert and Watson are evident from the latter's claim that the *Dendrologia* would provide an account of 'trees and shrubs that will live in the open air of Britain throughout the year' which would be 'useful to proprietors and possessors of estates, in selecting subjects for planting woods, parks and shrubberies; and also to all persons who cultivate trees and shrubs'.[36] It was clear to Lambert and Watson that Linnaean taxonomic studies on their own would never have satisfied such readers who required more comprehensive information concerning tree characteristics and qualities. On the other hand, the need to satisfy the demands of landowners, botanists and artists encouraged them to devote enormous cost and attention to the production of plates which made the *Dendrologia Britannica* and *Genus Pinus* prohibitively expensive. Despite Lambert and Watson's claims that their books would be useful for gardening and estate management, the fact that the plates consisted largely of sectional views would have limited their appeal for this market.

The plates of the *Genus Pinus* were painstakingly drawn and fully coloured. By the late eighteenth and early nineteenth centuries the standard of production of botanical illustrations had risen to a very high level allowing the publication of expensive and lavish hand-coloured books. Drawings were usually made from dried specimens in herbariums and field specimens thus combining artistic conventions and practices with botanical expertise. Knowledge of plant anatomy and physiology was required and the processes of reviving dried specimens using soaking to capture colours and form. These practices presented special problems for the production of arboricultural illustrations. It was not possible to preserve whole trees as dried specimens in herbaria so as many novel exotic types were initially only known in Britain from descriptions and small dried parts such as branches, twigs, leaves, cones or dried flowers, the application of botanical and artistic interpretations and conventions was required. This also explains why many arboricultural works featured sections rather than full tree illustrations which guided botanists and collectors in the field as to how they should record and interpret new trees. The production of botanical drawings characterized by single specimens usually with no backgrounds and frequently with dissections or depictions of cut sections, were intended to distinguish such works from more popular flower painting books which encouraged botanical plant illustration as a female drawing-room accomplishment. For Linnaean botanists they also helped to justify and stabilize orders and species, for instance by paying special attention to reproductive organs and idealized type specimens and removing supposedly incidental peculiarities. Rather than being objective representations of living trees, botanical illustrations embodied the distillation of conventions, operation of sets of rules and internalization of tacit knowledge by engravers, artists and editors. As Endersby has emphasized, illustrations both embodied and represented experimental practice and shaped the manner in which naturalists and others collected, classified and interpreted plants. They were therefore crucial to the production of knowledge and practices rather than merely the reproduction and dissemination of ideas. It was frequently asserted that the best practice for botanical illustration was to draw from life but in practice, like most introductory Linnaean texts on the market, many reproduced other illustrations from books and magazines, reinforcing rather than challenging conventions and traditions. However, new ideas in landscape design, drawing and painting did impact upon botanical drawings and contrariwise, as Thomas Sandby's incredibly detailed paintings of trees demonstrate through their attention to delineating individual botanical characteristics of different types of trees. Botanical illustrators such as James Sowerby utilized ideas and practices from landscape design, drawing and painting, recommending drawing from real life where possible, using single strokes and proceeding from outlines to define shape and then shading and colour, which it was claimed facilitated greater accuracy. Technological improvements enabled the production of cheaper mass-produced illustrated botanical works, opening up new markets and fostering new audiences, including the introduction of mechanical printing and later steam presses.

Copper plates were costly but provided unrivalled detail, especially when hand-coloured, whilst the application of woodcuts allowed botanical engravings to appear within the text of popular works, developments which John and Jane Loudon took advantage of in their publications.[37]

The Plates of the *Genus Pinus* were life size, but there were virtually no illustrations of full trees either here or in the *Dendrologia*, only sections, branches, cones, seeds, catkins and other reproductive parts. This underscores the importance placed upon Linnaean taxonomy and botanical practices and also the debt that both owed to collectors, travellers and dried herbariums. Only very occasionally a full tree was included, such as the 'Jersey Pine *Pinus inops*' drawn by Sowerby based upon a tree in the royal garden at Kew. The 1837 quarto edition included an engraving in black and white of *Pinus sylvestris* from Coombe Wood with fence and plantation in the background. The plates were generally drawn by Franz and Ferdinand Bauer with James Sowerby, George Dionysius Ehret and others. It is significant that despite the fact that only dissections were included, Goethe valued the plates tremendously and considered that they captured the essence of trees as living organisms more successfully than other illustrations. After the plates were descriptions providing details of the shape, colours, growth patterns, products and how to obtain the best results from planting, fructification, location of trees and shrubs and information on how to procure seeds. The lavish production quality of the first editions of the *Genus Pinus* restricted their circulation to the wealthiest individuals and institutions. The first two folio editions were praised for the quality of their plates and botanical descriptions and the quality of their production, which surely make them one of the most lavish arboricultural works ever published. In 1824 Watson, who made copious use of Lambert's 'very extensive library and herbarium', called for a cheaper octavo edition of this 'valuable and splendid work' which he thought would be 'gratefully received by the botanic public'.[38] Lambert eventually responded with an unsatisfactory quarto edition of 1832 this time dedicated to the King. This edition incorporated the plates used for the earlier folio edition but they were now awkwardly cut and folded to cram then into the reduced space, others had some material removed to accommodate them.

The 172 beautiful coloured plates of exotic and hardy trees and shrubs which followed the review of middle and northern European arboricultural literature and lengthy introduction to descriptive botany in the *Dendrologia* were also only sections of branches, leaves, flowers and fruit. The plates were numbered but bound alphabetically and each was also accompanied by a page of detailed description and analysis in tabular form more systematic than the information provided by Lambert.[39] Some plates seem to have been produced in a hurry and are partially coloured for no apparent reason, others were delayed. There are no complete paintings or drawings of any tree or shrub whatsoever, all the coloured and partially-coloured paintings featured leaves, flowers, and sections of trees with branches being the largest part ever reproduced. A reader who knew nothing of the appearance or shape of trees would have little idea from reading the work. Watson's concept of dendrology was in this sense different from Loudon's arboriculture which embraced trees and shrubs in their entirety as well as their component sections. Although the *Dendrologia Britannica* was original in bringing together much information concerning exotic varieties, and living specimens and private collections provided the sources, it was still restricted by dependency upon Linnaean botany and dried herbariums.

The information under the heading 'Observations' concerning the dissections with illustrations tended to increase over the course of publication. As a landscape painter, Watson took a keen interest in how the plates of trees and shrubs had been drawn and coloured, commenting and criticizing recent French and German arboricultural works in aesthetic and scientific terms, including the use of colour.[40] He produced the initial coloured drawings for the *Dendrologia* plates from trees and shrubs in the metropolitan vicinity. These were then painted for publication by J. Hart, E. D. Smith and Mrs F. Travis. Of the plates in the *Dendrologia*, 103 featured North American varieties that had been imported to Britain, the remainder including examples from southern Europe and the near East. Watson aimed to provide 'a full description and coloured drawing, from nature, of such hardy trees and shrubs as had not generally been figured by recent British botanists, at the time the drawings were made'. There were, however, important exceptions limiting the arboricultural usefulness and appeal of the work. The entire *Genus pinus* was omitted so as not to duplicate Lambert's book. Watson also ignored many trees and shrubs described in French and German works such as those by Schmidt, Wildenow, Hayne and the two Michaux on the basis that these authors 'had nearly the same object in view as himself '. However, he did include 'such exotic trees as will flower in Britain, which are not very numerous, to indicate the real state of their floration in our island'. Although these exclusions reflected the cost of commissioning new plates and the desire to avoid duplication, it seriously limited the appeal of the *Dendrologia*. The result was that the choice of trees and shrubs for the work appeared quite arbitrary. Those who required a fuller English-language overview of the subject when hundreds of exotics were being introduced needed to consult a series of European arboricultural books some of which, like Lambert's *Genus Pinus*, were only available in very expensive folio editions. This underlines why, as the only sources of non-private comprehensive tree collections, botanical society arboretums were so important. The incompleteness of Watson's coverage may also help to explain why sales were poor in both the part and completed editions, which discouraged him from continuing. Watson admitted he had presumed 'a more liberal support from the scientific public' but would continue the work as far as

160 plates even though 'its sale has by no means covered the heavy expense of publication'. He sought 'reimbursement only; profit, personal expenses, and his own labour' he would willingly sacrifice for the good of science.[41]

The *Dendrologia Britannica* provided a model for Loudon's *Arboretum Britannicum*. Loudon acknowledged that it was 'the most scientific work devoted exclusively to trees which has hitherto been published in England' and praised the quality of the 170 plates. This was at a time when 'the only work hitherto published in England which contains a description of all the hardy trees and shrubs in the country, in addition to that of all the others plants, ligneous and herbaceous, described by European botanists' was Thomas Maryn's voluminous edition of Philip Miller's *Gardener's Dictionary* (1807). Although, as Loudon observed, with the exception of the introductory summary on tree publications, the letterpress of the *Dendrologia* consisted 'solely of technical descriptions of the figures, arranged in tabular form under a given number of heads', he considered that this presented a 'very effectual mode of preventing any point, necessary to be attended to in the description of a plant, from escaping the notice of the describer'. This meant that 'in this respect', the work was 'superior to some of its contemporaries, in which the descriptions are sometimes rather disorderly if complete; and are often incomplete, apparently from want of being taken in some fixed and comprehensive order'.[42]

The impact of global botanizing and new tree discoveries is very evident in the *Genus Pinus* and the *Dendrologia*, particularly the taxonomic problems that this presented. Watson and Lambert owed a considerable debt to explorers, merchants and company officials, foreign botanists, naval officers and other individuals based around the world for the seeds, cones and other specimens that they were prepared to send. Lambert obtained considerable information from officials in the East India Company to whom he acknowledged 'many obligations', and the oriental scholar Charles Wilkins provided access and advice concerning the rich collections in the museum at India House. Lambert emphasized that 'botanical science' was 'so much indebted' to the court of directors of the East India Company for 'the great zeal and liberality they have shown in making known the valuable treasures of their extended Asiatic possessions'. Through the appointment of collectors in parts of India, and the establishment of 'one of the most magnificent botanic gardens in the world, at Calcutta' they had 'furnished to the gardens of this country their chief ornaments'.[43] The importance of economic botany is reflected in Lambert's inclusion of an appendix containing details of trees and shrubs that were not necessarily connected with the *Genus Pinus* but which he judged 'remarkable' in their own right. These included a *Quercus grandifolia* from Nepal, the *Maclura aurantiaca* or Osage orange, *Ilex paraguensis* or tea-tree of Paraguay and plates of two species of Nepenthes or pitcher plants discovered in Singapore by his 'highly valued and lamented friend' the late Sir Thomas Stamford Raffles. Lambert also included an account of the potato and an 'account of the medicinal and other uses of various substances, prepared from trees of the Genus pinus' by William George Maton MD, FRCP, FRA, LSS.[44]

The interpretation of information concerning novel trees from traveller's accounts, cones, seeds and other dried parts returned to Britain before they could be successfully grown presented considerable challenges which are reflected in the alterations made to the *Dendrologia* and the successive editions of the *Genus Pinus*. The fragmentary nature of initial information concerning foreign varieties also encouraged supplementation of Linnaean-style taxonomies and the adoption of the natural system. Even the vice-president of the Linnean Society found that he could not wait lengthy periods for newly imported varieties to fruit or flower before attempting classification. After learning of recent French botanical work, Watson made some alterations to the 'usual established genera' in his *Dendrologia*. For instance, he regarded Planera as distinct from Ulmus 'from its bearing a capsule (not a samara)', Lyonia was 'distinguished from Andromeda, by its very different male aments and flowers' whilst 'Spartium was merged into Genista' 'after the example of the French'. Similarly, Oxyzaccos was (after Pursh) separated from Vaccinium as bearing only eight stamens, Cydonia, Malus and Sorbus were included in Pyrus, 'having the spermoderm (seed-coat), Cartilaginous' and Crataegus was merged into Mespilus, 'from the seed being contained in Pyrenes'.[45]

The production of Lambert's *Genus Pinus* also illustrates very clearly how international networks impacted upon all aspects of arboricultural study. Botanists and authors needed to assess the veracity of tree descriptions and the status of explorers helped to inculcate trust in their accounts. Even so, it was sometimes hard to establish the source for particular seeds or cones or to be sure that particular descriptions referred to the same kind underlining the importance of co-operation between scientific institutions and their respective explorers in sharing information and specimens with each other. Lambert emphasized that 'more frequent intercourse with foreign parts', and 'greater facilities' of 'scientific travellers' in visiting 'remote regions', had 'tended infinitely to enlarge ... every branch of natural history'. This was through the 'vast increase of new species' and the accumulation of a 'rich fund of interesting facts and observations'.[46] The impact of these discoveries on British arboriculture is demonstrated clearly when we compare different editions of the *Genus Pinus* between 1803 and 1837. In the second edition (1824), for instance, Lambert noted that through the kindness of his 'excellent friend' Carl Friedrich Philipp von Martius (1794–1868), he had been able to incorporate the fruits of the latter's five year study of Brazilian botany patronized by Maximilian Joseph I, King of Bavaria. Through Martius he was able to provide a portrait of the *Araucaria brasiliana* and a fully-grown

cone of the same species. Lambert also received materials from the celebrated
East India Company explorer William Moorcroft (1767–1825) which furnished ample information for the plate and description of the *Pinus deodara* in
the *Genus Pinus*. Similarly, cones from the *Pinus orientalis* (*Picea orientalis*) were
collected from near Teflis by the diplomat Sir Gore Ouseley (1770–1844) who
also provided information. Specimens of a pine collected near Montivedeo were
supplied by Lambert's 'distinguished friend', the royal navy captain and explorer
Phillip Parker King (1791–1856), who was engaged surveying the South American coast. King was told that the tree had been raised from seeds originating
in Tenerife, but as it was 'quite unlike the *Pinus canariensis*' but more similar
to *Pinus massoniana*, Lambert was inclined to consider that it may have been
imported by seed from China instead.[47]

Tantalizing reports of new trees were received with considerable excitement,
but the fragmentary nature of the information is apparent in the different editions of the *Genus Pinus*. Lambert and Don paid considerable attention to newly
published arboricultural works internationally and used these where necessary
to supplement their analysis. Lambert utilized Thomas Nuttall's *Genera of North
American Plants* (1818) for instance, and obtained a description of *Maclura
aurantiaca* from him enclosed in a letter from Liverpool dated 12 April 1824.[48]
Similarly, he obtained information concerning the beautiful *Pinus spectabilis*, or
purple-coned fir, which he regarded as 'undoubtedly the finest of the fir tribe',
from his friend the East India Company surgeon and botanist Nathaniel Wallich (1786–1854), who provided specimens and seeds. Although they knew of
no example that had been successfully grown in Britain, Sowerby and Weddell
were able to produce a colourful, vibrant and spectacular plate based upon these,
one of the most striking in the *Genus Pinus*.[49] When Lambert had almost finished the second edition in 1824, he received intelligence from his friend Joseph
Sabine (1770–1837), secretary of the Horticultural Society, that a 'most remarkable pine of extraordinary dimensions and bearing large cones with edible nuts'
had been discovered in California. David Douglas (1799–1834), the Society's
botanical collector, had just returned from a three-year expedition to the north-
west coast of America where he had made the discovery which he reported in a
paper to the *Transactions*, naming the tree *Pinus lambertiana* in honour of the
vice-president. The tremendous size of the 'remarkable tree' and its cones provided a heroic element to the discovery as it was to do later with the *Sequoia
sempervirens* and Wellingtonia. Lambert excitedly reported that 'the trunk of
the largest ... measured was 215 feet high, perfectly straight, with a diameter of
19 feet ... cones 16 inches long, and 9 in circumference'.[50] His pride at the naming of the *Pinus lambertiana* is demonstrated by the decision to have a drawing
of himself with a plate of the tree as the frontispiece for the second edition of
the *Genus Pinus*. The tributes to Lambert – such as the naming of Douglas's

discovery – were an acknowledgement of the debt that plant collectors owed to Lambert's patronage and his status in the botanical world as vice-president of the Linnean Society. Lambert's work sanctioned and confirmed the value, novelty and status of their collecting efforts, helping to secure economic benefits, such as sales to nursery companies, and reputations in the field of natural history.

The fragmentary knowledge of oriental pines during the first decades of the nineteenth century presented Lambert and Don with special difficulties and they were sometimes forced to undertake revisions in response to new American and European arboricultural works. Louis Claude Marie Richard's study of Coniferae published by his son Jean Michel Claude Richard led Don to re-examine the *Dammora australis* and *orientalis*, as well as *Taxodium sempervirens*, and to 'correct some important errors into which he had fallen in his description of them' in the *Genus Pinus*.[51] Lambert considered that a Chinese tree known only from cones and a drawing at the Horticultural Society in London could be called '*Pinus sinensis*', suggesting a failure to appreciate the size and botanical variety of Chinese flora. Some novel oriental trees were being raised in Britain but this required him to wait until they were sufficiently mature to fruit or flower before a definitive name could be given. Similarly 'having lately seen drawings by Japanese artists of the *Pinus abies* and *larix*, noticed by Carl Peter Thunberg in the *Flora Japonica*', Lambert became 'fully satisfied of their being perfectly distinct from the European species with which Thunberg has confounded them'. He suggested that '*Pinus thunbergia*' would be the better name for the former and '*Pinus kaempferi*' for the latter. Of other Thunberg designations, Lambert thought that '*Pinus strobus*' was 'evidently the same with my *Pinus excelsa*' whilst of '*Pinus cembra* and *sylvestris*', he thought it likely that they would prove distinct species. Likewise, a single plant of a species '*Pinus foliis ternis*' suspected to be from Timor which had grown to 16 ft and borne male catkins annually, he suspected to be close to '*Pinus longifolia*'.[52] In all these cases identification and classification remained contingent and tentative whilst more information was gathered and attempts were made to grow specimens to maturity in British gardens or hot houses. In these circumstances gardeners, curators, nurserymen and plant collectors were as likely to first fully appreciate arboricultural characteristics as the metropolitan elite of the Linnean Society.

Conclusions

It was generally in the British semi-public botanical society gardens that tens of thousands of patrons, subscribers and visitors were introduced to systematic planted tree and shrub collections and it was through the requirements of management and presentation for these audiences that problems facing Linnaean taxonomy in arboriculture were confronted. Presented starkly as differentiated and organized spaces, different arrangements demonstrated the merits of the natural system and the inadequacies of the Linnaean scheme for anything other than elementary botanical education. Botanical society tree collections also provided an opportunity for study of physiology and structure in addition to taxonomy, reflecting the inspiration and requirements of horticulture, agriculture and industry and the economic opportunities that arboriculture was thought to offer. They had a major impact upon early nineteenth-century British natural history and underscored the importance of trees and shrubs for celebration and study. Although Kew gardens continued to hold a pre-eminent position in British and imperial botany (as the royally-patronized focus of international plant exchanges demonstrate) and the Horticultural Society gardens also claimed some pre-eminence, their position was strongly challenged by the major Irish, Scottish and English provincial botanical society tree collections. As we shall see, under William Jackson Hooker and William Andrews Nesfield, with the major expansion of Kew and its arboretum from the 1850s, metropolitan institutions again strove for pre-eminence in experimental arboriculture and international plant exchange networks, yet the greatest innovations had occurred in the context of non-metropolitan botanical gardens. This inspiration was reflected in the schemes adopted in early nineteenth-century botanical and arboricultural treatises such as the presentation of Linnaean and natural arrangements in successive editions of Loudon and Lindley's *Encyclopaedia of Plants*.

4 JOHN CLAUDIUS LOUDON'S ARBORETUM

Loudon's *Arboretum Britannicum* is the most important systematic study of hardy British trees and shrubs to have been published in the past two centuries, and arguably ever. Its greatest significance lies in the fact that it combined a comprehensive study of trees and shrubs with geographically founded histories of arboriculture, analysis of the importance of trees and shrubs in landscape gardening and full-length portraits of trees at different growth stages. Loudon's ideas concerning arboretums developed in various ways during the 1820s and 1830s in response to his experience of trees in various contexts, including private gardens such as his Bayswater villa, botanical and horticultural society gardens, and country estate collections. The significance of the *Arboretum Britannicum* and arboretum concept, manifest in various changing forms, can only be fully appreciated if examined in the context of British scientific culture and particularly cultures of natural history. The book must also be judged in terms of Loudon's other efforts to promote botany within gardening and horticulture, encouraging new audiences and practitioners amongst all social classes in the British Isles and Ireland. Scientifically-informed arboriculture and landscape gardening should not be merely the preserve of aristocracy, gentry and their agents as in the days of Humphry Repton. This combination of elements, the role of the *Arboretum Britannicum* in promoting a vision of natural history in British culture and society, the attention to arboriculture in landscape gardening, systematic tree botany, arboricultural history and extensive engravings based upon carefully identified growing specimens rather than dried sections, explains why the work was of political as well as cultural and scientific significance.

However, there are some striking omissions from the *Arboretum Britannicum* which are also very revealing, such as a lack of information concerning the physiological and anatomical botany of trees. The ghost of a book that Loudon had begun to compose entitled the *Encyclopaedia of Arboriculture* – written in manuscript but never published – haunts the pages, helping to explain these omissions. There is arguably more information on the physiology and anatomy of trees in the *Encyclopaedia of Agriculture* than there is the *Arboretum Britannicum* which focuses on external characteristics of trees, habits of growth and cultural historical significance rather than internal physiological matters. The tree sections in the former as well as the *Encyclopaedia of Gardening* give some idea of the material that might have been included in the projected *Encyclopaedia of Arboriculture*. As we shall see, the *Arboretum Britannicum* had a major impact on British and international arboriculture, inspiring the creation of many private and public arboretums and tree collections and the composition of other arboricultural works, including studies that sought to emulate Loudon's comprehensiveness.[1]

Loudon's Arboretums and British Scientific Culture

Although the *Arboretum Britannicum* was easily the most important delineation of Loudon's arboretum ideas, these ideas underwent considerable change during the 1820s and 1830s. Before examining the *Arboretum Britannicum*, this chapter considers how these ideas were modified during the period. It argues that the significance of Loudon's arboretum conception can only be understood if placed within the context of British late-Georgian and early-Victorian science and scientific cultures as well as landscape-gardening and horticulture. Loudon's arboretums were inspired by enlightenment progressivism and systematic botany, but they were also a response to rapidly changing British society experiencing population expansion, urbanization and industrialization, which challenged establishments and reformers alike. Loudon's belief in the efficacy of practical scientific education and his political philosophy was shaped and reinforced by his arboriculture. Various strands in British science in the period have been identified which moulded Loudon's arboretums. Inspired by his Scottish enlightenment background and education mediated through Andrew Coventry at Edinburgh University and others, the need for scientifically-informed agricultural and horticultural improvement was an important objective. Secondly, Loudon aimed to foster scientific education amongst middle and labouring classes to improve gardening and horticulture using arboretums and Candollean arboriculture.[2] There are important parallels with contemporary attempts to open established science to the middle classes and challenge aristocratic, elite and metropolitan domination of national associations such as the Royal Society, Royal Institution and Royal Botanic Gardens at Kew. Just as the mechanics' institutes and the British Association for the Advancement of Science were trying to foster provincial urban public science, so Loudon was striving to bring the botanical sciences into gardens, parks and homes.[3]

Loudon's development and promotion of botanical gardens and arboretums was part of this general reforming scientific agenda, although his role in British natural history has received surprisingly little attention. In his classic history of botany for instance, John Reynolds Green included Loudon in the chapter on 'Subordinate Features of the Period', describing him as 'only of minor importance' although an 'eminent landscape gardener and writer on horticultural subjects'.[4] During the 1820s and 1830s, Loudon used the publications he composed and edited, especially the *Gardener's Magazine* (1826–44), *Encyclopaedia of Plants* (1829), successive editions of the *Encyclopaedia of Gardening* (from 1822), *Magazine of Natural History* (1828–36) and *Arboretum Britannicum* (1838) to promote natural history for all social classes and the provision of systematic plant collections in public and private gardens. Subsequently this work was undertaken in collaboration with his wife, Jane, who learned botany with her husband's encouragement and became a successful author in her own right, publishing a variety of works on gardening and natural history including some intended for women and children. Often literally crammed with blocks of frequently miniscule text, profusely illustrated and regularly updated with voluminous supplements and additions, the pages of Loudon's works were intended to convey, economically and accessibly, as much information as possible to enhance individual education and attract new audiences. The latest publishing and printing techniques were employed to intersperse numerous illustrations within text and in separate plates in a technique only applied previously to children's books. Dinner parties were held at Loudon's Bayswater house in which naturalists, explorers and writers met to converse on all manner of subjects in literature, the arts and sciences and enjoy the beautiful botanically-informed systematic collections of front and rear gardens.[5]

John and Jane Loudon tried to entice botanical novices, demonstrating the interconnections between the sciences, horticulture, arboriculture, agriculture and landscape gardening. As we shall see, Loudon's fascination with the scientific, aesthetic and landscape gardening significance of trees and shrubs encouraged him to undertake systematic botanical observation and study and to develop his notion of the gardenesque. *The Magazine of Natural History and Journal of Zoology, Botany, Mineralogy, Geology and Meteorology* which Loudon edited between 1828 and 1836 is generally regarded as the first popular periodical exclusively devoted to the subject, its ambitions evident from the full title. The magazine proved to be so successful that Loudon decided to publish it monthly instead of bi-monthly, and like the *Gardener's Magazine* it developed an extensive network of national and international correspondents becoming the most important regular popular and professional natural history forum for the exchange of ideas and practices during the late 1820s and early 1830s. It was Loudon's promotion of the Wardian case in the *Magazine of Natural History* and *Gardener's Magazine* in 1834 for instance, that popularized its British use long before Ward himself published *On the Growth of Plants in Closely Glazed Cases* (1842). In addition to middle-class men with botanical interests, much encouragement was given to women and the labouring classes to pursue botany and natural history for enjoyment and improvement in their leisure time which also expanded the commercial market for botanical literature. The *Magazine of Natural History* included work by distinguished botanists and naturalists such as John James Audubon, Charles Waterton and John Stevens Henslow and launched the writing career of John Ruskin, who also contributed a major early series of articles to Loudon's *Architectural Magazine* during the 1830s. The *Magazine for Natural History* lasted longer than most similar contemporary ventures, providing a model for Victorian natural history periodicals which came to rival it, although few were able to so successfully transcend the growing divisions between popular and professional science and become as commercially successful as Loudon's publication had done.[6]

Loudon's closest scientific collaborator was probably the botanist John Lindley (1799–1865) whose work challenged the scientific establishment dominated by an Anglican, aristocratic and clerical amateur Oxbridge-educated elite until at least the 1830s. Initially under the patronage of Sir Joseph Banks, Lindley became associated with the Horticultural Society and was the first professor of botany at the new University College, London, after William Hooker had rejected the position. He collaborated with Joseph Paxton to direct the Hudson Bay Company exploration in North America and worked with Loudon on the *Encyclopaedia of Plants*. Like Loudon, Hooker and Lindley sought patronage from aristocracy and gentry with horticultural interests, especially the dukes of Bedford and Devonshire, and supplemented their salaries by writing and expanding the audiences for botany, gardening and natural history. Lindley and Hooker also strove, with support from the Whig aristocracy, as Drayton has shown, to reform the Royal Botanic Gardens at Kew as a national British imperial botanizing institution on the model of British provincial, Irish and Scottish botanical gardens and the Parisian Jardin des Plantes and Museum d'Histoire Naturelle. In the *Gardener's Magazine* and *Magazine of Natural History*, Loudon had been calling for such a national institution throughout the 1820s and 1830s. Although he strove to professionalize botany on the model of the French academic sciences, as editor of Paxton's *Gardener's Gazette*, Lindley also shared Loudon's view that gardeners and landowners needed an improved botanical education to take full advantage of the global plant riches being brought back to Britain. Drayton and Shteir have contended that in so doing, he was masculinizing botany, striving to transform it from the polite Linnaean female pastime into a serious professional discipline. As we shall see in the following chapter, Lindley shared and encouraged Loudon's belief in the superiority of de Candolle's version of the Jussieuian natural system of botanical classification.[7]

Loudon believed that de Candolle's improvements simplified and rationalized arboricultural taxonomy, making the planting, presentation and understanding of systematic collections easier for a wider spectrum of British society. With his

highly spatialized perception of taxa and economical mode of species differen-
tiation, de Candolle encouraged the application and presentation of systematic
collections in gardening. Loudon saw public and semi-public botanical gardens
and arboretums and the introduction of systematic collections into public parks
and suburban gardens as a means of promoting botanical education to all social
classes. He also hoped that they would solve problems created by the growing
tension between specialization and professionalization on the one hand and
the need for popularization and dissemination of knowledge on the other. Loudon's efforts to professionalize and formalize the education and qualifications
of the gardening profession and to undermine the perception of gardeners as
merely unthinking low-status servants, partly paralleled the efforts of scientific
reformers to formalize, institutionalize and extend scientific education, but also
challenged notions of professional elitism. This was also reflected in Lindley's
interest in the relationship between botany and gardening, including his promotion and editorship of the *Gardener's Chronicle*. Loudon's final major work was
the *Self Instruction for Young Gardeners* (1844), which provided a broad course
of education including attention to mathematics, practical geography, natural
history, meteorology and geology, thereby challenging the continuing perception of gardeners as menial servants.[8]

On the other hand, whilst Loudon sympathized with many of the objectives
of Lindley and the scientific reformers, there were some important differences
which, as we shall see, the development and impact of arboretums demonstrate
well. Loudon's arboretum conception was as much inspired and shaped by the
practical requirements of gardening, landscape gardening and horticulture as it
was by abstract scientific ideas, although he shared Lindley's belief in the efficacy of the Candollean system and the practical inadequacy of Linnaean botany.
This meant that whilst botanists such as Lindley continued to make modifications to their taxonomies over the decades, Loudon required greater certainty
and needed to present a fixed version of the natural system as a pre-requisite for
the practical design and management of arboretums by planters, gardeners and
landowners. Loudon was therefore drawn towards the Candollean simplicity
and fixity of taxa. Despite the apparent egalitarianism of Loudon's arboretum
concept, which he claimed could be employed in all kinds of gardens from public parks to suburban spaces, in many respects it was still much more suitable
for aristocratic and wealthy landowners with the staff and resources to plant
ambitious systematic collections. We can also see how the presentation and
experience of systematic collections promoted an idealized and in some respects
quite oppressive, regulated, hierarchical and ordered Edenic vision, whilst classifying and exploiting the treasures hoarded by the most powerful imperial nation
and attempting to 'equalize' global flora. This helps to explain why attempts to
apply arboretums to public gardens experienced only mixed success, although

Loudon's dry and unemotional facts and statistics were always intermixed with numerous literary, poetical and artistic arboricultural allusions and quotations.

Loudon's emphasis on estate improvement was very appealing for Victorian aristocracy, gentry and wealthy landowners who were able to adopt many of his arboretum schemes without subscribing to his social and political views. Yet just as Loudon's magazines provided forums for all social classes from gardeners to aristocrats and clergy to engage in discussion and debate with an equality unusual in early-Victorian society, so arboricultural enthusiasm was also able to transcend social strata and geographical and cultural divisions such as those between town and country, metropolis and provinces, and English and Celtic countries. Similarly, whilst the evidence concerning plant adaptations and the changes induced by horticultural breeding was fully documented in works such as the *Arboretum Britannicum* and *Encyclopaedia of Gardening* and utilized in support of evolutionary ideas, not least by Charles Darwin in the *Origin of Species* (1859), clerical naturalists and creationists continued to celebrate horticultural improvement and the promote natural history as support for natural theology.

Arboretum Concepts and Contexts

Loudon was interested in the problem of balancing systematic botanical collections with landscape gardening aesthetics and the importance of the distinctions between the Linnaean and Jussieuian systems from the earliest stages of his career. Loudon's interest in the formation of arboretums grew during the 1820s. One of the most important inspirations came from the Loddiges nursery arboretum at Hackney, and in 1831 Loudon had portraits of all the trees in this garden taken along with observations of arboretums at Kew and other places in preparation for the book (Figure 4.1). Reminiscent of Loudon's spiral Linnaean botanical garden published in 1812 (Figure I.1), the Loddiges arboretum (i) was approached across a bridge over a public lane (k) and carried on along the right side of the path before terminating in an American garden (m). The Horticultural Society's arboretum at Chiswick provided inspiration for Loudon's arboretum conception which he developed in a series of critical articles in the *Gardener's Magazine* and the *Encyclopaedia of Gardening* during the 1820s and early 1830s. Loudon was also unsympathetic to the elitist amateurism of the Horticultural Society having described their publications as 'costly quartos for the rich' with little practical value and he temporarily fell out with their president, Thomas Andrew Knight.[9]

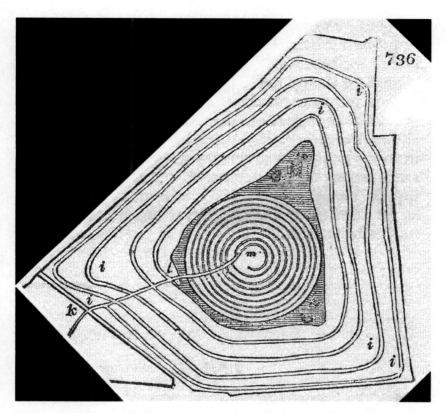

Figure 4.1: Loddiges' spiral arboretum at Hackney from J. C. Loudon, *Encylopaedia of Gardening* (1830), p. 801. Reproduced from the authors' personal copy

In the second edition of the *Encylopaedia of Gardening* (1824), Loudon argued that the plan was 'defective in general arrangement' and shows a 'want of grandeur and unity of effect as a whole, and of connection and convenience in the parts'. One of the most glaring errors which according to Loudon would strike any visitor, was 'forming the arboretum in a large rectilinear clump' when it should have been arranged around the circumference so that visitors could have driven round the whole (see Figure 3.5). Subsequently, Loudon was more strident and provided more detail of how he would change the gardens, which were 'so bad' that they could not be improved and ought to be 'totally obliterated'. After sale of much of the stock, his first suggestion was that the arboretum 'should be formed as a belt, combined with hardy herbaceous plants, and arranged in the natural manner'. The walk within the belt ought to be perfectly circular, but it could be a square or wavy.[10] Loudon softened his criticisms of the Horticultural Society arboretum during the 1830s, which was useful as he intended to

obtain drawings of all the trees there for a projected work on landscape gardening and garden architecture. In 1833 he noted that the trees in the arboretum
were 'thriving', many of them having 'now attained a considerable size' though
this was so much that 'in most of the clumps they are crowding each other, so
that the characteristic forms of the individual species will soon be lost'. He could
not resist reminding readers of his ten years of criticisms regretting that the trees
and shrubs were not spaced and distributed round the circumference where they
would have demonstrated individual shapes and provided 'a really useful knowledge' of 'forms, colours, and effect in the landscape' for the public. Instead of the
public taking delight in constantly changing arboricultural novelty there was 'no
thinking gardener' who could not foresee that in a couple of years 'this arboretum will become nearly useless for every purpose of the garden artist'. As we shall
see, it was his experience of cramped planting in the Horticultural Society collection that later encouraged Loudon to situate specimens in their own spaces
in the Derby Arboretum and to recommend the drastic action of periodic total
removal once they had attained too large a size.[11]

For Loudon, the Loddiges' arboretum – especially when compared to the
Chiswick garden – looked better all the time. After walking around the two
outer spirals of the coil again in 1833, Loudon commented that there was 'no
garden scene about London so interesting to us as this arboretum', and he only
wished that it would 'induce the planters of parks, shrubberies, and pleasure
grounds ... to examine it'. He expressed himself astonished that 'thinking men
amongst gardeners' who knew its characteristics 'should continue to plant the
commonplace and monotonous mixtures in shrubberies which they do now'.
He wanted ladies from gentlemanly families and mothers who educated future
generations to come and experience it so that they could instil a taste for exotic
arboriculture in the young. It was only the panoramas being exhibited in London, such as one of Niagara being shown in 1833, that allowed many a glimpse
of these exotics through the 'rich colours' of the trees constituting the woods in
the background of that magnificent picture. He singled out the American acers
being exhibited at Loddiges for their luxuriant colours, birches and crataegus.
The problem was that so few had access to such arboretums given that most did
not have 'sufficient ground to allow the different roots to attain their full size
and to display their separate natures'. In the wake of the Great Reform Act of
1832 and the reforming Whig administration, Loudon argued that the time was
'just commencing for the establishment of public parks and gardens adjoining
towns' where such arboricultural delight could be witnessed 'at the expense of
all for the enjoyment of all'. He urged that 'the whole of Regent's Park would
be required to plant one of each of the species and varieties' in the Loddiges
arboretum 'at proper distances varied by suitable glades'. Similarly, Hyde Park,
Kensington Gardens and Greenwich Park could also accommodate such col-

lections and become 'sciences of rural enjoyment independently of their beauty to general observers' and so become 'schools of botany and gardening', spreading good taste to the rising generations of all classes. When this had been done 'what a paradise this island will become, displaying, as it will do, all the trees and shrubs in the world' that would grow in temperate climates.[12]

By 1830 Loudon was contemplating some kind of 'Arboretum Britannicum' as a natural successor to the *Encyclopaedia of Gardening* and the *Encyclopaedia of Plants*. In many respects, responding to queries raised by correspondents in the *Gardener's Magazine* in addition to his own experiences as the landscape gardener, the whole *raison d'etre* of the *Hortus Britannicus* (1826) was to facilitate the creation of arboretums and other systematic garden collections. Introducing the natural arrangement, Loudon stated that species were placed under their various classes, orders and tribes 'with reference to their garden culture', using gardening to preset a spatial visualization of the system. The space of ground to be occupied by the plants in each order was estimated 'with a view to the formation of arboretums and herbaceous grounds' arranged according to the natural system. A system of symbols distinguished in shorthand form between trees, herbs and water plants. Hardy plants under one foot in height were allowed one square foot, those of 6 to 30 ft per square (or circle formed from the radius) of 10 ft and those above this of 20 ft. In theory Loudon's scheme therefore allowed for planters or gardeners to combine these squares (or less easily, circles) to estimate the total area likely to be occupied by the presentation of an entire tribe or order within an arboretum or other systematic collection. Although his scheme encouraged the presentation of systematic collections in geometric forms, Loudon argued that as 'every gardener' knew how to modify squares to parallelograms, triangles, or circles 'of the same capacity', so the system would easily allow collections to be accommodated within different designs or shapes of land. The 'best method' for laying out such arrangements according to the natural system was 'by circular groups, on a plane, or on a regularly convex or concave surface with a circular boundary', the spaces between being covered by grass, gravel or pavement, the groups being edged by box. To facilitate the translation of this rather two-dimensional plan into three dimensions, Loudon suggested a method of arranging the groups systematically and throwing 'the numerous circles into agreeable figures'. This entailed drawing circles representing each tribe on circles cut from coloured paper which were then attached to larger circles representing orders and pasted onto larger shapes representing the intended surface in outline, also drawn to scale. Once arranged satisfactorily, the same plan stuck together with gum or paste could be used as the working plan for arranging different kinds of arboretums and other systematic collections for pleasure grounds. Loudon recommended that one of the best methods of planting such arboretums was to place the trees and shrubs 'along one or both

sides of a winding shrubbery walk', with no tree or shrub being closer to the
walk or other specimens than half its own height. He claimed that a 'complete
arboretum' following this plan would 'extend along two miles of walk'. In his
Illustrations of Landscape Gardening (1830), he provided examples of systematic collections formed according to this method of demonstrating the natural
system which he thought appropriate for residences in addition to institutional
collections. These included an 'Elementary diagram for the composition of arboretums in lines along the margins of walks', showing a tree collection distributed
along a long walk divided into sixty-four zones and containing 2,512 species and
varieties presented in 'natural orders and tribes'.[13]

From Suburban to Urban Arboretums

Another important stimulus for Loudon's arboretums was the problems of planning the gardens of the Bayswater villa, which he constructed in 1823. Prior
to this he had moved with his mother and sisters to a rented house in Bayswater in 1816 called the Hermitage which had a large garden.[14] A prototype for
Loudon's idealized middle-class villa, the front and rear gardens of the house
allowed him to experiment with systematic collections in suburbia as he strove
to accommodate the largest variety of plants in confined enclosures and emulate landed elite estate arboretums in microcosm. Many of the ideas originally
developed for the Bayswater gardens were later applied to the Derby Arboretum
which also occupied a small site on the urban periphery. Loudon presented a
strong case for the general importance of trees and shrubs in villa gardens in his
Suburban Gardener and Villa Companion (1838) emphasizing how maximum
aesthetic and botanical impact could be achieved with even the smallest com-
pass. One design for a one and a half acre villa crammed in around the house
and offices, a kitchen garden, reserve garden, botanical ground for herbaceous
plants, American ground, orchard and shrubbery. Even for small villas plantations were as important in achieving general effects and in demonstrating the
'taste and means' of the intended occupier as the house architecture. Garden
trees connected objects with each other, united the house with offices, helped
to conceal the latter, and united 'the place as a whole with other places in the
neighbourhood with adjoining scenery'. Trees and shrubs provided vital shelter
and shade, screens in thickened masses, and sheltered other more delicate plants,
whilst 'subdividing compartments' in the garden. Of course, the choice of varieties depended on the taste of owners, but Loudon considered that the choicer
species and varieties, especially evergreens, ought to be placed nearer the house,
as the 'centre of art and refinement'. Some might choose to present a 'complete
arboretum' in their gardens, others 'showy trees and shrubs', large rapid growing varieties or fruit trees, in addition to the beauties of all kinds including their

'individual expression and character as pictorial objects, variety and intricacy in combination, and botanical interest'. Just as domestic design and interior divisions and functions distinguished private and family life from public and business life, so in Loudon's vision trees and shrubs played important social and cultural roles in addition to demonstrating the taste and discernment of owners and skills of architects and gardeners. Where business was conducted in the house, they could screen the offices from domestic quarters. They were not just barriers dividing public from private spaces, but defined the terms of middle-class family engagement with social urban life beyond, masking houses from each other and screening the private spaces of dwellings from the outside world. Inviting attention and encouraging the advance of botany through the example of displayed exotic varieties, they linked houses with immediate neighbours and other street villas if systematically planted through urban gardens. Family education and private botanizing aside, Loudon's suburban arboretums would have been largely wasted if not observed by others. Although they helped to define private domestic life they also invited display, observation and emulation. In this respect, although Loudon's idealized villas and gardens were tremendously influential for nineteenth-century suburban living, they were also at the cusp between enlightenment public progressivism and Victorian private domesticity.[15]

In his *Suburban Gardener* Loudon described the various changes made to his garden, presenting it as a model of an ideal aesthetically pleasing and botanically significant fourth-rate suburban garden. He recognized that whilst a grand national arboretum was desirable, it was also possible and equally important to develop one by combining multiple smaller arboretums. The front and rear gardens of the semi-detached Bayswater villa demonstrated how an ambitious suburban garden might be laid out with a botanically significant collection which if it consisted of hardy plants, required 'little skill in management and allowed owners to procure specimens easily and cheaply'.[16] Presenting various designs for the front garden, Loudon noted that planting 'all the more ornamental species' would allow the garden to become 'a school of practical botany' to the house and family. The combination of significant collections in each garden with 'trees along the front of these gardens next the road or street' would then form part of a series, which, if the street is long enough' might include 'the whole of the popular species of the British arboretum'. Furthermore, he suggested that a kind of symmetry could be achieved by adopting such a series of trees in the front gardens on both sides of long streets, so presenting an organized collection to members of the public 'in walking or driving' past.[17]

Whilst Loudon provided plans for a seven-acre arboretum of numbered trees and shrubs for a second-rate villa and another for the garden of a larger first-rate villa, the most detailed account of a suburban garden tree collection was for the Bayswater villa.[18] In 1831 Loudon changed the planting in his rear garden beds from a selection of flowers to 'a representative system of hardy herbaceous plants', adding to the numerous other gardening features and collections crammed into this small space. In 1826 another border in the back garden was divided into 464 divisions in order to represent the orders and tribes of the 'whole vegetable kingdom, hardy or tender, indigenous or exotic, cultivated in Britain'. In 1830 all the flints, vitrified bricks and plants of the 'universal representative system' were removed and replaced with a 'miniature arboretum' of 'representatives of hardy trees and shrubs cultivated in Britain' which extended to fifty-eight species. The 'stronger growing' trees and shrubs in this collection were taken up annually and their roots were reduced 'in order to keep them of moderate dimensions'.[19] Loudon's attention to the maintenance of his garden tree collections even extended to trying to police how his neighbours, tenants in the adjoining villa, behaved towards their trees. Although the north house was let to 'parties who care little about gardening', he made sure they were 'precluded in the lease from ... taking down or planting trees' or making other alterations without his permission so that the 'general appearance' of both front gardens remained the same from the road. Furthermore, a list of trees planted in the front and back garden was provided to prevent the tenant cutting any down without permission to 'preserve the unity of appearance' between the gardens and villas.[20] According to Loudon, the 'numerous experiments' that he undertook in his garden aimed to 'try everything that could be tried in so small a space', primarily for the sake of 'scientific knowledge rather than enjoyment'; to become 'intimately acquainted with all kinds of hardy trees and shrubs adapted for enriching pleasure grounds'. The trees and shrubs grew so rapidly that he was forced to remove many larger ones so that of the maximum 2,000 or so species of plants accommodated in the garden only a few hundred remained. Ironically, in 1835 or 1836 Loudon also had to lay off his gardener because he could not afford to keep him as a result of the economies occasioned by the cost of producing the *Arboretum Britannicum*. However, he was confident that the effect of the garden remained the same, owing to the relatively little care required by trees and shrubs beyond pruning and thinning compared to annuals.[21]

The adoption of arboretums for suburban villas was not just something for owners and tenants to develop but an objective that Loudon also hoped to further through architectural education – part of the formalization of the profession occurring in Britain during the first half of the nineteenth century. In his *Encyclopaedia of Cottage, Farm and Villa Architecture* (1833) Loudon recommended that architects intending to become landscape gardeners serve one year in a botanic or horticultural-society garden. This would enable them to 'acquire a correct knowledge of the names, heights, characteristic forms' and foliage colour of hardy trees and shrubs and of the 'principal families of herbaceous plants'. He argued that this could only be achieved if architectural students produced individual and group sketches of many of these species and varieties. This was to be supplemented by reading about 'the native countries of trees' and their associations with particular soils, surfaces and topographies, such as the associations of willows with damp conditions, which ought to prevent them being planted near houses which needed to be associated with dryness. The year would also enable students to get to know 'the routine practices of gardening', the construction and uses of hot houses and other gardening structures and to further their scientific education. Students would also benefit from an apprenticeship with a country architect where they could learn to sketch landscapes and garden scenery 'from nature' which would facilitate arboricultural education. This would provide enough education and experience to allow the qualified architect to obtain sufficient knowledge of landscape gardening to enable him to lay out grounds with the assistance of a gardener similar to the help he would receive on the house from builders. Through developing architectural education, Loudon strove to open another front in his urban arboretum campaign supplementing the arguments presented to tenants and property owners in his *Suburban Gardener and Villa Companion*.[22]

Trainee architects' knowledge of landscape gardening would be enhanced by 'access to some botanic garden containing a rich arboretum' which meant in London the Loddiges collection at Hackney, Loudon's favourite arboretum, or the Horticultural Society. Using these collections, Loudon suggested that architectural students went through the process he was undergoing in preparing for the *Arboretum Britannicum* by making 'portraits of all the trees and shrubs'. Simultaneously students could acquire a general knowledge of gardening by observing villa gardens and mastering gardening works including the *Gardener's Magazine*. This was because for Loudon, imported trees and shrubs served the vital function of distinguishing the landscape gardeners art. Far from landscape gardening being merely a matter of imitating 'natural scenery ... in artificial grounds', this encapsulation had to be undertaken 'in the spirit of art' which it could only be in the 'native or modern style' through the placement of exotic trees. It would then transcend the kind of 'mere mimicry of nature' or 'tame monotony' that characterized 'half the villa landscape or park scenery of Britain'.[23] The idea that gardening was inspired but not imitative of nature had been maintained by British writers and philosophers for decades including Alexander Pope, Joshua Reynolds and Walter Scott. However, the work of the influential octogenarian French architectural theorist Antoine-Chrysostome Quatremere de Quincy (1755–1849), particularly his treatise on imitation in the fine arts (1823), seems to have most stimulated Loudon's notion of 'recognition by art' in gardenesque landscape gardening during the 1830s.[24]

The impact of Loudon's experience in designing, planting and changing the Bayswater garden inspired him to encourage landowners and gardeners to plant arboretums in his other publications. In response to a list of prizes issued by the Caledonian Horticultural Society in 1827 Loudon suggested a set of prizes to encourage landowners and employers of gardeners to participate 'to a greater extent than they have hitherto done' in advancing horticulture. It is striking that apart from the erection of gardeners' houses and offices and a garden library, all of these prizes were primarily for the formation of systematic collections rather than general landscape gardening achievement. There was a prize for the 'formation of the most complete arboretum arranged according to the natural system', for 'the most complete collection of herbaceous plants' similarly arranged and for 'the most complete systema vegetabilium of living hardy trees and shrubs' arranged either according to the Linnaean or natural systems. The latter was to be laid out either in groups on a lawn, in beds with gravel or in rows. Another two prizes more explicitly encouraged the union of systematic collections with aesthetics. One was for the formation of a country seat laid out to demonstrate 'in the woods, shrubbery, and flower-garden, a complete system of hardy trees and plants arranged according to their natural affinities'. The highest value was to be accorded the garden of smallest extent as 'the most difficult and the most expensive to execute' relative to intended effect. Finally, inspired by the panoramas on display in metropolitan pleasure gardens such as Vauxhall, Loudon recommended another award for living floras representing localities, countries, continents, 'temperate parts of the world' or the entire globe, arranged 'geographically on a plot of ground' laid out like a map of both hemispheres'. The torrid zone was to be a belt of hot houses and the temperate zone partly covered by moveable grass. Large rivers, seas and mountains were to be reproduced to scale with even their geology being 'correctly imitated' and the situation of larger cities was to be marked by garden seats or buildings.[25] Arboretums such as those of Messrs Loddiges did serve the important practical function of providing information for landscape gardeners, garden owners and amateur gardeners concerning tree characteristics. Loudon devoted the fourth part of his *Hortus Britannicus*, which provided a list of all British hardy and half hardy plants, to providing a classification of hardy trees and shrubs according to the time of flowering, the height they grow to and the colour of the flower.[26]

The *Arboretum Britannicum*

Loudon began collecting material for the *Arboretum Britannicum* in 1830, publicly signifying his intentions in the *Gardener's Magazine*.[27] The engraving and printing of the book commenced in August 1834, the first number being issued in January 1835 and the final sixty-third number appearing on 1 July 1838.[28] The extent to which the work grew beyond Loudon's original expectations is evident from references remaining in the first edition to tree portraits from a second or final volume of the originally projected two-volume work which had grown to eight volumes. The deterioration of Loudon's health by the 1820s seriously impacted upon the composition of the *Arboretum Britannicum* whilst the expanded size greatly increased production costs. There is no need to describe these problems in detail which are recounted in Jane Loudon's 'Life and Writings' of her husband, except to explore the impact they had upon the composition of the *Arboretum Britannicum*. Although some relief was received through the ministrations of London surgeon William Lawrence, Loudon lost movement in his legs, the use of one arm, and only retained limited use of the other. Finally, he suffered from such poor general health that all movement became difficult and painful. One serious implication was that he was initially obliged to give up practical landscape-gardening commissions, although he was forced to return to this to make money towards the end of his life, always driven by a Calvinistic determination to see every unoccupied minute as wasted. The second was that, as a result of the arm amputation and rheumatism, he could barely write after 1825. For the composition of all major works from this date including the *Arboretum Britannicum* he was forced to rely upon an amanuensis, and after his marriage to Jane Webb in 1831 he relied on his wife for the completion of manuscripts, whilst two of his sisters took up wood engraving. Loudon dictated the text to his amanuensis and to Jane employing artists to produce most of the tree portraits from living specimens. Given his relative immobility and the difficulty he faced drawing and writing with just two fully active fingers on one arm, the benefits of a virtual book arboretum with good quality tree portraits must have been all the more apparent.[29]

In 1834, after deciding to return to the project, Loudon sought information by sending out numerous questionnaires to landowners, an example of these being reproduced as an appendix at the end of the fourth volume. Printed questionnaires were a well-established means of gaining natural historical, meteorological, antiquarian and other forms of information for county historical and chorographical studies over the previous two centuries. However, this was the first time they had been employed to exclusively solicit local information concerning British trees on a comprehensive basis. Loudon posted approximately 3,000 printed lists of trees and shrubs and received numerous replies from individuals listed as contributors in the first volume. Over seventy of these replies survive in a bound volume and the importance of the information provided is evident at many places in the text.[30] His greatest debts were to the Loddiges' arboretum at Hackney and to the Horticultural Society arboretum despite his criticisms of the latter. Many tree portraits were undertaken in these arboretums whilst Donald Munro (*c.* 1789–1853), head gardener at the Horticultural Society, and George Gordon (1801–93), foreman of the arboretum who later published *Gordon's Pinetum* (1858), supplied information and the institutional herbarium provided drawings of botanical specimens sent home by David Douglas and others. Loudon was 'deeply indebted' to Loddiges for the 'kind and liberal manner' in which over a decade they had provided free access for his artists to make drawings, supplied specimens and opened 'their unrivalled collection of hardy trees and shrubs as if it had been his own'. The staff of other botanical institution gardens, to a lesser extent, also provided books, specimens and information, especially the Linnean Society and their librarian, George Don, whilst Aylmer Bourke Lambert opened his 'magnificent library', 'unique herbarium' and valuable living conifer collection. The Royal Botanic Garden at Kew was also of some use. The Director William Townsend Aiton (1766–1849) supplied specimens of trees and shrubs from the arboretum, loans of manuscripts and other 'valuable information and assistance', some of which had been through the offices of John Smith (1798–1888) the 'scientific and assiduous botanist', cultivator and foreman of the garden. Aristocracy, gentry and landowners had also rendered significant assistance. The most important of these was Hugh Percy, third Duke of Northumberland (1785–1847), who allowed Loudon's artists to come and take portraits of the 'splendid exotic trees at Syon' and provided, at his own expense, drawings of over 100 of the 'largest and rarest of those trees' by G. R. Lewis. Mrs Lawrence of Studley Royal, Yorkshire, employed a London artist, H. W. Jukes, to take portraits of many of her trees, the Countess of Bridgewater provided portraits of her gigantic beeches at Ashridge. Similarly, through Philip Frost, the 'intelligent and most industrious' head gardener at Dropmore, Lady Grenville allowed portraits and provided much information concerning her 'magnificent pinetum'.[31] The Duke of Bedford and Mr Ireland, the gardener, supplied much valuable information concerning the oaks, Scotch pines and Cedars of Lebanon at Woburn, offering full use of 'that magnificent work', the *Salictum Woburnense*.[32]

The *Arboretum Britannicum* comprised eight volumes: four devoted to text and the remaining four consisting of plates. The first volume included an analysis of the history of trees and shrubs from antiquity to 1834. The most detailed section concerned the history of the indigenous and foreign trees and shrubs of the British Isles. This was followed by a similar analysis of the 'indigenous and introduced trees and shrubs' of other countries with similar temperate climates, drawing extensively on European and North American arboricultural literature.[33] The second part was devoted to the science of trees and examined them 'in all their various relations to nature and art', as 'component parts of the general scenery' and with 'regard to the expression and character of particular kinds', especially in relation to pictorial delineation. Trees were then examined with respect to uncultivated nature, cultivated nature, and to humankind.[34] The third part or main body of the text contained histories and description of various species and varieties of trees and shrubs, native, indigenous, useful and ornamental in British cultivation.[35] In describing each species, Loudon followed a scheme that began with botanical identification followed by 'synonymes, botanical and vernacular', specific characters, popular descriptions, geography and history and poetical and legendary allusions. He then turned towards properties and uses of each species and variety, soil and situation under which they flourished best, propagation and culture, susceptibility to parasites and diseases, examples of British and Continental growth, and finally price in some British, Continental and American nurseries.[36] The third and fourth chapters of the first volume provided lists of the trees and shrubs described and classified 'according to their different capacities for fulfilling' various purposes required by planters and landscape gardeners.[37]

After providing a historical account of the progress of British and then global arboriculture in the first volume, Loudon devoted a set of chapters to the study of trees from botanical and aesthetic perspectives. These explored the botany of arboriculture, artistic representations of trees considered as pictorial objects and as parts of general scenery and status of trees 'as organised beings, or botanical objects and forests or plantations' impacting upon the 'physical circumstances of a country' and the human condition. These chapters examined the 'kind of expression' produced by individual tree forms, their 'attitudes', 'other pictorial qualities' and history in which Loudon claimed to record all facts respecting each species old or young 'which may lead to interesting associations'.[38] The main body of text provided scientific, cultural and commercial information concerning each species with inset botanical illustrations (Figure 4.2). The section on each species and variety generally followed a template in which the 'specific character, description, history, and uses of each individual species, race, or variety' were examined in order. So a description of the specific characters of the *Abies excelsa* or Norway spruce fir for instance, was followed by a list and description of different varieties, a general description including analysis of cultural and economic history, geography, properties and uses, propagation and culture and statistics. Similar information is then provided for the white spruce fir, Douglas fir and others in the same genus, drawing upon a range of correspondents. Illustrations within the text depict leaves, branches, cones and other sections and different examples of the Norway spruce on its own, in groups or in landscape scenery in Norway, Scotland and other countries.[39]

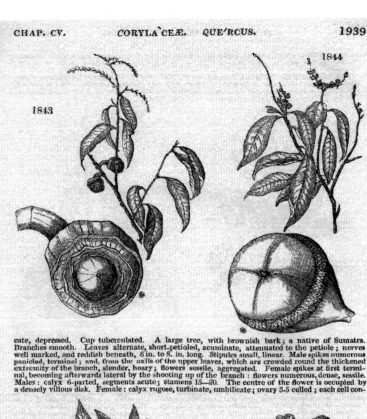

cate, depressed. Cup tuberculated. A large tree, with brownish bark; a native of Sumatra. Branches smooth. Leaves alternate, short-petioled, acuminate, attenuated to the petiole; nerves well marked, and reddish beneath, 6 in. to 8. in, long. Stipules small, linear. Male spikes numerous panicled, terminal; and, from the axils of the upper leaves, which are crowded round the thickened extremity of the branch, slender, hoary; flowers sessile, aggregated. Female spikes at first terminal, becoming afterwards lateral by the shooting up of the branch: flowers numerous, dense, sessile. Males: calyx 6-parted, segments acute; stamens 15—20. The centre of the flower is occupied by a densely villous disk. Female: calyx rugose, turbinate, umbilicate; ovary 3-5 celled; each cell con-

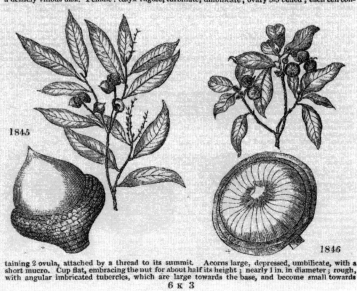

taining 2 ovula, attached by a thread to its summit. Acorns large, depressed, umbilicate, with a short mucro. Cup flat, embracing the nut for about half its height; nearly 1 in. in diameter; rough, with angular imbricated tubercles, which are large towards the base, and become small towards

6 K 3

Figure 4.2: Acorns and oak leaves from J. C. Loudon, *Arboretum Britannicum*, vol. 3, 2nd edn (1844), p. 1939. Reproduced from the authors' personal copy.
1843: Q. costata; 1844: Q. rotundata; 1845: Q. daphnoidea; 1846: Q. platycarpa

The definitions of orders and tribes usually followed those of de Candolle's *Prodromos* and Lindley's *Introduction to the Natural System* (1830) and *Key to Structural, Physiological and Systematic Botany* (1835). Genera of each order or tribe with characters were usually provided straight after the general character of the order or tribe following de Candolle and Thomas Martyn in his edition of Philip Miller's *Gardener's Dictionary*. The species, races, or varieties of each genus followed enumeration of the genera, distinctive characters being given with English name, habit, flower colour, season of flowering and year of British introduction. Where possible, these were correlated with descriptions in author-itative works, recognizing historical or geographical synonyms with derivations of generic and specific names and synonyms where 'instructive or interesting'.[40] Descriptions of each specimen also generally followed an order beginning with the roots and then moving upwards through the stem, leaves, stipules, inflores-cence, bracteas, flowers, fruit and seeds. Particular characters or qualities were noticed above others. For instance, roots were considered in respect of their 'figure, quality, substance, bark, duration, direction, rootlets, fibres, spongioles', susceptibility to produce buds from cuttings, soil, native habitat, and other qualities. Leaves were noticed with regard to 'vernation, internal structure, fig-ure, articulation, insertion, circumscription, colour, texture' and other qualities whilst flowers were described in regard to appearance, colour, structure, size and other general characteristics and component sexual parts.[41]

Following analysis of the distinctive characters of each species, race or vari-ety, each was examined at greater length paying attention to a large range of matters including general characteristics in both natural and artificial states of growth, habit, bulk, figure and duration. The main ethos behind the descrip-tions of species, races and varieties (and mules, hybrids and variations) was to convey information to cultivators concerning the general nature of trees, their roots, branches, wood and seeds, so that they could appreciate how they were best propagated and applied for various purposes. Loudon appreciated that it was important to convey information concerning rates of growth, anatomical structure, physiology, species affinities and susceptibility to diseases, especially with regard to propagation and culture. Inspired by Alexander von Humboldt (1764–1859) and de Candolle, details of geographical distribution and history were regarded as crucial, especially concerning countries of origin, botanical dis-covery and global spread and conditions under which they flourished such as soil and elevation.[42] This was supplemented with considerable information con-cerning the different means by which trees and shrubs were propagated naturally through seeds and suckers and artificially by seeds, cuttings, layers, grafting and other means. The applicability of these varied considerably according to grow-ing conditions and soil, situation or nursery treatment. It was also desirable to imitate natural processes of succession as far as possible for greatest success as exhibited by forests recovering from fire depredation with different kinds of species from those originally destroyed. Sometimes where species were very well known this information was extremely detailed. The analysis of the propagation, culture and diseases of the elm (Ulmus) for instance, ran to seven pages.[43]

Loudon's symbols were a kind of arboricultural shorthand intended to convey information rapidly without practical gardeners having to wade through pages of text (Figure 4.3). The system employed signs under the titles of chapters and in some sections which were intended to show at a glance the general habit of each tree or shrub; these provided information as to whether the tree or shrub was large, small or middle sized, deciduous or evergreen, other aspects of the shape likely to be attained and growing habits. Loudon adopted the signs placed before each individual species and variety from the *Gardener's Magazine* and *Hortus Britannicus* which allowed readers to distinguish immediately between deciduous and evergreen trees and shrubs, climbers, trailers, creepers and those with other growth characteristics. Other signs appended to these in the table of contents were intended to signify the degree of protection that each might require in the London climate. Pronunciation conventions were indicated with the use of symbols and botanical names were all accented with their origins being indicated by various means. Besides names derived from Latin or Greek, classical botanical names were given with the first letter in italics whilst commemorative names were also indicated by italics around the original name and aboriginal or names of uncertain origin were italicized in full.

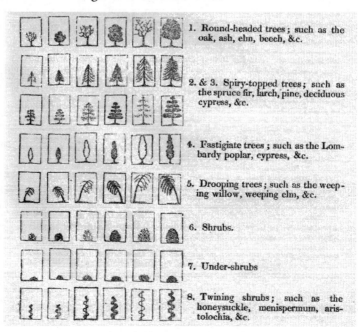

Figure 4.3: Symbols of tree and shrub shapes from J. C. Loudon,
Arboretum Britannicum, vol. 1, 2nd edn(1844), p. xiii.
Reproduced from the authors' personal copy

Inspired by Loudon's experiences of living tree collections especially the arboretums at Hackney, Chiswick and his Bayswater villa, there were many different ways in which readers were encouraged to wander around Loudon's virtual British arboretum. Short 'descriptive notices of species and varieties' that Loudon hoped would be introduced were printed in small type providing a virtual and idealized element to the tour. These were supplemented with accounts of others that had been introduced but lost, kinds not seen by him but described in catalogues and half hardy varieties requiring special treatment. Statistics or accounts of the dimensions of trees received from around the British Isles or Europe were also printed in small type to save room. Like print in newspapers, differences in the size of type in each entry denoted the relative importance, reliability and specialized interest of the information. Judged of 'less interest to the general reader', statistical information was provided on smaller typeface. Equally, all material relating to species unseen by Loudon and concerning half-hardy varieties that required special winter protection also appeared in smaller type, even though he recognized that this was 'likely to prove one of the most interesting parts of our work to many gardeners and amateurs'. Examples of these included half hardy types of cacti that Loudon suspected might grow in unusual situations around London such as behind walls.[44]

There were therefore different levels in which the *Arboretum Britannicum* could be 'read' following the methods already pursued in the *Encyclopaedia of Gardening*. Various indexes and sections signified by different type sizes were utilized to impart maximum information and present hierarchies of interest and importance guiding readers from professional gardeners and botanists to mildly interested amateurs through text and illustrations. The contents pages and indexes served as ground plans for the different routes that walkers through Loudon's magisterial virtual collection could follow guided by divisions akin to sectional tallies. The straightest and easiest path to the description and history of particular species or varieties was to go immediately to the alphabetical index at the end of the fourth volume. Similarly, the most direct path to tree and shrub portraits was to go straight to the illustrations index at the end of the eighth volume. The general text and descriptions were best approached through the epitome of the contents and full contents tables in the first volume, although each of the first four volumes were prefaced by their own tables of contents. The tables of contents, indexes, lists of works cited, lists of engravings, lists of contributors and other parts of the scholarly apparatus totalled some 220 pages and had much the same relationship to Loudon's virtual arboretum as printed catalogues had to living collections.

Picturing and Placing Trees

Descriptions of trees and shrubs in the *Arboretum Britannicum* were intended
to impress the importance of arboriculture upon 'the minds of proprietors and
their families', especially the 'rising generation'. Loudon also strove to have them
drawn in a spatially and temporally standardized manner. Illustrations derived
from portraits of trees and shrubs in the London area in specific 'gardens or
grounds within a limited distance' of the city. To promote the spread of these
trees, Loudon also provided details of the nurseries where they were cultivated
and sold, catalogues of various nursery companies and details of the prices that
trees could be purchased from English, French, German and North American
nurseries. The number and quality of full-length tree portraits set the *Arboretum Britannicum* apart from previous British arboricultural works. These were
not only of botanical sections as had been customary in recent works such as
Lambert's *Genus Pinus* and Watson's *Dendrologia Britannica*, but also full length
views of 'the greater number of species of trees'. The first edition of Loudon's
Encyclopaedia of Gardening in 1822 was one of the earliest books to provide
hundreds of numbered finished wood engravings interspersed in the text rather
than placed upon separate pages, closely integrating text with illustrations. One
of the models seems to have been the use of wood block illustrations within the
text in children's books. Loudon employed the technique in all of his popular
gardening and horticultural works including the magazines and the *Arboretum
Britannicum*, where comprehensive verbal and technical descriptions were a poor
substitute for sensory experience of living specimens. It provided a new kind of
visual immediacy enveloping images within textual descriptions encouraged by
the need to simulate sensory and aesthetic experience, inform and stimulate the
intended audience.

Firstly there were depictions of ten-year-old trees from the London area
drawn to a scale of a fourth of an inch to a foot. Secondly, on the same plate
there were engravings of botanical specimens in flower, in fruit and with winter's wood in the case of deciduous trees, depicted to natural size or on a scale
of two inches for every foot if larger than this. Thirdly, there were illustrations
of fully grown versions with leaf specimens also on the scale of two inches to
the foot. The engravings were intended to demonstrate the varied progress of
trees in different soils and situations and the size, form and character they took
when mature, providing information for gardeners and collectors. Fully grown
trees were presented on a scale of a twelfth of an inch to a foot. Within the main
text of each species, pictorial signs were placed at the beginning of each genus,
or at sectional divisions described whether these were trees or shrubs, deciduous or evergreen, climbers, twiners, trailers, creepers and other characteristics.[45]
Engravings of botanical specimens within the text and accompanying the tree

portraits in the final four volumes were drawn to a scale of two inches to a foot, although some of the dissections, where signified, were either magnified or drawn life size. The botanical drawings were mostly by James de Carle Sowerby (1787–1871), Loudon's trusted friend, who had already produced many of the drawings for the *Encyclopaedia of Plants*. These showed young tree branches in flower, the winter's wood when the tree was deciduous, a branch with ripe fruit, with autumnal leaves and dissections of flowers and fruit. Other details at the base of the full-length portraits showed groups of leaves designated as 'artist's foliage' rather than 'botanical specimens' because they were drawn by the artist who made the full study and who recognized artistic 'differences' rather than botanical characteristics, providing a 'more correct idea of what is called the "touch"'.[46]

The full-length portraits were intended to provide a 'palpable representation of their forms and magnitudes' to make a 'stronger impression' in the minds of readers. Different stages in tree life cycles were shown with portraits of ten- or twelve-year-old specimens from within ten miles of London taken between 1834 and 1837, all drawn to a scale of 1 in to 4 ft. A second set of portraits depicted fully grown trees mostly from the London area drawn at 1 in to 12 ft.[47] Where reasonable fully grown examples were not available within ten miles of London portraits were taken from other places in the British Isles, particularly Studley Park and Dropmore or in a couple of cases, from Continental examples. The set of younger tree portraits were intended to reveal the 'general magnitude, form and character' and varied growth rates of different species and varieties even 'when growing in the same soil and climate' after ten or twelve years growth. Loudon claimed that this would provide 'valuable assistance' in landscape gardening by suggesting species best capable of producing 'any desired effect of wood or of trees' in given localities and times. Readers just had to 'turn over' the portraits within the final four volumes in order to select those species that seemed to exhibit the 'forms and magnitudes' that would 'produce the effect desired'. The fully grown tree portraits were intended to demonstrate the 'magnitude and character' of some mature species and to guide planters with regards to 'pictorial effect', timber production, shelter and shade. Engravings of some species and varieties, to a scale of two inches to a foot, with the flowers and other parts less than an inch in diameter of natural size were given along with the text. The tree portrait volumes and engraved botanical sections were intended to foment an interest in arboriculture for those who had previously given it little attention, the 'grand object' of the *Arboretum Britannicum*.[48]

Most of the tree portraits were taken between August and October, although some were done in winter to illustrate the 'skeleton tree without its foliage', as familiar to many readers as when fully covered. Loudon emphasized that all the engravings had been 'drawn from nature' by 'competent artists', names and tree places being listed in part of the table of contents.[49] The extensive artwork throughout the *Arboretum Britannicum* caused greater difficulties than Loudon originally envisaged, particularly because of the degree of standardization that he sought from using recognizable living trees, which helped to make them very expensive. As the work originally appeared in numbers, one result was that blanks had to be left when good quality specimens of the right age could not be found. Although these spaces were, in most cases, subsequently filled, confusion of names, changes in identification and radical differences between trees at various stages of their lives posed additional challenges to Loudon's system, resulting in some 'inaccuracies' that had to be corrected in the second edition.[50]

Conclusion

Whilst the immediate impact of the *Arboretum Britannicum* was great, its importance, originality and comprehensiveness ensured that it became the standard work on British arboriculture. Until the end of the nineteenth and early twentieth centuries, no other study strove to provide a comprehensive analysis of hardy British trees and shrubs, subsequent Victorian arboricultural works examined only specialized parts of the subject such as forestry or individual genera or species. No book sought to rival the *Arboretum Britannicum* until Henry Elwes and Augustine Henry's *Trees of Great Britain and Ireland* (1906–13) and William Jackson Bean's *Trees and Shrubs Hardy in the British Isles* (1914). Loudon's *Arboretum Britannicum* was rooted in study and representation of trees and tree collections in places from Loddiges' Hackney nursery arboretum to the Bayswater villa back garden arboretum and the vast semi-public and private collections such as those of Woburn and Kew.

The cost of producing the *Arboretum Britannicum* was not measured in financial terms alone. As we have seen, Loudon's health was already seriously compromised before he began the work and his wife believed that her husband's physical problems were deepened by the debts that he laboured under from the 1830s, largely due to the *Arboretum Britannicum*. Looking back, she suggested her husband had 'not contemplated the expenses which he should incur' in the production of the work and having begun, found that it was 'impossible to compress it into the limits originally intended'. In Jane Loudon's view, 'in his determination to make the work as perfect as possible, he involved himself in the difficulties which hastened his death', doggedly bringing out each number and launching the *Architectural Magazine* and the *Suburban Gardener* whilst continuing to produce the *Gardener's Magazine* and other publications.[51] However, to her husband, constantly reminded of his own mortality, the *Arboretum Britannicum* must have presented his greatest opportunity to leave a lasting monument in contrast to the everyday concerns of the *Gardener's Magazine*, like the longevity of its tree subjects compared to transitory annuals. Having lost much of his wealth from farming through unwise financial speculations, Loudon gained a steady income from the *Gardener's Magazine* and other publications, although this declined as competition appeared such as Paxton's *Horticultural Register*. The costs of producing the *Arboretum Britannicum* mushroomed as it grew in scope and ambition. Large sums were paid to artists and other assistants and by the time of publication in 1838 Loudon owed £10,000 to the printer, stationer and wood-engraver. Although the creditors did not press for money but gave him 'a chance of reaping the benefit of his labours' by agreeing reimbursement through his publishers and much was paid off, further debts were incurred through the production of the *Encyclopaedia of Trees and Shrubs* and other books. This would have been fine had not the wood-engraver become bankrupt; it was harassment from his assignees threatening bankruptcy and possible imprisonment that made Loudon's final days so miserable. It is significant that the *Arboretum Britannicum*, which in Jane Loudon's view 'hastened his death', should have been the source of some relief after he died, copies bought by the nobility, landowners and businessmen after a special appeal raising hundreds of pounds for his family.[52]

5 THE BOTANY OF THE
ARBORETUM BRITANNICUM

Introduction

The most important scientific message of Loudon's *Arboretum Britannicum* was that botany was essential to all aspects of arboriculture from the economic and cultural uses of trees and tree products to effective artistic depictions of trees and landscape gardening. This paralleled Loudon's friend and collaborator John Lindley's attempt to place botany on a more rigorous, serious 'masculinized' and scientific institutional footing and raise its status in relation to physics and chemistry. With relatively few indigenous species and a flourishing commercial nursery sector, hardy and half hardy trees and shrubs provided a manageable subset of plants and demonstrated the efficacy of the Candollean natural system. The extensive scientific, economic, cultural and statistical information in the *Arboretum Britannicum* conveyed a sense of weighty learning, authority, professionalism and comprehensiveness whilst satisfying the Victorian desire for natural historical facts, objects and specimens. Loudon was criticized for a lack of moral and religious feeling and content in his applications of science to nature for instance, in relation to garden cemeteries. However, the unrivalled and daunting enumeration of facts that characterized all his works generated a secular form of wonder partially replacing the celebration of divine nature and natural theological assumptions underpinning most early and mid-Victorian natural historical works. The relentless accumulation of information added weight to the core scientific message that modern physical, anatomical and taxonomic botany was essential to modern arboriculture. The post-Enlightenment image of fixity and order that characterized Loudon's arboretums, both virtually in text and planted in the ground, demonstrated the efficacy of the Candollean system. It provided a sense of order in response to the chaos and uncertainty still prevailing in early nineteenth-century British botany in the face of rival natural and numerically-founded systems and the continuing appeal of Linnaean botany, especially to amateurs.

By striving to professionalize gardening and formalize gardening education and training, and extolling the benefits of scientific education for gardeners of all social classes, Loudon encouraged contemporary scientific reformers, although his perception of botany was less hierarchical and elitist than William Hooker and Lindley's interpretation of academic institutionalized botany. Like Hooker and Lindley however, Loudon was striving to expand audiences for botany through the publication of popular works whilst working closely and sometimes uneasily with influential and wealthy aristocratic amateurs and patrons, leavening suggestions and criticisms with inducements and flattery. The *Arboretum Britannicum* demonstrated the interconnection of physiology, taxonomy and anatomy in modern arboriculture as a serious science differentiated from the gathering, listing and naming of Linnaean natural history, underscoring the difficulties of employing the latter for identification of trees and shrubs. Furthermore, Loudon's application of the Candollean system was encouraged by the practical demands of tree planting in landscape gardening, providing apparent order, regulation and simplicity in the face of the bewildering array of specimens. As a highly successful farmer, landscape gardener and proprietor of the *Gardener's Magazine* and other publications with their extensive correspondence networks from all social classes, Loudon was highly attuned to the practical needs of his farming, gardening and popular science audiences. This helps to explain why he followed Candolle even more closely than Lindley in simplifying and reducing taxonomic distinctions, becoming more of a 'lumper' than a 'splitter'. Conventionally Linnaean botany was still perceived as ideal for amateurs and novices. Loudon, however, believed that the Candollean natural system, which as we have seen, emphasized distinctions between species more than Jussieu, was more intuitively meaningful to non-professional botanists. In applying the Candollean natural system to arboriculture, Loudon aligned himself with most leading contemporary Victorian botanists including John Lindley, Joseph Hooker, George Bentham, Charles Darwin and John Stevens Henslow.[1]

This chapter examines how Loudon drew upon the systematic and physiological study of trees and shrubs to demonstrate the importance of arboriculture and the application of these ideas and practices in landscape gardening and landscape art.[2]

The *Arboretum Britannicum* and the Natural System

Loudon followed the principles of the natural system in the *Arboretum Britannicum*, which he believed would demonstrate its effectiveness for arboricultural taxonomy. However, although required to fulfil various objectives, the application of the natural system in the rapidly changing state of international arboriculture was not a straightforward process in the face of competing taxonomies. The need for a British arboricultural work founded upon the natural system was underscored by the fact that John Lindley, Loudon's close collaborator in the *Encyclopaedia of Gardening*, had 'long had a similar' work to the *Arboretum Britannicum* 'in contemplation'. Loudon 'entirely' gave up the idea of proceeding with his projected arboricultural encyclopaedia until 1834, when Lindley seems to have informed him that he was no longer interested in the project. During the 1830s and 1840s as professor of botany at University College and editor of the *Gardener's Chronicle*, Lindley remained preoccupied with combating supporters of Linnaean botany and improving the natural system. These efforts are evident in his *Introduction to the Natural System of Botany* (1830), the much enlarged *Natural System of Botany* (1836) and *Vegetable Kingdom* (1846).[3] By 1846 Lindley regarded the battle to have been won, although in fact it took another decade or two for the natural system to become fully established. Opponents used Lindley's suggestions for improving the Jussieuan system in the *Vegetable Kingdom* as an argument for the efficacy of Linnaean botany. Even though Lindley never completed his work on British arboriculture and was content to leave the field for Loudon, the fact that the leading crusader for the natural system regarded such a book as necessary is highly significant and no special justification was regarded as necessary for organizing the *Arboretum Britannicum* according to the Candollean system.

To examine the large number of hardy trees and shrubs covered in the *Arboretum Britannicum*, Loudon marshalled a formidable range of national and international evidence, practical experience and scholarship from Europe, North America, Australia and beyond. Although the book was primarily intended as a study of British arboriculture it was, as he emphasized, in many ways an international study reflecting the fact that as microcosms of the botanical world, arboretums were inherently global. Loudon made full use of the international network of botanists, professional and amateur gardeners and landowners fostered by his European tours and publications. These international contacts supplied him with specimens, seeds and other tree parts and products, portraits of trees and shrubs, whole specimens, and a large amount of information. A number also assisted by providing the geographical histories of international arboriculture in the first volume which included separate sections on France, German, the United States, Australia, China and Turkey provided by de Candolle at Geneva, Francois André Michaux (1770–1855) and Philippe-Andre de Vilmorin (1776–1862) of Paris, Joseph Franz Jacquin (1766–1839) and M. Charles Rauch of Vienna and the Prussian-Dutch botanist Professor Caspar Georg Carl Reinwardt (1773–1854) of Leyden. The French naturalist and entomologist Jean Victoire Audouin (1797–1841), a member of the Institute of France, for example, provided information concerning diseases of the elm, de Candolle read through the proof sheets of the article on Salisbùrie and the sec-

tions on French and Swiss arboricultural history along with Vilmorin. Christian
Abraham Fischer (1785–1836) Inspector of the Göttingen Botanic Garden and
Christoph Friedrich Otto (1783–1856), Inspector of the Berlin Botanical Gar-
den, corrected the section on German arboricultural history. In North America,
Loudon's most important contacts were in Philadelphia. These were the distinguished physician and natural philosopher Dr James Mease (1771–1846)
and the printer Colonel Robert Carr (1778–1866), who maintained Bartram's
Botanic Garden in the city on the banks of the River Schuylkill – now the oldest surviving botanic garden in North America – with his wife Ann Bartram
Carr (1779–1858). Information from Australia was provided by John Thompson (1800–61) of the Surveyor General's office in Sydney.[4] As this suggests,
although the *Arboretum Britannicum* was less of a collaborative work than the
Encyclopaedia of Plants, it still entailed considerable national and international
cooperation.

Information was also obtained from many British and Irish and some European and North American landowners and gardeners. More direct assistance
came from Rev. Miles Joseph Berkeley (1803–89), author of *Monograph of the
British Fungi* (1836) and often regarded as the father of British mycology, who
supplied the lists of tree fungi. The entomologists William Spence (1783–1860)
and John Obediah Westwood (1805–93), Secretary to the London Entomo-
logical Society, provided descriptions and drawings of tree insects. John Denson
(d. 1874), who had been a gardener and curator of the Botanic Garden, Bury
St Edmunds in Suffolk during the 1820s and one of Loudon's co-collaborators
on the *Magazine of Natural History*, drew up the characters of the orders and
genera whilst overseeing theoretical botanical aspects of the plan. David Don
who, as we have seen, was librarian to the Linnean Society and professor of botany at King's College London, helped to arrange and define the whole of the
generic characters, and read through sections such as that on Coniferae. Loudon
also utilized the special knowledge of others on individual genera. The Sussex
landowner and botanist William Borrer (1781–1862) for instance, generally
regarded as the father of British lichenology, used his experience developing a
large Salicetum to oversee the arrangement into groups of the numerous species
of the difficult genus Salix and read through the proof sheets of that section.[5]

Although Loudon claimed that the superiority of the natural system was
obvious and that it had been widely accepted by 'botanists and scientific cultivators', he still saw the need to justify its application in the *Arboretum Britannicum*.
Candolle's version of the natural system was just one version competing with
numerical forms of classification and the Linnaean system in British botany until
at least the 1850s. The work therefore provided an opportunity to showcase the
effectiveness of Candollean taxonomy to many readers who remained uncon-
verted and to present it as inherently logical, approachable and democratic when

applied to trees and shrubs compared to the scholarly elitism of Linnaean abstractions. There seemed little need to justify not using an alphabetical arrangement even though most arboricultural works published over the previous couple of centuries had adopted what Loudon regarded as such an unscientific scheme.

The Candollean natural system efficaciously grouped together 'objects which resemble one another in the greatest number of particulars' or were 'most alike in their qualities', whilst the various indexes facilitated rapid referencing. This meant that everything known concerning the properties, uses and cultures of tree and shrub groups could usually be inferred from individuals. Similarly, however much names might change, they would 'always be found associated together in the same group or in groups nearly adjoining' and when new plants were received their natures could 'be anticipated' by observing their 'resemblance to some group already known'.[6] Loudon had to admit that many, especially those who knew 'little of scientific botany', believed the Linnaean system superior for 'practical men' and 'amateurs'. This was mistaken because although mastery of the natural system required 'much study and perseverance', even an imperfect knowledge of it was of 'far greater use' to cultivators, medical man, travellers and amateurs 'than the most profound knowledge' of Linnaean botany. The underlying principles of the natural system also accorded more intuitively with those of human sensory, intellectual and educational development. Although they might not know any botanical names, 'to a certain extent' all children 'in the habit' of seeing trees, shrubs and plants understood the natural system. To 'preserve order' in their ideas and assist their memory, they were 'obliged to throw all the conspicuous plants' they encountered into 'groups palpably distinct'. They would therefore naturally form 'three grand classes of trees, shrubs, and herbaceous plants' and group together 'broad leaved and fir-leaved trees, deciduous and evergreen kinds, fruit-bearing, barren' and other categories. These accorded well with the principles underlying natural system divisions which operated by bringing things 'together', designating them under 'one common name on account of their general resemblance'. Arboricultural analysis informed by the natural system demonstrated how botany was approachable and essential for amateurs and scholars alike, serving Loudon's political agenda by breaking down barriers between elite and amateur science whilst apparently reinforcing the rigour of the latter. There was 'little difference' between the 'natural system of the most learned botanist' and the 'most ignorant country labourer', except the former had 'gone more profoundly into the subject' gaining knowledge more from 'principles deduced from ... facts accumulated by his predecessors' rather than 'personal experience' alone. The effectiveness of the natural system in arboriculture demonstrated that 'all sciences not purely abstract' were 'founded on some simple instinct of our nature, which is perceptible in the customs, not only of ignorant persons in civilised society, but of the rudest savages'.[7]

Taxonomy of the *Arboretum Britannicum*

Despite the fact that Loudon estimated there to be only 1,500 kinds of trees and shrubs in Britain, the large number of exotics being imported fomented considerable 'uncertainty in the application of names'. He claimed that in 'genera consisting of many species', there were 'scarcely two' London nurseries 'where the same names are applied to the same things'. What was regarded as a variety in one was elsewhere 'elevated ... to the rank of a species'. Similarly, arboricultural ignorance from travellers rendered their remarks on trees 'of little use' because it was 'impossible for botanists to ascertain' species from their generally unscientific descriptions.[8] Loudon contended that applying the Candollean system encouraged a reduction and simplification in the number of species, and sometimes varieties of trees and shrubs and would have reduced these further if he had been 'inclined to trust entirely' to his own opinion'.[9] He thought it 'well known' to 'cultivators of trees and shrubs', that many names of species enumerated in botanical works or even the *Hortus Britannicus* as having been introduced to Britain were in fact 'not to be found in any nursery or even botanic garden'. Some might have been introduced and lost, others could be names of plants already in the country reintroduced under new names, yet 'according to the present mode of compiling botanical catalogues' fostered by the Linnaean system, these would all be recorded whether they existed or not, which added to the confusion. When collectors ordered them from nurserymen, they either 'did not receive any plants at all' or got something they did not want or already had. Secondly, 'in order to supply the demand for novelties or appear to have large collections' nurserymen 'too frequently' introduced names into catalogues plants that could not be supplied or synonyms without making this clear. The effect was that 'gentlemen intending to form collections, finding their intentions frustrated, frequently give up the pursuit in disgust'.[10]

Given the imperfection of catalogue records and the 'erroneous ... nomenclature' of public nurseries, it was difficult for practical gardeners to know precisely which trees and shrubs were available, inducing arboricultural conservatism. Loudon claimed that only gardeners who had worked in arboretums at one of the major botanic gardens such as Kew, the Horticultural Society, Edinburgh or Loddiges could be said to 'know the names of one fourth of the trees and shrubs already in the country', underpinning the importance of national systematic tree collections. He intended that the *Arboretum Britannicum* would 'go far to remedy' the evil of botanical ignorance by enabling gardeners and their employers to ascertain exactly what trees were in the country and where they were growing and could be procured. Although Loudon relied upon some descriptions provided by other botanical authorities, the identification of species and varieties was primarily achieved through examination of living trees and shrubs and comparisons with existing engravings and descriptions. The attention to living trees and shrubs within the metropolitan vicinity allowed the *Arboretum Britannicum* to 'contribute' to his principal objective, that of 'diffusing a taste for planting collections of trees and shrubs'.[11] Nurserymen, 'by referring to these living trees' would have opportunities to verify catalogues of their own collections and to augment these with cuttings or plants. Using the nomenclature and illustrations of the *Arboretum Britannicum* and the references to living trees and shrubs, buyers could have greater confidence whilst nurserymen would be stimulated to improve accuracy and enrich their collections by propagation and purchase. Loudon claimed that after the publication of the *Arboretum Britannicum* it would be 'the fault of the nurseryman alone' if his collection did 'not contain plants of all the species and varieties which we have figured and described'.[12]

To achieve these objectives and provide a work useful to botanists, nurserymen and collectors, Loudon tried to utilize clear and consistent definitions of species, varieties and other taxa. In the face of multiple imports and confusion engendered by misunderstandings, commercial demand and taxonomic disagreements, contradictions and inconsistencies seemed inevitable. Much difficulty resulted from the problem of distinguishing between species or 'natural and permanent forms' and forms 'accidental' and the result of 'cross fecundation', 'culture, soil, situation, disease' and other factors that had led to a proliferation of species.[13] Loudon claimed that careful observations of metropolitan trees and shrub collections through the seasons over a number of years had guided him towards a more satisfactory arrangement of species, races, and varieties.[14] The problem was that 'modern' plant catalogues contained an astonishing 'number of specific names ... ranked under one generic name', yet many of the relationships between these were dubious and almost impossible to discover. One name denoted colour or another property others indicated native countries of species or commemorated individual or place names. Observations of the columns in catalogues denoting native species countries 'increased' rather than lessened 'the confusion by placing plants from the tropics with others from the frigid zone' and so on. The correct method according to Loudon was to proceed inductively by accumulating facts and generalizing upon them. However, it appeared that 'as far as regards species and varieties, the great object of botanists' had been to increase numbers 'without much regard' to relationship groupings. Although the *Arboretum Britannicum* could not solve all these problems, Loudon strove to make species, races, varieties, and variations clear and distinguishable from mules and hybrids.[15]

Loudon's scepticism concerning botanical varieties and perception of needless proliferation was informed by his knowledge of plant breeding, particularly in horticulture and agriculture. He placed much emphasis upon the importance of arboricultural improvement through cultivation, botanical tree studies, economic exploitation and landscape gardening, which underscored the importance of consistently reliable forms of classification. He emphasized that 'many species' could undergo improvement by 'cross-fecundation with other species nearly allied to them or by procuring new varieties through the selection of remarkable individuals from 'seedlings raised in the common way'. Novel and 'curious' varieties could also be produced through the 'selection of anomalous' shoots known as 'sports' which preserved their peculiarities through grafting. Loudon reminded cultivators that all the most valuable plants, 'whether in agriculture, horticulture, or floriculture', from fruit trees to herbaceous vegetables, had been 'more or less indebted for their excellence to art' such that even botanists couldn't distinguish between wild and cultivated plants of the same species. It was probable that methods used to improve fruit trees could be applied to create useful kinds of timber trees just as some types of oak, elm and magnolia had been improved through 'accidental crossing' and deliberate selection of varieties with variegated or drooping leaves, erect shoots or other qualities. Although arboricultural improvement through cross-fecundation was still 'in its infancy with respect to timber trees' and it was unclear how far it would go, Loudon saw no reason why there could not be a purple-leaved oak, elm, or ash as well as a purple-leaved beech or drooping ash. He praised work that promoted the importance of breeding to improve the quality of timber for naval and construction purposes such as that of the Scottish landowner Patrick Matthew. Oak trees demonstrated astonishing variation by culture and after the 'numerous American varieties' had been introduced into Britain and borne seed, there seemed 'no end to the fine hybrids that may be originated between them' and European species. Although the importation of foreign species as the 'raw material with which we are to operate' was still in its infancy, there seemed as great a potential to improve ornamental trees and shrubs as there was to improve fruit varieties.[16]

With some modifications, Loudon adopted John Lindley's definition of a species as a 'union of individuals agreeing with each other in all essential characters of vegetation and fructification'. They also had to be 'capable of reproduction by seed, without change; breeding freely together, and producing perfect seed, from which a fertile progeny can be reared'.[17] This 'general definition' maintained by botanists 'evidently' required 'some modification' given the fact that as in the case of many cultivated annual plants, varieties and races were 'reproduced from seed'. If it was the defining criterion then 'red, white, woolly-eared, and smooth-chaffed wheat' and the different varieties of cabbage, turnip and common lupin would be regarded as 'distinct species'. Similarly the 'common varieties' of cultivated fruit trees would be species given that 'when no cross fecundation has been effected' their seedlings always bore 'a nearer resemblance to the variety which produced the seeds' than to any other or to original species. With regard to reproduction of varieties from seed, Loudon argued that trees and shrubs were subject to the same laws as annual plants, although he accepted that lack of knowledge concerning varieties and tree longevity made it difficult to appreciate the 'identities of their natures'. Observation of examples raised from seed such as oaks in an oakwood or hawthorns in an uncut hedge suggested that no two individuals were 'exactly alike' in terms of foliage, flower, fruit, mode of growth, or timing of budding, flowering, fruit ripening, or leaf fall. He had 'no doubt' that if acorns or haws of varieties in such woods or hedges were sown they would be 'more like the parent variety' than any others in woods or hedges just like wheat, cabbage or fruit tree seedlings, unless artificial or accidental cross-fecundation had occurred.[18] Following de Candolle, he defined these as 'cultivated varieties' or 'races' to be distinguished from 'accidental varieties' that seemed less likely to be propagated by seed. Some weeping varieties of trees for instance, such as the common ash and fastigiated kinds like the 'Exeter elm' would only occasionally 'come true from seed'. It seemed probable that variegated trees and shrubs 'would not always come true from seed, any more than variegated animals or bulbs', although 'a certain proportion' of progeny would indeed be variegated. The 'raising of seedlings from such accidental varieties' demonstrated that they were not 'entitled to rank with cultivated varieties or races'.[19]

Underpinning Loudon's approach to taxonomy in the *Arboretum Britannicum* and encouraged by de Candolle was a belief in the potency and constancy of specially created aboriginal species from which all varieties had been derived through time. Loudon and Lindley believed this knowledge could be used to distinguish between aboriginal species and more ephemeral varieties, despite the considerable difficulties entailed. Lindley emphasized the problems of distinguishing varieties from species, noting that 'the manner in which individuals agree in their external characters' was 'the only guide which can be followed in the greater part of plants', given that the effects of sowing seed or mixing pollen remained largely unclear. Determining differences between species and varieties was 'wholly dependent upon external characters', which could only be appreciated through painstaking empirical work, although it remained 'in all cases, to a certain degree, arbitrary'. Lindley argued that it was probable 'in the beginning' that species alone had been formed and that each had, 'since the creation, sported into varieties' by which species limits had 'now become greatly confounded'. For instance, from just one or a few aboriginal rose species 'endless varieties' had been produced 'in the course of time', some of which, for a 'long series of ages', had become dependent upon 'permanent peculiarities of soil or climate'. These had been 'in a manner fixed, acquiring a constitution and physiognomy of their own'. Subsequently, these 'again intermixed with each other, producing other forms', a continuing process that rendered it virtually impossible to recognize originals. However, whether or not the hundreds of rose varieties were produced in this manner, the 'forms' into which they divided were 'so peculiar' that classification was pragmatically 'indispensable to accuracy of language', and therefore some had been regarded as species. Of these some were 'natural species' presumed to be aboriginal, whilst others were 'botanical species' identified by 'external characters'. The former were 'ascertained to a very limited extent', yet the latter dominated 'nearly the whole of systematic botany' which explained Lindley's broad definition of a species as 'an assemblage of individuals agreeing in all the essential characters of vegetation and fructification'. Fine in theory, but this meant in practice that, as Lindley noted, 'the idea of distinguishing' 'with anything like absolute certainty' between species, botanical species, or varieties seemed 'almost hopeless'. It was necessary to define the nature of 'essential' species characters which depended 'upon a proneness to vary' or be 'constant in particular characters'. One class of characters might be 'essential in one genus, another class in another genus', which was only to be determined by careful observation and experience. Hence species determination was 'in all respects arbitrary' depending upon 'the discretion or experience' of botanists, although Lindley suggested that 'decided differences' in leaf forms, stem figures, plant surfaces, inflorescence, proportion of parts, or forms of sepals and petals usually constituted 'good specific differences'.[20]

Whilst Loudon found Lindley's advice and experience instructive, this counsel of despair provided an insufficiently certain taxonomic framework to achieve the practical objectives of the *Arboretum Britannicum* or for planting arboretums. Informed by Lindley, Loudon sought a more positive definition of distinctions between taxa from his friend de Candolle whose *Theorie élémentaire* and *Physiologie végétale* Loudon believed presented the matter 'in the clearest light'. He followed de Candolle in defining species as 'aboriginal species'; the individuals of which all bore a 'sufficient degree of resemblance to each other' suggesting probable origination 'in one being or one pair of beings'. The criteria of resemblance used to judge species varied considerably within different families, and two individuals really belonging to the same species, such as fruit tree varieties, often differed 'more between themselves in appearance than others ... of distinct species'.[21] It was possible over time to distinguish 'with absolute certainty' aboriginal from variable features if all alleged 'species and varieties of any tree, shrub, or plant', however numerous and apparently remote from original species, could be 'collected together and cultivated in the same garden'. Despite the thousands of apple varieties 'in cultivation throughout the temperate regions of the world', including crabs, there was never any likelihood of one of these being mistaken for a pear. One general character of leaf, flower and fruit was common to all, though it might not necessarily be easy to define this 'essential character' so as to render it observable to any who had not seen many apple and crab varieties before. Equally, in the case of common hawthorn, though fruit colours varied from pale red to yellow and black and some had drooping branches whilst others were 'rigidly erect and fastigiate' and so on through cultivation, yet there was 'never ... any difficulty' in determining that they belonged to the same species. Loudon also suspected this applied to other varieties of genera listed in catalogues as distinct species including Elmus, Salix, Quercus and Pinus. If all native American ashes could be brought together into one plantation and observed for years, Loudon had 'no doubt' it would be much more difficult to assign each to one species than it had been to designate the 'different varieties of the European ash to the *Fráxinus excelsior*'. Furthermore, following de Candolle's principles, all European elms might be reduced to a mere three species and there would be 'a tithe of the names which are ranked as such under *Salix*, *Quercus* and other species'.[22]

Turning to races, Loudon also followed de Candolle in regarding these as 'such a modification of the species' produced either by 'exterior causes' or 'cross fecundation', that could be 'transmitted from one generation to another by seed'. Taking the case of cultivated vegetables and fruits again, the 'greater number' of these ought therefore to be regarded as races because they could be grown from seed, the 'culture given and other circumstances being the same'. If the culture was 'neglected for a series of generations', there was 'no doubt that the race would revert to the aboriginal species' given the tendency towards this 'found to take place both in plants and animals'.[23] The fact that variation induced by environmental stimuli was transitory and might be reversed upon a return to original conditions, as de Candolle maintained, further illustrated the importance of careful and conservative arboricultural species delimitation. Varieties differed from races, 'in not being susceptible of propagation by seed' without 'anything like constancy and certainty'. So the ordinary jargonelle pear could be continued by seed whilst the variegated foliage variety couldn't be 'propagated with certainty', although if many seeds were sown, some plants 'would probably show' foliage variegation. Seedlings of variegated hollies, fastigiate or pendulous-branched plants would also probably exhibit suchlike variegation in some cases, but to what extent it was 'impossible to say'. This would not be enough to cause the confounding of races with varieties or to 'render it desirable to propagate varieties in this way', so they would always be propagated 'by some modification of division', such as using cuttings, layers or grafting.[24] Of what Loudon perceived as the lesser categories, variations differed from varieties in 'not being transmittable by any mode of propagation' and being produced by 'local circumstances' such as changes in soil, water or light conditions, and would therefore disappear when these ceased to operate.[25] Finally, the term mule ought to be limited to hybrids 'raised between different aboriginal species' that appeared 'not susceptible of propagation by seed', whilst hybrids generally were 'the produce of cross fecundation of different races and varieties of the same species'. The latter was 'one of the most important elements' of horticulture and gardening, having 'given rise to the most valuable garden flowers, table fruits, culinary vegetables, and agricultural plants'.[26]

Simplifying the Categories

Loudon used the *Arboretum Britannicum* to demonstrate how botanical species could be simplified and distinctions between species and varieties improved using the Candollean system. The natural system also helped to promote his model of arboretums accessible to the middle classes and not merely the landed elite. Loudon emphasized that he was not undervaluing botanical species nor denying 'the distinctness of many' or the 'propriety of having different names for them, and keeping them distinct'. On the contrary, comparing plants with men, he suggested that aboriginal species were akin to 'mere savages' whilst botanical species were 'according to De Candolle's classification', races and varieties like 'civilised beings'. However, conflation of varieties or botanical varieties with aboriginal species created confusion for beginners and collectors who would be unable to make accurate selections. Secondly, he wanted to prevent collectors from having to puzzle themselves 'unnecessarily' concerning 'minute differences' between botanical species, many of which were highly subjective and misleadingly presented in botanical textbooks. Loudon asserted that the 'nicety' of these distinctions had deterred numerous individuals from studying 'practical botany' and prevented others from ever hoping to gain proficiency. Most importantly, it had created a kind of botanical and gardening elitism that 'prevented many' from 'forming collections of trees and shrubs' by creating the impression that they were the preserve of the rich. The *Arboretum Britannicum* would provide reassurance by demonstrating the small number of hardy trees and shrubs compared to the number of 'herbaceous plants or stove or green-house plants', showing that there was not 'the slightest difficulty' in coming to know all species, provided they were 'only seen together'. It would also show how small the cost of forming a complete collection of trees and shrubs from metropolitan nurseries actually was, placing it well 'within the reach of every planter' of single acre villa grounds.[27]

In his practical application of de Candolle's natural system, Loudon strove to retain as many botanical species and varieties as he believed to exist, placing them after those believed to be the aboriginal species to which they belonged. In the case of the genus *Fraxinus* or ash for instance which in the *Hortus Britannicus* had been described as consisting of forty-one species and twelve varieties, Loudon now ranked thirty of the species under '*Fraxinus americana*', two under '*F. lentiscifolia*', and the rest under '*F. excelsior*'. He did not make 'distinct genera of these three species' because all were 'so obviously of the same general appearance as evidently to belong to the same family'. Similarly be believed that the separation of the oak family into distinct genera could not be justified although there were 'not a dozen aboriginal species' of tree globally. These divisions facilitated generalization and particularization alike, and distinctions were much easier after 'throwing all the races or varieties of *Fraxinus americana* into one group' than they were by leaving them as distinct species.[28] Loudon did not attach much value to the 'specific characters of botanical species' defined as 'races or varieties' because it was often impracticable 'to discover a species or race by such characters alone'. However, specific characters of 'aboriginal species', insofar as they could be determined, needed to be cogently and economically expressed so 'tree names' could easily be discovered. With regard to the genera *Fraxinus* and *Crataegus*, for instance, Loudon challenged 'any botanist' to determine 'from the specific characters of the botanical species' of *F. americana* or of *Crataegus oxyacantha* the 'individuals' to which these applied 'without having recourse' to tortuous and impractical descriptions, dried specimens or engravings. A similar problem existed with regard to 'most of the alleged species of *Salix, Ulmus, Quercus, Pinus*' and 'more specifically the genera *Quercus, Salix, Ulmus, Rubus* and *Tilia*'.[29] Specific characters were of some use when applied to distinguish varieties of aboriginal species. For instance when considering '*Fraxinus pubescens, caroliniana* and *lancea*' as absolute species to be 'compared with different botanical species of the same aboriginal species', and *F. excelsior*, it was 'extremely difficult, if not impossible to apply them'. However, 'if it were known that these botanical species were only varieties of *F. americana*', then the 'difficulties of distinguishing' were 'greatly diminished'. For these reasons 'in many cases', Loudon provided the 'specific characters of botanical species given by botanists supplemented with 'descriptive particulars' from his own observations.[30]

Loudon suggested various reasons for the 'great imperfection' of 'specific characters of botanical species' and the proliferation in their number to an extent that he doubted existed 'in nature'. The first of these was the dependency upon descriptions of dried specimens and the second was the tendency amongst collectors to conflate varieties with species impelled by the desire to 'discover novelties' to satisfy and impress clients and patrons and establish botanical reputations. Although 'very natural' and the result more of youth and inexperience than intended deception, ardent collectors seized upon 'every variation produced by climate, soil, situation, age, or even accident' to add supposedly novel specimens to herbariums as botanists competed for the honour of discovering new species. Differences between foreign and British conditions encouraged the tendency because of the different appearances plants frequently adopted in 'wild' and 'cultivated' states exacerbated by the misleading qualities of many dried specimens sent home. Decisions of naturalists in studies still tended to be privileged above those of collectors in the field, who too frequently deferred to the former. Although a problem associated with all plant types, this was 'more particularly the case' with trees and shrubs given the length of time they often required to mature and their tendency to 'assume very different appearances under different circumstances'. Widespread acceptance of the natural system, systematic global provision of arboretums and botanical gardens and publication of works such as the *Arboretum Britannicum* would discourage these tendencies and help to distinguish species from varieties, encouraging international plant descriptions observed from nature. Loudon was confident that systematic planted collections of all tree and shrub species and genera concentrated together in different local climatic and geological conditions would eventually reveal genuine aboriginal species. The process was 'happily' well under way as botanists sprang up 'in every civilised country' or emigrated across the world to settle in 'newly discovered countries' as the 'wealthiest' European governments promoted botanical gardens in all climates. Loudon suggested that each botanical garden could assemble 'all the alleged species or varieties of at least two or three genera' which most suited their 'climate, situation, soil, and extent'. The widespread international informa-tion and specimen exchanges thus promoted would reveal the 'real' aboriginal species.[31] General understanding and dissemination of the natural system led by international arboriculture would discourage the extravagant species pro-liferation associated with Linnaean collecting, naming and describing, making it much easier to reverse errors. Re-ordering of the botanical hierarchy would encourage this process, as anatomical, physiological and chemical studies sup-planted 'comparatively mechanical' taxonomy as the 'highest departments', thus raising the status of arboriculture and botany as sciences.[32]

Despite Loudon's efforts in the *Arboretum Britannicum*, as he would have expected, arboriculural taxonomic disagreements concerning species delimi-tation continued throughout the nineteenth century. Faced with the task of producing a comprehensive arboricultural treatise on hardy British trees and shrubs, William Jackson Bean noted the 'enormous number of new species' that had become available for cultivation since Loudon's day through the activities of collectors such as William Lobb (1809–64) in Chile and California, and Robert Fortune (1812–80) in Japan and China. Chinese varieties proved particularly challenging given that many remained unclassified or unnamed, whilst the des-ignation of 'hardy' was difficult to apply as it depended upon gardening taste, economic value and experience. Systems of nomenclature provided no neces-sarily fixed framework either. In Bean's view, the tendency within botany had gone from reducing the number of genera and species in the middle of the cen-tury, which had been 'carried too far', towards the revival of older generic names. However, by the early 1900s, particularly in Europe and the United States of America, attempts to subdivide species, genera and natural orders had become so prevalent that Bean feared it would 'involve such confusion and readjustment of nomenclature as to render its acceptance by cultivators' in Britain highly unlikely.[33] Just as the international political scene deteriorated with the jarring of empires, so the arboricultural world seemed to be fragmenting nationally. This favoured the production of systematic arboricultural treatises and prestigious national arboretums for botanical experimentation and classification, which could then be utilized as authorities for national botanical publications.

The many international disagreements that occurred throughout the nineteenth century and the difficulties faced by the promoters of systematically planted arboretums are evident with respect to Coniferae. During the 1820s, Louis Claude Richard arranged the order into three tribes of Taxineae, Cupressineae and Abietineae, which was basically followed by Loudon and other botanists, although changes were made by some such as Heinrich Friedrich Link's separation of Picea (spruce) and Abies (silver firs) as distinct genera from Pinus. Loudon also adopted the suggestion of John Lindley that the order of Taxaceae be divided from Coniferae, but this was not generally adopted. Subsequently, Stephen Endlicher in Vienna defined Cupressineae, Abietineae, Podocarpeae and Taxineae as natural orders subdivided into tribes, which was generally accepted by Continental botanists. However, in 1881, George Bentham tried to simplify the systematic arrangement of Coniferae into six tribes, with various adjustments at the level of genera, whilst August Wilhelm Eichler in Berlin arranged the genera into two primary divisions of Pinoideae and Taxoideae, including conifers proper in the former and Taxads (the genus of yew) in the latter. In 1892, the Royal Horticultural Society tried to stabilize and standardize these international movements by holding a 'Conifer conference' at Chiswick which, it was claimed, brought together 'the most remarkable collection of specimens cut from Taxaceous and Coniferous trees and shrubs ever assembled' and resulted in the publication of a special volume of papers. The resulting systematic revision of the order was undertaken by Maxwell Masters and published by the Linnean Society, but, almost immediately, practical textbooks such as Veitch's *Manual of Coniferae* adopted 'deviations' from this.[34]

Disagreements concerning nomenclature were particularly evident where names celebrated political figures, nations, or empires which were not universally popular and there were 'numerous and perplexing' differences between 'European and American authorities' concerning the names of Coniferae. Taking the example of Sequoia again, the application by John Lindley in Britain of Wellingtonia and Albert Kellogg in California of Washingtonia to *Sequoia giganteum* on the basis that a Californian tree should not bear the name of a British soldier, caused international botanical controversy and resulted in the widespread use of *Sequoia Wellingtonia* and lengthy lists of names in books and labels as a compromise.[35]

Later in the century George Bentham and Joseph Hooker emphasized how even experienced botanists, provided with the most detailed and careful descrip-

tions in 'systematic works of the highest repute' were 'occasionally led into false
determinations' because species varied 'within limits' which was 'often very difficult to express in words'. There was, 'in a number of instances, great difference
of opinion as to whether certain plants differing from each other' in particular
ways were

> varieties of one species or belong to distinct species. Similarly, there was disagreement
> as to whether 'two or more groups of species should constitute sections of one genus,
> or distinct genera, or tribes of one family, or even distinct families.[36]

It was therefore sometimes necessary to compile analytical tables to divide genera and species using brackets to represent different alternatives where doubt
remained, although this was to be avoided unless absolutely necessary'. Anoma-
lies and aberrations were a further complication and might prevent a 'species
from being at once recognized by its technical characters' and these were exacerbated by hybrids 'frequently to be found in gardens through artificial production
or less commonly, in the natural world'. Bentham and Hooker provided descriptions of some of the most common aberrations with the most likely causes of
these in the numerous editions of their authoritative *British Flora*.[37] Botanists,
arboriculturists and natural historians were repeatedly urged to make their plant
descriptions as 'clear concise, accurate and characteristic' as possible so that 'each
one should be readily adapted to the plant it relates to' and no other. Descriptions were to be 'clear and readily intelligible' using ordinary 'well established
language' as much as possible' but giving 'precise technical meaning' to terms
used 'vaguely in common conversation'. Botanical terminology ought to be as
precise and clear as possible whilst avoiding 'that prolixity of detail and over-
loading with technical terms which tends rather to confusion than clearness'.
However, Bentham and Hooker admitted that botany was not an exact science
and the 'aptness of botanical description like the beauty of a work of imagination' always varied 'with the style and genius of the author'. The arrangement
according to 'natural divisions' facilitated comparison between closely allied
plants and if accompanied by 'an artificial key or index' would guide the student
to the correct identification.[38]

Trees in the Economy of Nature

Although the main focus of the *Arboretum Britannicum* was gardening, horticulture and arboriculture, it also demonstrates the degree to which Loudon
drew inspiration from 'uncultivated nature', which he interpreted in terms of
the economy of nature. He certainly did not perceive trees as merely utilitarian
entities of service to humankind for aesthetic and economic purposes. In a brief
but important chapter inspired by accounts provided by Joseph Banks, Alex-

ander von Humboldt and other botanists and travellers, Loudon emphasized the interdependency of humans, plants and animals in the 'economy of nature'. He argued that the high proportion of 'ligneous vegetation' globally and the 'immense extent' of forest compared to meadow, pasture or plain, demonstrated the importance of trees and shrubs in the natural economy. He contended that in uninhabited countries 'the influence of forests must be on the climate, on the soil, and on the number of wild animals and herbaceous vegetables'. The evidence of what later became known as palaeophytology, coal, early history and the exploration of the vast forests of North America demonstrated that the 'greater part of the surface of our globe has been, at one time, covered with wood'. Furthermore, the impact of trees in these uninhabited or uncultivated lands must have extended to animals, 'herbaceous vegetables', soil, waters and climate. Wild animals would have been furnished with shelter and sustenance, whilst birds, insects and reptiles were the most 'common inhabitants of forest scenery'. Although some herbaceous plants could not survive in the 'dense forests' of the torrid zone, yet some kinds, such as 'epiphytal lichens', mosses and some Orchideae were encouraged by the 'thickness of the shade and the moist heat which prevails among the trunks and branches of the trees'. Of most importance was the 'great influence of forest scenery in a wild state' upon the soil through decaying leaves, trunks and branches, which made them 'a provision of nature for preparing the earth's surface' for the cultivation of food crops for people and animals, such as corn. Forests also impacted upon the rivers and lakes of countries, acting 'more or less as a sponge in retaining the water which falls on them' and so supplying water more gradually than in countries where extensive clearances had taken place. Likewise, the 'influence of forest scenery in increasing the moisture of the atmosphere, and in preventing a climate from being so hot in summer and so cold in winter' was 'well understood'.[39]

Studies of the 'influence of trees' in uncultivated regions provided 'useful hints' for how best to plant and thin woodland in 'civilized' countries. Although such forests operated on a 'grand scale' of 'many thousand acres' they could still teach much to cultivators of smaller useful and ornamental plantations. British forests were no longer large enough 'to afford a shelter for wolves and hyenas' but they still harboured 'foxes, polecats, snakes, and other noxious animals, and several kinds of carnivorous birds, such as the hawk' whilst French and German forests contained wolves and wild boars as well as criminals fleeing from justice. Wild forests supplied food to birds and insects but also in 'civilised' countries, birds, insects and sometimes reptiles abounded wherever there were 'trees and shrubs to supply them with food and shelter'. In Belgium and parts of Holland, some moth caterpillars were so abundant in woodland for one season annually that it was 'a part of the business of the government police to see that they [were] destroyed'. This demonstrated the advantage of learning which trees and shrubs were 'obnoxious to particular insects' and which more vulnerable to insect depredation whilst the 'total destruction of herbaceous plants' in wilder dense forests demonstrated that they were more likely to thrive under relatively thin planting. Similarly, the 'influence' of decaying leaves, branches and trees in 'adding to the soil' demonstrated 'how barren soil may be improved by trees', a 'natural effect' imitated by 'trenching down entire plantations of Scotch pine grown on extremely poor soils' in parts of Scotland. Undrained woodland and especially copse woods retained rainwater much longer than 'open groves or plains' and, 'as increased exhalation and evaporation must be going on from such woods', so 'increased moisture must be thus produced in the atmosphere'. This suggested to Loudon that the impact of dampness engendered by putrefying leaves needed to be considered when planting extensive shrubberies near dwelling-houses, particularly those that were intended, 'by frequent digging, always to present a surface of naked loose soil'. Knowledge of the 'influence of trees in modifying both the temperature and moisture of the atmosphere in civilised countries, and in artificial scenery' would help determine the 'disposition of trees and shrubs' around houses, particularly in 'low situations'. Ignorance of this had led to 'naturally healthy' English country residences becoming damp and increased insect infestation due to a 'superabundance of trees and shrubs'.[40]

Finally, analysis of trees and shrubs in their 'wild state' provided information concerning their 'aboriginal natures' whose 'habits' had been changed, like domesticated animals and livestock, through years of cultivation. Loudon argued that the significance of this change needed to be recollected when 'speaking of the native soils and situations of different species'. To improve species it might be necessary or advantageous to 'place them in a different soil or situation from that in which they are found in a wild state'. Some situations or soils in which trees or shrubs were found in a wild state, however, could 'hardly be improved by art' such as peat bogs or peaty soils like those of America and Alpine situations.[41]

Towards Global Arboricultural Equalization

Of course by promoting the importation of numerous exotics into Britain the *Arboretum Britannicum* was a work of nationalistic and imperial botanizing. However, Loudon believed that this process of equalization by fostering the international exchange of species was a noble civilizing endeavour that would benefit all nations and, as we have seen, appreciated the importance of trees in the 'economy of nature', including in wild and 'uncivilised' countries. The *Arboretum Britannicum* did not only analyse hardy and half-hardy British trees and shrubs but presented a history connecting these with those of similar climates to demonstrate what had been done to introduce them and what might be 'anticipated from future exertions'. It was therefore a 'general history of the trees and shrubs of temperate climates, but more especially of those of Britain'.[42] Its most important purpose was to facilitate the spread of trees internationally so that, like the advance of progressive scientific and political ideas, climatic and geological factors permitting, there would be a global equalization of species. Although nineteenth-century international botanizing is often interpreted merely as a manifestation and function of imperialism, to Loudon this vision was supported by the notion of his friend Jeremy Bentham that it would contribute to universal progress by ensuring the greater happiness of 'the great human family'. The 'accumulated experience' of human tree encounters over millennia was considerable and given that the 'first characteristic instinct of civilised society' was to improve the 'natural productions' that surrounded them and appropriate these for commerce, so international tree exchanges would be facilitated for improvement, 'equalisation' and increasing enjoyment.

Despite the use, grandeur and beauty of timber trees, however, compared with 'herbaceous vegetables', only a small number of species had hitherto been distributed such as palms, bananas, pineapples and other popular or botanically significant varieties, most of which tended to be small. All timber trees except bamboo were dicotyledenous and there were probably not a thousand genera of these which grew to over 30 ft in height, most of which were in warm regions, equivalent genera in temperate regions numbered barely a hundred and in frozen areas were non-existent. This explained the small number of indigenous genera in temperate countries such as Britain, which Loudon estimated had scarcely more than twelve genera and thirty species normally attaining heights over 30 ft. Given the probable number in other temperate regions it was a 'beautiful work of civilisation, of patriotism, and of adventure' to bring these back to Britain and distribute them internationally. Britain could enjoy European, American and oriental trees whilst distributing her own 'and those which she has appropriated' globally, thus Loudon confidently predicted, 'contributing almost imperceptibly' but 'most powerfully to the progress and equalisation of civilisation and ... happiness'.[43] It was fascinating for philosophers and philanthropists to speculate how far this process of plant exchanges had come. Although much had been accomplished over the previous century, from the vast swathes of unexplored lands, there was reason to believe 'this department of the civilisation of the great human family is yet in its infancy'. The first part of the *Arboretum Britannicum* was therefore devoted to a general history of tree introductions to Britain, Europe and other countries within the European ambit, and those that might be reasonably expected.[44]

Besides hardy varieties, the *Arboretum Britannicm* also included ligneous plants already found to be half hardy in the metropolitan climate and others that Loudon considered likely to be so 'from their native countries and habits'. Many would be found 'quite hardy' from experimental introductions as '*Cydonia japonica, Hydrangea Hortensia, Aucuba japonica*' and others. The common passion flower, for instance, had lost its leaves in winter when initially introduced to the Edinburgh Botanic Garden but 'in the course of a few years, the same plant retained the greater part of them at that season'. Although the 'nature of species' could not be 'so far altered' that tropical trees could survive outdoors in cold climates, many plant habits did 'admit of considerable variation', and hot and cold climate plants had proved adaptable to British conditions. For example, Loudon believed it was likely that various newly discovered varieties of South and central American, African and Asian oak would be introduced to Britain, just as North American oaks had been able to join the native oak, hence he included full descriptions of these at the end of the lengthy *Quercus* section.[45] Exotics could be cultivated for their own sake as they added greatly to the interest of gardens, demonstrating the skill of gardeners and taste of owners. There were 'few scenes' in ornamental gardens or pleasure grounds of 'greater interest' to those with botanical knowledge than conservative walls 'covered with trees and shrubs, natives of foreign climates'. Though these might be killed during winter, they exhibited 'a degree of luxuriance' during the summer never displayed in green houses or conservatories. Even were all to be killed by frost every winter and 'a reserve obliged to be kept in greenhouses or pits to supply their place every spring', the 'splendour' of their summer appearance and 'novelty of their forms' would easily compensate for the trouble incurred. The ease with which conservative walls could be flued and borders heated by water pipes meant that they could be designed 'superior to our present green-houses and conservatories'.[46]

Even tender exotics such as cacti could be imported with sufficient observation and support. Every gardener had observed that 'common weeds' springing up in pots, hot beds or hothouses during the summer were killed or damaged when set outside in winter or spring whilst the same species was flourishing in the open air. The 'obvious conclusion' was that the 'habits of plants' admitted 'a certain degree of change with regard to the climate' which could be ascertained by experiments involving the movement of hothouse varieties outside. Trials suggested that many Australian and New Zealand trees and shrubs would 'ultimately become so habituated to the climate of London, as to live through the winter against a wall, with scarcely any protection'.[47] In addition to describing trees and shrubs already introduced, Loudon suggested other possibilities. For instance, whilst it had been virtually impossible to categorize or identify new species and varieties using the Linnaean system until they flowered, which delayed formal botanical recording, the natural system facilitated the process. In these cases the 'general habit' of the genus was discovered long before flowering, hence although not yet in catalogues, plants were distributed among cultivators and propagated for sale in nurseries 'under some provisional name'. Loudon provided details of 'expected introductions' which were species of genera so numerous and geographically extensive that 'from these circumstances alone' accommodation in British gardens could be reasonably expected. Other as yet unintroduced genera from temperate zones identified by botanists as 'suitable for our climate' were likely candidates to make travellers, 'wealthy individuals, or societies or associations' aware of so they could instruct their collectors.[48]

Tree Botany, Landscape Gardening and Art

According to Loudon the *Arboretum Britannicum* was primarily composed to diffuse among the landed gentry 'a taste for introducing a greater variety of trees and shrubs in their plantations and pleasure-grounds'. Whilst 'many new and beautiful' foreign trees and shrubs were annually introduced into botanical gardens and nurseries, it took longer for these to diffuse into gardens of country residences and 'many fine old specimens' suffered neglect. 'Numerous' British horticultural societies 'powerfully promoted the general taste for horticultural and floricultural productions' for fruits, culinary productions and flowers, yet neglected arboriculture. Trees and shrubs were next to buildings 'the most important ornaments' to be introduced, and 'their more general distribution and culture' was therefore crucial. The very little care required to manage fully grown trees and shrubs, their 'magnitude' and 'permanent influence' upon the 'general scenery of the country' made them 'greatly superior to herbaceous plants'.[49] Loudon originally intended to include all trees and shrubs in temperate lands in the *Arboretum Britannicum* but, finding it was already double the length originally envisaged, he thought it advisable to publish this 'generalisation' as a future single-volume separate work, which, however, never appeared.[50] Loudon did say, however, that the arboricultural encyclopaedia would feature a general study of trees and their culture. In other words, how they ought to be treated '*en masse,* whether as seedlings in the nursery, as useful and ornamental plantations, as yielding timber and other useful products, or as ornaments in the lawn and shrubbery'.[51]

Loudon argued that the *Arboretum Britannicum* would be of tremendous use to 'the gardening world' and landed proprietors. It would exert a major influence in 'promoting a taste for the culture and spread' of foreign trees, exciting a desire for the introduction of others and 'for originating new varieties' by art. Given that every landowner was already 'either a planter' or possessed trees, he would be encouraged to 'combine beauty with utility' by planting exotics on the 'outer margins' of woods, along open rides in the hedgerows of lanes or public roads which would be 'ornamental' and 'useful'. If the estate was already fully planted he would be encouraged to 'beautify his plantations' by 'heading down large trees of the common species and grafting on them foreign species of the same genus' which was a 'common practice' in orchards and could be so in parks and pleasure grounds as well. Hawthorn hedges, for instance, were 'common everywhere' and there seemed no reason why the 'twenty or thirty beautiful species and varieties of thorn' in nurseries could not be grafted onto them or allowed to grow as trees along roadsides. Similarly, Loudon suggested that scarlet oaks and acers could be grafted onto more common species of the genera 'along the margins of woods and plantations' at very little expense but to immediate effect. As 'every gardener' could 'graft and bud' and as all landed proprietors could procure nursery stock from which grafts could be taken or obtained scions from horticultural society or botanical gardens, so every nondescript hedge or plantation could become colourful and varied mini arboretums.[52] Similarly from the countryside to suburbs and towns, 'amateur landscape-gardeners and architects' laying out villa grounds would be encouraged to select the most aesthetically effective trees and shrubs. There were opportunities for tree planting everywhere from streets and squares to hedges and even the new railway cuttings that Loudon sped through with his wife Jane on their garden tours.[53]

The *Arboretum Britannicum* exploited the customary importance of trees in British landscape gardening since the mid-eighteenth century as well as analysis of the picturesque qualities of trees by Joshua Reynolds, William Gilpin, Uvedale Price and Humphry Repton. Repeating and extending some of the arguments made in his earliest major work, the *Observations on ... Useful and Ornamental Plantations* (1804), Loudon argued that trees were 'in appearance, the most striking and grand objects of the vegetable creation', contributing 'the most to human comfort and improvement'. In the *Arboretum Britannicum*, however, he demonstrated a much more profound knowledge and experience of the systematic and physiological botanical study of trees which he argued was essential for their effective deployment in landscape gardening. The utility of trees in providing shelter or shade in 'improving the climate and general appearance of whole tracts of country', in 'forming avenues to public or private roads, and in ornamenting ... parks and pleasure-grounds' was well known.[54] It was universally acknowledged that they were 'among the grandest and most ornamental objects of natural scenery' and that landscapes were 'utterly barren' without them. Loudon asked what the 'charm of hills, plains, valleys, rocks, rivers, cascades, lakes, or islands' would be without the 'hanging wood, the widely extended forest, the open grove, the scattered groups, the varied clothing, the shade and intricacy', contrast and varieties of form and colour 'conferred by trees and shrubs?' Whilst shrubs such as roses had some similar qualities, trees were grand objects in themselves, bold perpendicular elevations with a 'commanding attitude' and character 'greatly heightened' by knowledge of their 'age, stability, and duration' that rendered them 'sublime'. However, their 'characteristic beauties' and 'general forms' were 'as various as their species' just as the 'beauty and variety of the ramifications of their branches, spray, buds, leaves, flowers, and fruit' and changes in colour over the seasons, a source of 'perpetual enjoyment' to nature lovers. There could be little more fascinating than observing the development of spring buds or the 'daily changes which take place in the colour' of autumn foliage.[55] Although many trees took long to mature, many examples in the Loddiges' and Chiswick arboretums, for instance, had attained 30 or 40 ft after ten years and would 'practically speaking' afford shelter and shade, displaying individual beauty and character and conferring 'expression on landscape scenery'.[56]

Only fully appreciable through botanical study, Loudon went so far as to state that 'all the more important beauties and effects' of 'modern landscape-gardening' depended upon the use of 'foreign trees and shrubs'. It is not therefore an exaggeration to say that Loudon's conception hinged upon botanically inspired arboriculture, which explains why he regarded the *Arboretum Britannicum* as his most important book and the Derby Arboretum as his most vital landscape-gardening commission. It also underscores the extent to which Loudon's arboriculture was a practical concern, founded on experience of trees in a multiplicity of contexts rather than simply the application of abstract theories. The presence and treatment of exotic trees was pivotal because recognition of all arts from music to poetry, painting and sculpture, required them to be 'avowed'. Similarly in 'modern landscape gardening' it was unnecessary to 'conceal art' and the apparent contradiction of the fact that landscape painting imitated nature yet was clearly distinguished as art was to be explained by the difference between imitation and repetition. In the 'imitative arts', such as landscape painting and sculpture, the intention was never only to copy but to imitate 'in such a manner as to produce a totally distinct work from the thing imitated', to 'show the original through the medium' of art such as canvas or stone. The analogous process in landscape gardening was immediately apparent in the geometric style with its 'artificial surfaces, forms and lines', formal gardens displayed taste and discernment and were easily distinguishable from the 'woody scenery of the surrounding country'. Yet 'modern style' villa gardens laid out with grounds 'disposed in imitation of the undulations of nature' and trees scattered 'in groups and masses', would be 'mistaken for nature' save for the configuration of exotic and unusual trees distinctive from local types. For Loudon the gardens of 'no residence in the modern style' could be laid out in 'good taste' when 'all the trees and shrubs employed' were neither foreign nor 'improved varieties of indigenous ones'. So the 'grounds of every country seat from the cottage to the mansion' could become arboretums, differing only in the number of species accommodated. Botanically-informed arboriculture defined modern landscape gardening.[57]

Although established over centuries, British arboricultural taste remained, according to Loudon, 'in its infancy' and there was great neglect of the study of individual trees. Great respect was usually accorded them and there was general appreciation for their beauty which was misdirected leading to an 'unwillingness' to cut them down or thin plantations even where trees injured each other or spoiled gardening aesthetic effects. This 'indiscriminate regard for trees' and the 'morbid' disinclination to fell when 'wrongly placed or too thick' resulted from ignorance of botany, concerning the relative beauties of different species and lack of landscape-gardening taste. Despite the sale of American trees in British nurseries for at least a century, landowners, gardeners and planters had 'few opportunities' of seeing – and acquiring a taste for – exotic specimens. Similarly, want of botanically-informed arboriculture had prevented artists from portraying them accurately or tastefully. Loudon intended for the *Arboretum Britannicum* to remedy this problem whilst encouraging the spread, popularity and accessibility of private arboretums, public botanical and horticultural gardens by 'exhibiting living specimens'.[58]

Just as he strove to unite botany with aesthetics in his landscape gardening, living arboretums and the virtual collection of the *Arboretum Britannicum*, so Loudon argued that scientific arboricultural analysis would enable landscape artists to improve and differentiate tree representations. Although some artists such as Thomas Hearne and Paul Sandby (1731–1809) had produced striking picturesque tree depictions showing acute sensitivity to individual growth patterns such as 'Morning' (1794) and 'The Rainbow' (*c.* 1800), Loudon adjudged that urban-centred artists had too frequently copied trees from each other with little botanical knowledge, experience of living collections and studies of taxonomy, foliation and other aspects of arboriculture. With the possible exceptions of some studies of the oak, ash and weeping willow, no books hitherto published contained accurate depictions of tree species with 'characteristic expression' and scientific accuracy. Even illustrations in admirable works such as the oaks in Edward Kennion's *Essay on Trees in Landscape* (1816) bore 'not the slightest resemblance' to growing examples. Loudon asserted that even 'ladies who reside in the country and have studied botany' if they had a taste for landscape, would 'imitate the touch of trees better than professional landscape-painters'. The lack of 'scientific knowledge of the touch of trees' was partly attributable to lack of British literature on the subject, Gilpin's *Remarks on Forest Scenery* (1791) and Kennion's book being honourable exceptions. Loudon claimed that the combination of full-length and sectional tree portraits with detailed botanical descriptions in the *Arboretum Britannicum* would facilitate artistic depictions of seasonal foliage changes and other aspects of the 'difference of touch' from nature in individual species. It would also encourage London artists to forsake the studio and travel to visit trees and use this experiential knowledge gained 'in their walks, or professional excursions' to recognize individual species and varieties and represent them accurately in art. John Ruskin was the most influential writer to take up Loudon's challenge and after criticizing centuries of arboricultural art in *Modern Painters*, he also urged that painters subject trees to lengthy minute and painstaking study.[59]

Conclusions

The *Arboretum Britannicum* was much more than a description of the present state of British arboriculture; it was a means of transforming tree collections for the future and hastening changes already occurring. In addition to hardy and half-hardy trees and shrubs in the metropolitan vicinity, Loudon sought to include others that might be suitable for introduction, others more recently introduced and available in British nurseries, and types which had once grown but since disappeared. This introduced an element of indeterminacy into the book as the inclusion of these required generalizations about British climate and conditions, assessment of the individual qualities of sometimes hardly understood varieties, and predictions about the growth of trees wrenched from natural contexts. Loudon was striving to promote the process of global botanizing rather than merely presenting a summary of the *status quo*. Attention was given to half-hardy trees and shrubs that might survive in the London climate in mild winters and favourable conditions or if protection were used, but would not be suitable for planting outdoors in most of the rest of Britain. These trees and shrubs were of special importance because there was 'perhaps no scene in a British garden more interesting' than one in which exotics from warm countries, usually in greenhouses, appeared 'in a flourishing state in the open air'. Half-hardy exotics also appealed because their 'culture and management' necessitated a 'higher degree of scientific knowledge' and attention from gardeners, contributing to their 'improvement and ... usefulness'. Conservative walls offered protection from frost and elements and might be flued or support a roof. Although not yet 'very generally in use for ornamental exotics', they provided such a great 'interest and beauty' when 'properly designed, planted, and managed', that Loudon predicted they would become as popular as fruit walls and accorded equal status with conservatories and greenhouses.[60]

Although a remarkably thorough study, like the *Magazine of Natural History*, the *Arboretum Britannicum* was intended to be a forum for ongoing work and Loudon promised to publish a separate encyclopaedia of arboriculture from notes already completed on world trees. He also urged that readers send corrections, additions and suggestions to be published as an annual report in the *Gardener's Magazine*. This would include descriptions of all annual British trees and shrub introductions either through 'intentional hybridisation or accident' or from abroad, improvements in 'arboricultural nomenclature, such as species re-arrangements within each genus and everything necessary to keep the *Arboretum Britannicum* up to date'. Like the substantial supplements to the *Encyclopaedia of Plants* and *Encyclopaedia of Agriculture*, so the annual reports would be published simultaneously as pamphlets to allow owners of the *Arboretum Britannicum* to augment their knowledge without being forced to purchase a new edition. Despite its monumental comprehensiveness and the value of the tree engravings, the *Arboretum Britannicum* could only ever be a poor substitute compared to experiencing living trees and shrubs which it was supposed to encourage. Loudon also intended that it would encourage landowners to allow the general public to view their tree collections. He tried to make the book more accessible by publishing the much cheaper, abridged *Encyclopaedia of Trees and Shrubs* in 1842, which included most original text and sufficient sectional engravings for nurserymen, gardeners and foresters to use for identification purposes but excluded the historical and geographical studies of arboriculture and full length plates. Although praised for its accessibility and encouragement to new audiences, the differences between the two publications seemed to reinforce social divisions in arboricultural appreciation by suggesting that only the middle classes and landed gentry needed full length tree portraits whilst the general public could make do without. Loudon and his publishers were constrained by the crippling cost of producing the *Arboretum Britannicum*, especially the painstaking tree portraits, and needed to recover money through sales of the full price work. Institutional arboretums offered an opportunity to reach a wider public, especially if they offered gratuitous entry. Although Loudon had been urging the creation of a national arboretum since the 1820s it was only after the publication of the *Arboretum Britannicum* that he embarked upon the creation of a semi-public arboretum. This was the Derby Arboretum which he came to regard as his most important landscape-gardening commission.

6 THE DERBY ARBORETUM

Introduction

The popularity of botanical gardens and private tree collections during the 1820s and 1830s helps to explain why the latter played an important role in the provision of Victorian public parks providing both a general model and an inspiration for more varied planting schemes. Instructed by the donor, the industrialist Joseph Strutt (1766–1844), Loudon intended the Derby Arboretum to promote his vision of public arboretums and provide a living spatial realization of the *Arboretum Britannicum* (1838) and the Candollean natural system. In so doing they created one of the first specially designed Victorian public parks. But there were tensions between the need to preserve a botanically-significant tree collection and the demands of public access mirroring those experienced by botanical society gardens, which impacted upon the management of all public arboretums. However, for about thirty years the Derby Arboretum management committee was able to reconcile these demands whilst regularly holding some of the largest festivals in the region. The Derby Arboretum also demonstrates the importance of provincial scientific culture and rational recreation, and provides an example of how local initiatives in provincial towns, as well as government legislation, fostered Victorian municipal innovation. It was supported by a local group of scientific activists known as the Derby philosophers because of their association with the Derby Philosophical Society and other scientific institutions. In encouraging the foundation of other public parks and arboretums, the Derby Arboretum helped to set the pattern for Victorian public urban parks, although it remained only a semi-public institution.[1]

Strutt and Loudon's Derby Arboretum design attracted much national and international interest because it was taken to encapsulate the latter's theories of landscape design and arboriculture whilst providing a model for a self-sufficient institution funded and managed through subscriptions, donations and commercial events. Loudon's initial design and management recommendations fundamentally impacted upon the character of the institution and there were many changes in response to local demand such as a rapid increase in size by

1845. Although not officially managed by the corporation for forty years, the arboretum was largely a liberal municipal project inspired by a group of reformers who were primarily industrialists, nonconformists, professionals and urban gentry who sympathized with Loudon's political and scientific objectives.

Focussing on the part played by Loudon, the Strutt family and the Derby philosophers, this chapter examines the design, development and impact of the Derby Arboretum as botanical institution and public park. It considers why the arboretum was founded, how it was intended to facilitate social, moral and intellectual progress and how it was utilized, experienced and appropriated in novel ways, from botanical institution to urban leisure park signifying local civic and national pride. The Derby Arboretum was in many ways a physical planted embodiment of the *Arboretum Britannicum* as well as Loudon's gardenesque landscape-gardening philosophy and needs to be examined with the book that inspired it. Critical responses to the two melded. Visitors came to the park with preconceptions formed by the book and contrariwise. The Derby Arboretum demonstrates how Loudon's conception of public arboretum was applied, modified and experienced in the provincial urban context, the role of scientific ideas in rational recreational institutions and the varied impact of natural history in early and mid-Victorian industrial society.[2]

Industry, Science and Horticulture: Loudon and the Derby Philosophers

The Derby Arboretum was inspired by the shared interests of Loudon and the Derby philosophers, particularly the relationship between Loudon and the Strutt family, which long predated the park. The Derby philosophers were a group of professionals such as medical men, industrialists, manufacturers and urban gentry, many of whom, although not all, were Whigs and nonconformists. They had grown to a position of political prominence in the county, as the rise of the Strutts demonstrates, taking advantage of the fact that restrictions against dissenters holding political office tended to be ignored in Georgian Derby. Meeting in institutional forums such as the Derby Philosophical Society (1783), the Derby *savants* promoted the benefits of rational recreation and scientific education for social progress by supporting institutions such as the Derby Literary and Philosophical Society (1808–16) and the Derby Mechanics' Institute (1825).[3]

Brothers William (1756–1830) and Joseph and William's son Edward (1805–80), were major figures in local urban science and culture. The family's rise to business prominence was largely the result of Jedediah Strutt (1730–97), father of William and Joseph, who made a fortune in the textile industry in partnership with Richard Arkwright, establishing major manufactories at Belper and

Derby. The Strutts used their wealth to embark upon paternalistic campaigns for social improvement and political reform whilst improving their Derbyshire estates as landed gentry and fostering the growth of Belper into the second largest county town. William and Joseph were leading figures in the unreformed Derby Corporation, whilst the latter served as first mayor of the reformed body in 1836 and his nephew Edward was elected MP. William in particular utilized industrial technology and forms of organization to reform public institutions including the town watch and Derbyshire General Infirmary (1810). The Strutts' Unitarianism helped to inspire their public and political activities. In a sermon of 1820, Edward Higginson, minister of Derby Friargate Chapel, urged his congregation to 'extend the most unfeigned and comprehensive charity to all who differ from you', which must 'consist in something more than candid expressions: it must shew itself in kind thoughts and benevolent actions'.[4] Similarly Noah Jones, another minister at the chapel, noted after Joseph Strutt's death that 'as a protestant dissenter' he had been 'the consistent supporter of civil and religious liberty' and of 'all public institutions, which have for their object the promotion of human virtue and happiness, especially such as aim at improving the social state of the great mass of the community'. With other local philosophers, the Strutts were largely responsible for the foundation of the Derby Lancasterian School and Mechanics' Institute and supported the new University College, London. Joseph and George Benson Strutt also served as presidents of Manchester College, York.[5]

Loudon and the Derby philosophers had a shared enlightenment faith in social progress through the application of scientific and technological innovations, particularly in agriculture and horticulture. Although the industrial interests of pioneering manufacturing families along the Derwent Valley such as the Arkwrights, Strutts and Evanses secured their fortunes, their attention to horticulture and estate improvement was almost as significant. The Strutts and Arkwrights quickly purchased titles and estates and took a keen interest in their improvement. Joseph and Edward Strutt's promotion of the arboretum was partly an extension of agricultural, horticultural and arboricultural improvements undertaken on their extensive estates. These were patriotic acts which symbolized their status as manufacturers and landowners simultaneously challenging and embracing the lives of the landed gentry. These activities are fully detailed in John Farey's study of Derbyshire agriculture, topography and geology which Loudon regarded as one of the best Board of Agriculture county reports. Farey received considerable help from the Strutts and their philosophical friends and took an interest in their promotion of horticulture, noting for example that the Strutts had created gardens at Belper which utilized specially designed sewers to collect waste for manure and grow vegetables superintended by tenants. The family's arboricultural interests are evident from the number of plantations they created and Farey enthusiastically described their planting methods. George Henry Strutt had 'very laudably' applied 'himself to the super-intending of the planting of about 100,000 Larch, Scotch and other trees' on the Belper estate of his father, George Benson Strutt, who shared control of the business with his brothers William and Joseph Strutt. George Henry Strutt also oversaw the 'pruning and management of the extensive plantations previously made there' and kept 'accurate and systematic accounts of the expense and time of planting', pruning, thinning, the value of cut timber, and 'the measure and value of those trees standing in his father's plantations.'[6]

Encouraged by his own work on hothouses and the problems of applying cast iron and glass to hothouses, Loudon was also keenly interested in how William Strutt and his family had utilized industrial technology in horticulture.[7] William Strutt pioneered the use of iron as a fire-resistant material for the construction of manufactories, public buildings (such as the Derbyshire General Infirmary) and domestic architecture. He approached industrial and horticultural problems from an industrial perspective. Heating systems and fire-resistant construction methods from the cotton manufactories were applied to gardens and hothouses and introduced by the Strutts and their friend, the chemist and engineer Charles Sylvester, into semi-public institutions like the Infirmary and Manchester College, York. Loudon admired William Strutt's work, meeting him and his brother Joseph during his garden tours and advocating many of William's social and engineering innovations. Loudon was particularly interested in the social applications of industrial technology employed in the Strutt factory settlements and included very detailed descriptions of cottage window staybars, door staybars, lodge gate fastenings and trussed iron rod girders and rafters utilized at Derby, Milford and Belper in the *Gardener's Magazine*. In his *Cottage, Farm and Villa Architecture* (1833) Loudon stated that 'no oven has equalled that of Count Rumford as modified by W. Strutt esq. a man of most extraordinary genius', and included a detailed description of the Strutt kitchen. Loudon also recommended Strutt and Sylvester's *Philosophy of Domestic Economy* (1819) which described the system of heating and ventilation employed at the Derbyshire Infirmary and its many applications, claiming that it formed the basis for what was 'now in general use throughout Britain for large buildings.'[8]

William Strutt and Sylvester also investigated how heating systems could be applied to hothouses. In 1823, William wrote to Anne Strutt describing how Sylvester, who had moved to London and whose heating and ventilation installation business was growing, had returned to Derby where they obsessively discussed hothouses 'so that I shall have neither ennui nor solitude while he remains.'[9] Loudon enthusiastically described William Strutt's hothouses at St Helen's House, which became the residence of his son Edward after his death in 1830, and the hothouse at Joseph Strutt's residence in St Peter's Street featured in the *Encyclopaedia of Gardening* and *Gardener's Magazine*. William's hothouse was heated by hot water pipes under the soil and by one of his cockles which recycled warm air in winter to prevent heat loss, whilst there was much else 'in these gardens to observe and commend' including the system of training fruit trees with wires. Loudon first visited Joseph Strutt's house in 1826 and praised the moveable front sashes which allowed vines to be admitted to the open air in winter whilst remaining dry. Other plants commanding Loudon's attention including hothouse varieties such as 'well grown' orchids and an *Aristolochia sipho* on the wall of the residence, with large shoots one 50 ft long. He also approvingly described Joseph Strutt's domestic picture gallery, which served as a kind of town gallery, often open to the public. The 'handsome' central space was 'remarkably well lighted ... through double sashes set at an oblique angle and, inevitably, also heated by hot air and small tubes filled with water. The efficiency and economy rendered it, in Loudon's opinion, 'one of the most complete picture galleries that we know of' requiring remarkably little artificial heat even 'during the most severe weather'.

Rational Recreation and Early-Victorian Public Parks

Although the public had access to common lands, walks and pleasure gardens in the Georgian town, many of these were increasingly enclosed and available only to paying subscribers. Access to open land for recreational purposes was not too much of a problem because of the relatively small size of most pre-industrial towns, but with industrialization and the loss or enclosure of many common lands, such access to green spaces was curtailed.[10] In towns such as Nottingham, retention of large surrounding estates and enclosures hemmed in the urban poor, forcing up the price of land and cost of living.[11] There were attacks on the enclosure movement in Parliament during the 1830s and some encouragement was given to the parks campaign by the report of the Select Committee on Public Walks in 1833 which emphasized their value in terms of rational recreation and public health. The 1836 Enclosure Act backed by Joseph Hume exempted common fields from enclosure if they fell within a certain radius of large towns following the recommendations of campaigners such as Loudon. Joseph Strutt approved of the measure and acknowledged that it had helped to inspire the creation of the Derby Arboretum.[12]

At Derby, William Strutt, Erasmus Darwin and others of the Derby philosophers had been amongst the primary instigators of the sale of the Nun's Green common land to pay for urban improvements and establish an improvement commission. Other common lands disappeared, including parts of the Siddals used in 1839 and 1840 for the new railway station and engine sheds. Surrounding estates such as those at Markeaton and Darley, associated with the Mundy and Evans families, respectively, had informally allowed the public to wander and children to play in their parks. However by the 1820s these were rapidly disappearing along with the surrounding country walks owing to industry and residential development. At around the same time other forms of leisure enjoyed by all social classes, such as the Shrovetide football and horse racing, came under pressure from moral campaigners including some middle-class evangelicals and working-class radicals, and despite much public protest the former was suppressed in 1845.[13]

Loudon used the *Gardener's Magazine* and other publications to support the establishment of urban parks and gardens for all social classes as a utilitarian rational recreational reform using enclosed common lands and other means. He also proposed interlinked green spaces for the metropolis, suggesting that burial grounds and other places could become urban walks if correctly planted and maintained and joined the Metropolitan Improvements Society in 1842 where he met Joseph Hume and Edwin Chadwick. Although Loudon was the foremost British landscape gardener by the 1830s, these political views also attracted him to Joseph Strutt. He praised the Derby Mechanics' Institute exhibition of 1839 as an event that could not 'fail to have an excellent effect' whilst noting 'on good authority' – presumably the Strutts – that after only a fortnight it had been visited by more than 20,000 people. Strutt became one of Loudon's most important benefactors as debilitating disease took its toll during the early 1840s and work became almost impossible. As we have seen, the production of the *Arboretum Britannicum* dragged the Loudons into severe debt and Strutt provided financial assistance, tried to negotiate a deal with all the creditors and purchased multiple copies of the book to help the widowed Jane repay some of the loans.[14]

As part of his campaign to promote the education of gardeners and working-class gardening Loudon visited and described such efforts in the *Gardener's Magazine*. In 1839, for instance, with Jane and probably Edward or Joseph Strutt he was shown one remarkable garden on the outskirts of Derby which belonged to 'Mr. Bonham' who enjoyed 'excellent health and spirits' and was 'enthusiastically attached to his garden'. Although not more than twenty yards square, it was 'a work to wonder at', especially given that Bonham was over seventy years of age, had taken to gardening late in life, and still worked in a Derby brewery during the day. Bonham's garden was an interesting miniature working-class version of Elvaston which included ground 'thrown into hill and pits, varied by rockwork, roots, seats' and other objects combined with 'many curious and beautiful plants, shrubs' and trees all effected 'by his own personal labour in the evenings ... without the aid of money'.[15] Here was one solution to the lack of access to beautiful aristocratic gardens in the region and the fact that Elvaston and Chatsworth, two of the most important British arboretums, remained aristocratic preserves accessible only at the whim of wealthy families.

Loudon praised the Duke of Devonshire for his 'degree of liberty and impartiality' in allowing visitors both rich and poor into Chatsworth and thought that there was 'little doubt' the Earl of Harrington would do likewise once Elvaston had reached maturity, but the fact remained that access was conditional and in the case of the latter non-existent. As the wealthiest nonconformist manufacturers in the region the Strutts were mounting the strongest political challenge to the local aristocracy, especially the Cavendish family. They led campaigns for political reform for half a century, advocating the progressive application of modern industrial organization in society, opposing state interference in industry and supporting the establishment of a London university and other liberal ventures. It is thus not hard to see why Strutt asked Loudon to design the new public park and why he agreed so readily to assist. Despite the fact that there was only free access for two days each week, the Derby Arboretum was intended to be a popular counterpart to aristocratic arboretums like Chatsworth and Elvaston.[16]

The political, rational and recreational objectives of the Derby Arboretum were emphasized by Joseph Strutt in his speech at the opening ceremony in September 1840. The arboretum was essentially a utilitarian venture, providing instruction, improving the environment and giving pleasure to the citizens of Derby and elsewhere. The thousand specimens of trees and shrubs were intended to offer pleasure, education and moral improvement. As the sun had shone on him in life so he wanted to give something back to those whose toil had helped to create his wealth.[17] Just as the Strutts and their philosophical friends had supported the Mechanics' Institute, schools, scientific societies, libraries and museum, so they helped to create Loudon's 'living museum'. The arboretum and the mechanics' institutions were responses to the problem of rational recreation that had occupied many writers from the 1820s to the 1840s and resulted in the banning of Shrovetide football. This asked how the strains of the rapidly evolving industrial society could be prevented from resulting in immorality and licentiousness amongst the labouring population. Thus the arboretum can be interpreted as an attempt to regulate the leisure activities of the middle and working classes, though, as we have noted, such an explanation in terms of social control only partly explains the motivation of the Strutts and the Derby *savants*.

The Design of the Derby Arboretum

Using Loudon's original guide, the Derby Arboretum can be reconstructed just as it would have been in the early 1840s. By the 1830s, Loudon defined three basic gardening styles: the geometric, picturesque and gardenesque. The geometric style was the formal garden, the picturesque was 'characterised by that irregularity in forms, lines and general composition, which we see in natural landscape'. The gardenesque, developed and publicized in Loudon's many books and journal articles, was designed to show 'distinctness in the separate parts when closely examined' whilst being generally 'governed by the same general principles of composition as the picturesque style the parts, though blended, being yet connected'. In the gardenesque, each tree and shrub was 'planted and managed in such a way as that each may arrive at perfection, and display its beauties to as great advantage as if it were cultivated for that purpose alone'. There was some flexibility and trees and shrubs could be positioned in groups or following geometric lines if desired. In 1835, Loudon expanded these ideas in an article on the laying out of public gardens which provided a manifesto for public arboretum design.[18] Loudon's first principle followed Repton and held that 'every garden is a work of art' so that, though nature was to be imitated, it was not to be imitated 'in such a manner as that the result shall be a mistake for nature itself'. His second principle was 'unity of expression' by which he meant that there was to be overall purpose and meaning in all the features of the park. Walks and entrances, for example, should proceed logically and obviously, so that lesser paths were narrower than the main paths or that the same scene did not have to be viewed twice. Other principles were those of variety and of relation. A large variety of trees and shrubs had greater educational value, created more interest and were more aesthetically pleasing. According to the principle of relation, scenes in a garden should not follow each other at random but according to an order which 'should be recognisable from the first by the spectator'.[19]

As the *Arboretum Britannicum* made clear, the expression and character of trees meant that they were fundamental to landscape gardening, culture and history, as confirmed by enlightenment associational psychophysiology. Beauty arose through the character and individuality of trees (including geometry and shape) and their historical and cultural associations. Association gave rise to the 'moral and historical expression' of trees and was evident in the anthropomorphic and utilitarian qualities ascribed to them, such as the perception of oaks as strong, dignified or vigorous or the tendency to associate weeping trees with sorrow. The numerous 'historical and geographical associations' of trees such as the English oak, Cedar of Lebanon and yew and shrubs were a major component of the *Arboretum Britannicum*. Loudon distinguished between his own gardenesque and the two kinds of beautiful or smooth picturesque beauty as applied to trees by Uvedale Price and William Gilpin. The former described 'cultivated scenery' or scenery suitable for painting, the latter 'wild and forest scenery', also appropriate for a different kind of artistic representation. As interpreted by Loudon, picturesque trees could be mutilated although they had to be 'capable of readily grouping with another tree, or with any building, object, or animal, so that the combination may form a satisfactory whole'.[20] Gardenesque trees on the other hand were 'at all times, regular, or symmetrical, planted singly in 'favourable situations', 'not pressed on during their growth, by any other objects' and 'allowed to throw out their branches equally on every side, uninjured by cattle or other animals'. Gardening interventions had to be slight and intended to improve 'regularity and symmetry'. A 'truly gardenesque tree' when fully grown, always had 'some of its branches depending on the ground' to 'mark it as a tree of the garden or lawn and not one of the park', where lower branches would be separated from the ground by the line produced by grazing cattle. Nor was it similar to a forest tree pressed on all sides by others or growing to maturity under shade, which would render it picturesque or 'peculiar', the latter being, according to Joshua Reynolds, an object of 'deformity rather than beauty' in trees.[21]

Rejecting trees considered architectural or 'sculpturesque' as unsuitable for most contexts, Loudon emphasized the importance of perceiving them 'with reference to their beauty as organic forms' and the beauty and interest they generated through stimulating associated ideas. These were founded upon local circumstances which explained the perception of the superiority of oaks in commercial and maritime nations. Local associations greatly added to the pleasure of those beholding each species, such as the 'antiquity of the celebrated chestnut at Tortworth' or that on Mount Etna or the elm under which William Penn had signed his first treaty with the Indians. Young trees could therefore be as interesting as older ones because of the potential for growth and development that they embodied. From a picturesque perspective it could not be denied that older trees were more interesting, but from the gardenesque perspective young trees promised future increases in size and variation to delight the informed possessor with a 'historical and gardening knowledge of trees'. They embodied the multiple improvements 'contemplated or in progress' in global 'civilised society', which helps to explain Loudon's drastic recommendations that large trees in the Derby Arboretum be completely removed periodically. Independently of all 'moral, historical, and economical considerations', he claimed that so great is the botanical and horticultural interest connected' with young trees and 'so delightful' the job of preparing optimum growing conditions, if presented 'with a timbered estate' he would cut down all old trees to plant young ones.[22]

The design of the Derby Arboretum allowed Loudon to combine his social and landscape concerns, in harmony with those of the Strutts, to create what, according to Robert Chambers, he regarded as the most important landscape-gardening commission of his life. What was originally an eleven-acre site just beyond the boundary of Derby was carefully shaped with long undulations to create an impression of size following a pattern already established by Loudon at the Terrace Garden, Gravesend, Kent, where the walks were made as long as possible within restricted space and boundaries were carefully hidden with trees and shrubs (Figures 6.1 and 6.2). Loudon complained about the lack of purpose and direction in many parks by the 1830s, and the walks at Derby were designed to have meaning, terminating in lodges, fountains or benches. Strutt wanted the existing garden, plantation belt and tool shed to be preserved to minimize maintenance expenses but most of the arboretum featured a collection of 1,000 foreign and native trees and shrubs with the names and source of each displayed and detailed in a guide book for educational purposes. Thus pleasure and education, botany and geography were neatly combined. 350 seats were provided, each being carefully positioned – mostly by Strutt – to give views away from the boundaries for interest, security and comfort. The formally positioned lodges followed picturesque Elizabethan and Tudor styles with steeply pitched roofs and tall chimneys, thus reflecting both neoclassical and picturesque tendencies. A major secret of the arboretum was hidden beneath the ground. Loudon positioned nearly one mile of underdrains, consisting of semicylindraical tiles laid on flat tiles, running along the centre of each walk. Other drains led from the central drain to the edges of the walk, through cast iron gratings set in stone. This, with the carefully gravelled surfaces of the walks, kept the paths dry and prevented accumulations of water from forming in the mud.[23]

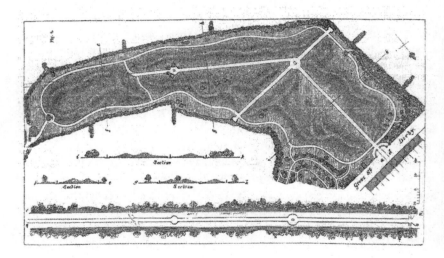

Figure 6.1: Relief plan of Derby Arboretum from J. C. Loudon, *Derby Arboretum* (1840), p. 6. Reproduced by permission of the Derby Local Studies Library

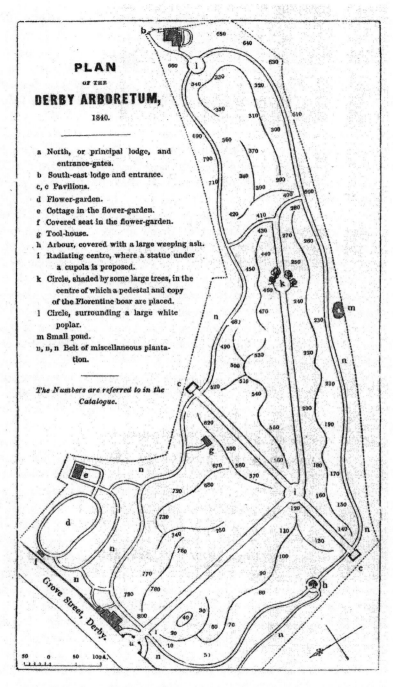

Figure 6.2: Plan of Derby Arboretum from J. C. Loudon, *Derby Arboretum* (1840), p. 75. Reproduced by permission of the Derby Local Studies Library

The arboretum was also, of course, a physical manifestation of Loudon's *Arboretum Britannicum* with its four volumes of plates and four of text.[24] The prohibitive cost of Loudon's arboricultural tome placed it beyond the means of most gardeners and underscored the need for publicly accessible planted arboretums as influential commentators such as John Lindley, professor of botany at University College, London, emphasized. Reviewing Loudon's *Encyclopaedia of Trees*, the relatively cheap abridgement of the *Arboretum Britannicum* without the full-length illustrations, Lindley argued that the latter was 'one of the many extraordinary instances of industry which Mr. Loudon's career as an author has produced'. Although this gave him 'great claims to the gratitude of gardeners', the price of the book 'placed it beyond the means of ordinary buyers', as ten pounds for a single volume was 'a sum not to be thought of by the mass of garden lovers'. Despite 'its merit', the *Arboretum Britannicum* would only ever therefore be seen 'in the libraries of the wealthy'. At not more than a quarter the price of the original work, the invaluable abridgement placed Loudon's work 'within the reach of thousands who could not before hope to obtain it' and could be recommended for 'universal patronage'.[25]

Given that Loudon followed the Candollean arrangement in the *Arboretum Britannicum*, it was only to be expected that he would pursue this in the Derby Arboretum.[26] Specimens were obtained from commercial and institutional collections including those of the Horticultural Society at Chiswick and nurseries at Exeter, Hereford and Canterbury. Following Loudon's principle of relation, the arrangement was intended to combine order with variety, representing global climatic zones without allowing glaring unnatural juxtapositions.[27] Turning the pages of the *Arboretum Britannicum*, readers passed through a virtual arboretum which, like the Derby Arboretum, allowed them to experience global trees and shrubs. The relationship between written textual and planted arboretums was underscored by the fact that copies of the book were placed in the north lodge for the perusal of visitors whilst entertaining extracts were reproduced in Loudon's catalogue and guide. As visitors and groups navigated the planted arboretum reading and discussing the book entries they observed each seasonally changing specimen, pausing as their curiosity was aroused to engage with Loudon's ideas and compare them with living forms. Virtual and planted arboretums were combined in multi-sensory, multi-dimensional, perpetually changing visitor experiences. Like most of the plates in the *Arboretum Britannicum*, only one specimen of each tree or shrub was displayed and planting was done so that growth was uninhibited, however, the planting of flowers in vases around the park and the mixture of deciduous and coniferous varieties provided colour and interest all year round. The order of Coniferae, for example, was represented by various kinds of pine trees, fir trees, cedars, cypresses and junipers. These included: the common Scots pine, the Norway spruce, 'the American arbor vitae', the still rare araucaria or Chilian pine, 'the pendulous branched deciduous cypress' and the Swedish juniper. Each plant was numbered and named on glass-covered brick tallies with botanical name, common name, place of origin, height in native habitat, and the date of introduction into Britain.[28]

Erasmus Darwin, founder of the Derby Philosophical Society, had popular-ized the image of the sublime and the beneficial value of nature and translated Linnaeus into English. As Loudon knew, Darwin's Lichfield botanic garden was arranged according to the Linnaean scheme which he presented poetically in the popular *Botanic Garden* (1791).[29] Darwin's *Phytologia: or the Philosophy of Agriculture and Gardening* (1800) contained extensive discussions of plant phys-iology and reproduction, drainage and watering, manures, the aeration of soil and the growing of fruits, roots, wood and flowers. Darwin and other members of the Philosophical Society, including William Brooks Johnson and Brooke Boothby, had been members of the Botanical Society of London. Richard For-ester, Joseph Strutt and Thomas Bent became fellows of the Linnean Society and Thomas Bent was a vice-patron of the Derby and Derbyshire Horticultural and Floral Society.[30] In addition, Bent and other promoters of the arboretum were involved in the Midland Horticultural Society whilst, not surprisingly, editions of Loudon's works were stocked in the libraries of the Philosophical Society and the Mechanics' Institute.[31]

In Darwin's *Botanic Garden* the beauty of the garden could be admired for its own sake, serving as an entry into a greater understanding of the natural world, science and human achievement. Loudon's gardenesque enthusiastically rec-ommended the use of natural taxonomic collections in gardens or arboretums which became 'living museums'. But whilst mixed Linnaean collections had to be represented in groups or linear strips as in nurseries or botanical gardens with the tallest examples at the back, natural arrangements could more successfully marry taxonomy with aesthetics (Figures 2.1 and 2.2). At the Derby Arbore-tum, each tree and shrub was carefully situated in its own space. Like one of the Derby philosopher's medical specimens, each plant was positioned, labelled, cat-alogued, and artificially isolated so that it could be studied. The trees and shrubs could have been exhibits in the Derby Town and County Museum, except that they were alive. The Derby Arboretum was a living spatial representation of the Candollean natural system with its simplified taxonomic scheme, emphasis on interconnected conservative species delimitation, plant geographies and physi-ology. A Jussieuian Derby Arboretum might perhaps have represented more loosely-defined species boundaries by planting some trees and shrubs as inter-mediary groups. Though he chose the Candollean natural system rather than the Linnaean, Loudon was following the type of scientific scheme favoured by Eras-mus Darwin, in preference to the more unscientific and aesthetically founded picturesque. For both writers a scientific scheme emphasized order and meaning in the natural world, enhancing understanding through the presentation of scientific collections, which both believed would facilitate social progress. In fact, the Derby Arboretum was originally to have been a botanic garden, but the projected cost of upkeep limited the venture to trees and shrubs.[32] The Darwin connection was noted by contemporaries. In 1843, one visitor wrote in the visitor's book that he would like to see 'a bust of the late excellent old physician' Dr Darwin in the arboretum who had, 'so sweetly tuned his poetic lyre, to sing the praises of and to inspire the taste for, a "Botanic Garden"'.

Simo has argued that Loudon 'would have liked to evoke something of the spirit of Derbyshire' in his design. However, although he had wanted to incorporate 'hollows and winding hollow valleys' and 'hills and winding ridges' but was hindered by the nature of the soil and topography, it is striking how little Strutt and Loudon's creation had anything to do with evoking the picturesque spirit of the Derbyshire Peak. Although the mounds were supposed to be from 6 to 10 ft high in Loudon's plan, and some illustrations suggest that some attained this height, most never seem to have been this tall. Some settling due to subsidence, natural compacting and washing of soil down with rain water, visitors walking over the mounds, the renewal of gravel paths and the use of the park in the dig for victory campaign during the Second World War might be expected. However after a visit in 1844 Charles Mason Hovey expressed disappointment with the mounds, which he thought were not large enough to look 'natural' and therefore not much use in obscuring walking individuals.[33] Similarly, whilst the *Derby Mercury* followed Loudon in describing them as 6 to 9 ft tall in 1840, according to an observer who visited in October 1847 they were 'almost everywhere from four to six feet in height'.[34] The mounds were supposed to foster an illusion of size by obscuring visitors from each other on adjacent paths and helping to hide boundaries. However, given that they never seem to have been quite tall enough to do this their primary function seems to have been exhibiting tree forms and root systems.

When they toured the region, the Loudons were most impressed with what they interpreted as evidence for interrelated progress in agriculture, planting, architecture and industry. Hence much of Loudon's description of his midland garden tour was dominated by detailed accounts of the construction of railroads, railway plantations and the applications of technology in architecture and domestic comfort by the Strutts at their manufactories and workers' houses and others. The Derby Arboretum was not supposed to evoke the Derbyshire Peak and did not provide a representative collection of local flora, which would have been far easier to accomplish than procuring and managing hundreds of rare species of trees and shrubs from national collections. Rather, it was primarily intended to celebrate and encourage the progress of arboriculture and botany in the world's leading industrial nation.[35]

The practical requirements of visitors to the Derby Arboretum were considered and Strutt stipulated that refreshment rooms were to be provided for nothing or a small charge with separate toilets. Following Strutt's instructions the park was designed with easy maintenance in mind, and Loudon thought one person would be sufficient for the upkeep.[36] Indeed the institution was a great success. The three-day opening ceremonies in September 1840 included processions from the corporation, trades and societies and children's day celebrations. The festivities included a balloon ascent, dancing and a fireworks display all of which received much attention in the local and national press.[37] However, many changes were made, the most important of which were the acquisition of a large extension during the 1840s and the construction of a new entrance with lodge and commemorative statue of Joseph Strutt during the 1850s. The unexpected popularity of the park was such that in 1841 the committee were already looking for extra ground to rent. An extension fund was created in 1843, which accumulated interest through investment in government stock and a six-acre extension to the south west of the original site was purchased in February 1845. The landscaping was undertaken by William Pontey junior, nephew of William Pontey senior who kept a nursery near Huddersfield and author of various arboricultural works including the *Forest Pruner* which was recommended by Loudon. Pontey junior succeeded his uncle when the latter became too ill and infirm to continue the business and the arboretum committee clearly employed him as an acknowledged expert whose later commissions included the Ipswich Upper Arboretum. Pontey senior had previously undertaken tree planting for Francis Jessop, mayor of Derby in 1840 and a member of the Derbyshire Horticultural Society and Derby Philosophical Society, probably at Loudon's instigation.[38] The extension was partly funded through a large legacy provided by Richard Forester, president of the Philosophical Society who died in 1843. His successor Dr Thomas Bent also contributed towards the building of the new lodge and statue in 1851, donated money towards a new glasshouse called the crystal palace and bequeathed a further £200 towards the enlargement or improvement of the arboretum.[39]

Critical Reaction to Loudon's Arboretums

The publication of the *Arboretum Britannicum* had an immediate impact on the British and international horticultural and gardening world. The *Quarterly Review* 'warmly' congratulated Loudon on the publication which it considered to be the work to be 'of solid value, worthy of a place in the library of every landed gentleman, as well as of every student of botanical, arboricultural, and horticultural science'. The accomplishment was a 'Herculean task; a task which few men, except himself, would have had the courage to begin, and still fewer the perseverance to complete' and 'must become' the 'standard book of reference on all subjects connected with trees'. The *Edinburgh Review* noted the considerable cost of the laborious and complex work by the 'well known' editor of the *Gardener's Magazine* which it considered 'a publication of the highest value in the class to which it belongs' and which will be essential 'for every person interested in the important subject of planting, whether as an amateur or in a professional capacity'. The *Times* considered the book a 'gigantic' and 'successful' undertaking bringing together, for the first time, 'such a mass of information on the subject of trees as was never before collected together' which would 'render it indispensable to every country gentleman and landed proprietor'. For the *Athenaeum* speaking on behalf of landowners, 'so useful a book upon trees and shrubs' was never 'to be found in any language' and it was 'not too much to say' that it deserved to 'form a part of the library of every country gentleman'. Similarly, the *Quarterly Journal of Agriculture* believed it to be 'by far the most complete work on the interesting subject of arboriculture that has hitherto appeared in this, or, as far as we know, in any other language'. The *Arboretum Britannicum* would be 'indispensable to every land proprietor, land steward, and forester' who would find 'much that is original, combined with all that has been written worthy of perusal up to the present time' on the subject. The *Edinburgh New Philosophical Journal* considered it to be the first great analysis of the 'natural and economical history' of trees and shrubs since John Evelyn's *Sylva*, noting how useful the 'beautifully illustrated volumes' would be to professional foresters, improvers and landowners, and emphasizing that it ought to be in 'all our public libraries' and the library of 'every country gentleman'.[40]

The international reception was mostly equally as enthusiastic. According to Professor Dietrich Franz Leonhard von Schlechtendal (1794–1866), the distinguished German botanist, it was manifestly a 'work of the most laborious assiduity, and persevering steadiness' from a talented and active man which could only have been completed in a country with 'so much capital and so great a taste for parks and pleasure grounds on a large and small scale'. Professor Alphonse Pyramus de Candolle (1806–93), son of the man whose system Loudon had followed in the *Arboretum Britannicum* and who succeeded to his father's chair at the University of Geneva, noted that it contained 'an immense quantity of matter in a very little space', managing to combine systematic study with entertaining facts. He considered that it would appeal equally to amateur and professional gardeners, whilst botanists would 'consult it for its descriptions, its synonyms, its numerous and beautiful engravings, and for the economical history which it contains of each species'.[41]

The judgement of the two most prestigious British botanists, Lindley and William Jackson Hooker, was very important. Lindley, who as we have seen had been contemplating a similar study himself until the early 1830s, noted that whilst it was partly a compilation, it was 'to a great extent original', providing a 'most valuable mass of information'. In a detailed analysis in the *Annals of Natural History*, Hooker judged the book to be of the 'highest importance and of the 'greatest utility' to arboriculturists all 'nobleman and gentleman of landed estate' who intended to improve their estates and enlarge 'the resources of his country' and to 'every botanist and cultivator who wishes to become acquainted with the trees and shrubs, whether indigenous or exotic' to the British climate. Everything Loudon promised had been amply and skilfully realized and there was 'not a naturalist in Europe who could have executed the task with anything like the talent, and judgment, and accuracy' displayed by Loudon. After describing the content of each volume, Hooker emphasized the importance of the 'very numerous wood cuts' that 'so beautifully' illustrated the text and included landscapes and scenic views as well as sections, illustrations of insects that preyed upon trees and much else. He believed that 'nothing' had been 'omitted, either in the descriptive or pictorial matter' that would illustrate the history and uses of trees and shrubs. Trees and shrubs supplied fruits and numerous other products, were 'the most valuable materials' for ships, dwellings and implements, and the 'greatest ornaments' of parks, gardens and pleasure-grounds. Regretting that he had not space for copious extracts, Hooker adjudged Loudon's *Arboretum Britannicum* to be a work of 'vast importance' to Britain and 'to every part of Europe, and the temperate parts of North America', even to 'all the temperate parts of the civilized world'.[42]

As probably the foremost living realization of Loudon's arboriculture, the Derby Arboretum attracted considerable national and international attention. The value of the arboretum to the town of Derby was made clear by the Unitarian minister, Noah Jones, when preaching Joseph Strutt's funeral oration in 1844. The arboretum was a 'noble gift' which would stimulate the 'rational social pleasures' of many, including the 'toilworn artisan' who would then supposedly work with greater diligence.[43] The health commissioners, in reply to Martin's report on the health of Derby in 1845, concluded that the park with its trees and shrubs had already had a 'perceptible effect' in improving working-class 'appearance and demeanour' and 'conferred an equal benefit upon their health'.[44] Martin was much more explicit and impassioned in a later report arguing that 'the most cursory reference to the condition and habits of life of the working classes' clearly demonstrated 'the immense advantages, moral and physical, that must accrue to the inhabitants of closely-built towns by the establishment of public parks' such as the Derby Arboretum.[45] This was 'a noble example', though one that 'few persons have the means of imitating' and he hoped that national government intervention would follow. For without this, owing to the 'rapid increase of towns', many of the labouring people would be 'reared from infancy to mature age without once breathing in an open space, or once enjoying the refreshment to be derived from exercise in the green fields'. This was an 'unnatural and cruel' though not uncommon position. A *Derby Mercury* editorial of 1851 agreed, arguing that Derby was 'privileged beyond most provincial towns' in having this 'beautiful property'. It was 'scarcely possible' to overestimate the benefit which so large a 'well managed' 'space of ground for air and exercise' was able to confer upon a 'dense population' who would readily appreciate the 'glimpse of the country' from amongst their 'crowded habitations'. This was 'constantly becoming more apparent' as the town increased. Derby had 'rapidly extended itself on all sides, and its population 'acquired an immense increase'. Green fields were giving way 'to vast masses of building', and the townspeople were becoming further 'removed from the beautiful walks and the pure air of the country'. Faced with the bewildering changes of Britain's industrial revolution, here was one response that all could support with enthusiasm.[46]

Most visitors were equally enamoured. The physician Augustus Bozzi Granville (1783–1872) wrote of Loudon's 'noblest combinations of artificial gardening' and suggested he take charge of all royal public parks and gardens.[47] William Edward Hickson (1803–70), editor of the *Westminster Review* and educational campaigner, thought that Strutt's munificence was the finest aspect and looked forward to the day when many parks would be founded in towns, which could be legally acquired and supported by the rates.[48] Hickson hoped for a new national policy of garden creation. In 1842 Edwin Chadwick, the sanitary reformer and poor law commissioner, stated that 'much evidence might be adduced from the experience of the effects of the parks and other places of public resort in the metropolis, to prove the importance of such provision for recreation'. This was not only for the pleasure they gave, but as a 'rivalry to pleasures that are expensive, demoralizing, and injurious to the health'. He recommended the arboretum as 'deserving of particular attention' and included extensive extracts from Loudon's guide, with a plan of the park in an appendix to his report.[49] The Scottish journalist William Jerdan (1782–1869), editor of the *Literary Gazette* thought Loudon's 'little Derby domain' to be 'the very treasury and epitome of the wide world's natural wealth'.[50] The US nursery proprietor and editor of the *Magazine of Horticulture*, Charles Mason Hovey (1810–87), of Cambridge, Massachusetts, who came to the arboretum in the Autumn of 1844, thought it superior to other British and French parks and even to Paxton's work at Chatsworth. He noted that there were no weeds, the flower garden was flourishing and the brick tallies were now in full order and he considered that Loudon and Strutt's example should be followed as 'we know of no object so well deserving the attention of men of wealth than the formation of public gardens free to all in crowded towns or cities'.[51]

Perhaps the most important and detailed North American assessment of the Derby Arboretum came from the distinguished landscape gardener Andrew

Jackson Downing (1815–52). Downing believed that the picturesque natural style, developed in Britain by Repton, Loudon and others, was the perfect form of landscape gardening to be adopted and adapted in the United States of America. He agitated to promote the benefits of public gardens as places for the recreation and intercourse between social classes, visiting examples in Britain and Europe that might be adopted in the United States. Shortly after his return to the United States in 1850 Downing immediately set to work on a design for extensive public grounds in Washington which incorporated many European ideas and included a garden of American trees and a living 'museum' of evergreens, and also urged for the creation of a large park in New York. During a tour that included Woburn Abbey, Haddon Hall and a stay at Chatsworth, Downing visited Derby which he found to be an 'interesting old town'. However, he had really come to see the arboretum which interested him because it had been 'especially formed for, and presented to the inhabitants' by Joseph Strutt. Furthermore, it contained 'a specimen of most of the hardy trees that will grow in Britain' laid out by Loudon. In Downing's judgement it was a 'noble bequest' that was 'in beautiful order' and 'evidently much enjoyed' not only by locals but by strangers from the country around. He met 'numbers of young people strolling about and enjoying the promenade' with nurses and children gaining strength in the fresh air whilst amateur botanists would carefully read the labels of the various trees and shrubs (which he described) and make notes in memorandum books. In Downing's view, 'the most perfect novice in trees' could, by walking around the arboretum, obtain in a short time 'a very considerable knowledge of arboriculture whilst problems of identification and classification could be solved by observation of living specimens'.[52]

Downing regarded the Derby Arboretum as 'one of the most useful and instructive public gardens in the world', agreeing with Strutt and Loudon that it combined 'the greatest possible amount of instruction' with pleasure for all classes, encouraged by the grass field extension where games such as skittles were played. However, whilst he accepted the botanical value of placing isolated specimens around, he confessed to being disappointed with the lack of taste displayed by the arboretum. Downing was not necessarily against the ridges which he accepted shielded individuals on the paths from view, however, he considered that groups and masses of trees and shrubs would have been preferable, 'dotted as they are with scattered single trees and shrubs' the result was 'a little harsh' with 'neither the ease of nature nor the symmetry of art'. In contrast, Robert Marnock's botanic garden in Regent's Park, 'in point of tasteful arrangement and beauty of effect', was very different with 'none of the stiff and hard look of the surface of the arboretum at Derby'. In mitigation, Downing recognized that Loudon had been aiming at 'a garden for instructing the British public in arboriculture' rather than primarily 'a specimen of public pleasure grounds'. The fact that the arboretum was the bequest of a 'private citizen to his townsmen' and to the country made it 'a magnificent donation' which would provide a source of recreation, enjoyment and local pride to all classes. Downing urged that his countrymen should follow Strutt's example and form public parks or gardens in large towns which would keep their memories 'more lovingly fresh in the minds of his fellow townsmen, and their descendents, than any other bequest it is possible to conceive'.[53]

Conclusions

The Derby Arboretum is significant in a number of ways. Firstly, it provided an opportunity for Loudon to promote his vision of public parks and gardens in towns and to demonstrate that arboretums were an appropriate model for such institutions. Secondly, as national and international contemporary observers repeatedly emphasized, the arboretum provided an influential example of how the problems of funding urban parks could be overcome. Strutt's donation demonstrated how the munificence of wealthy benefactors could play a major role in changes in taste and recreational practices. In that sense it was a triumphant success, as the popularity of the festivals and requirement for an extension demonstrates. Thirdly, the Derby Arboretum was intended to be a counterpart to museums, galleries and mechanics' institutes in the encouragement it gave to natural history and arboriculture. It was hoped that it would presage the development of similar authoritative provincial foundations by providing a systematic collection readily available for all social classes so that living specimens could be compared with examples in herbariums and textbooks. For its first four decades, the Derby Arboretum encouraged the provision of similar collections elsewhere, an increase in the richness and variety of trees and shrubs planted in public parks and provided an accessible and authoritative planted realization of the *Arboretum Britannicum*.

7 ESTATE ARBORETUMS

Introduction

Estate arboretums carried the impulse for tree collecting amongst the landed classes into the nineteenth century and were of considerable importance in the development of arboretums. The economic, botanical and iconographic significance of trees in British society, underscored by the prizes for planting awarded by the Society of Arts, encouraged landowners and their agents to undertake large-scale planting. Special kinds of tree collections or plantations were nurtured on many estates, usually associated with landscape gardens or their boundaries, although some came to occupy large areas or were spread along drives and rides. Estates also managed the remains of much older tree collections of great commercial, botanical or antiquarian interest, such as the sections of ancient Sherwood Forest lying within the Nottinghamshire Dukeries. Grand tree collections provided visible evidence of the magnificence, power, leisurely discernment and cultural pre-eminence of aristocracy and gentry – suggesting continuity and permanence just as their authority was beginning to erode. Frequently associated with the improvement, construction or reconstruction of houses, grand arboretums also provided an opportunity for the nouveau riche to assert their status and gain admittance to the landed classes. Larger estates had extensive resources and manpower, providing an important forum where landscape gardeners, gardeners and woodsmen could train and conduct experiments. Some gardeners and landscape gardeners such as Joseph Paxton, William Barron and William Coleman were launched into professional pre-eminence by great estates and aristocratic encouragement, their success enhancing their patron's reputation. Although access to estates and private gardens was often restricted, especially for working and middle-class visitors, public visits and practices publicized in the burgeoning horticultural press encouraged many to emulate the practices of the landed elites. Estate arboretums also played a major role in the development of professional and scientific forestry encouraged by forestry in Europe and the British colonies. Many aristocrats and gentry gained great wealth through the empire and served in colonial government. Trade and ownership links fostered the exportation of estate management practices whilst aristocrats and gentry brought back ideas and practices as well as exotic specimens from the colonies which they attempted to utilize in British estate management. They also encouraged and sponsored plant hunting exhibitions for their own gardens or institutions. As we shall see, it was working for the Duke of Devonshire at Chatsworth which enabled Joseph Paxton to become the pre-eminent Victorian landscape gardener and these experiences inspired the Great Exhibition of 1851 whilst William Barron's reputation was forged at Elvaston under the Earls of Harrington.

There were tensions between Loudon's enlightenment concept of an ordered botanically representative arboretum and estate arboretum requirements. Estate management and horticulture required plantations for economic and other uses. These varied demands were usually resolved by divisions between kitchen and pleasure gardens around the house and separate plantations for ornamentation, orchards, commercial exploitation, hunting, shooting and other purposes. Acquiring trees and shrubs for the purposes of pleasure, landscape ornamentation and ostentatious display as well as science, private collectors were not always primarily interested in botanical significance or representativeness. There was a cultural hierarchy of trees and shrubs, apparent in professedly objective nineteenth-century botanical works such as Loudon's *Encyclopaedia of Plants*, where those regarded as boring and insignificant were pejoratively described. In terms of estate landscape gardening and gentlemanly collecting, exotic, colourful, striking and unusual trees and shrubs were especially valued and placed in significant locations. As Humphry Repton emphasized, house and gardens were unified. Just as the magnificence of the former depended on its rooms and furniture so the 'diversity and succession of interesting objects' in the latter 'required to be enriched and finished' to demonstrate conspicuous good taste.[1]

Although landed gentry retained considerable power and influence during the Victorian period, the relative importance of landed power declined in the face of urban expansion and political reforms that increased the size of the electorate after 1832. Some British landowners had always, of course, taken a keen interest in industry and mining but the shift of wealth from land to business and other forms of investment became more pronounced during the second half of the nineteenth century. This was especially true after the 1870s when a period of long term agricultural depression began. Whilst landed families continued to present an appearance of wealth and gained considerable sums from agriculture and rents, much more income came from business ventures. The Cavendish family and their agents developed the port of Barrow in Cumbria, for instance, whilst much of the income of the Holford family, who owned Westonbirt, came from investments in the London Water Company. Although it should not be exaggerated and the landed elites remained easily the wealthiest British individuals, the development of systematic estate tree collections after 1850 took place against a background of relative agricultural decline. Arboretums were a symbol of the increasing tendency for estates to be sites of consumption rather than economic production. Systematic estate arboretums always attracted some criticism, particularly towards the end of the nineteenth century when landscape gardeners such as William Robinson and others discouraged the cultivation of exotics in favour of so-called 'native' varieties of trees and shrubs.[2]

The Collecting Impulse: Rivalry and Display

Some nineteenth-century aristocrats and gentry continued to collect trees and shrubs as enthusiastically as their Georgian forebears. The tree collection acquired by George, fourth Duke of Marlborough as the Marquis of Bland-ford from 1798 for his estate at Whiteknights near Reading in Berkshire set the fashion for grand estate arboretums. To celebrate landscape improvements and tree collections, the Marquis commissioned the artist Thomas Hofland to paint views of the house and grounds and these were published with a cloyingly celebratory text by his wife, the novelist Barbara Hofland. He collected plants of 'every description, built numerous hot houses for the exotics' and, with the help of Jones his gardener and another twenty four staff, developed a large walled garden with hardy herbaceous plants and 'the more choice trees and shrubs'. He formed an arboretum of several acres and distributed throughout the park 'a collection, as extensive as could then be procured' of the genus *Crataegus*. At great expense, the Marquis procured the latest examples of American trees and shrubs, magnolias, rhododendrons and azaleas despite 'enormously high prices' due to their rarity and recent introduction. If only one or two examples were known, he would try to procure all of them and Lee of the Hammersmith Nurs-ery informed Loudon that he had sold several examples of rare plants to the Marquis for twenty or thirty guineas each.[3]

The rarity, cost and display of the Whiteknights gardens was intended to demonstrate the superior taste and discernment of their owner whose political status was a pale shadow of that enjoyed by his illustrious predecessor. Accord-ing to Hofland, the 'beautiful walks, velvet lawns, exotic plantations, flowery arcades, rural bowers and gay pavilions' demonstrated the 'taste and spirit of their noble possessor' who had provided a 'rich plantation' of forest trees.[4] The botanic garden was an 'unrivalled storehouse of flora ... guarded and adorned by four lofty cedars of Lebanon'. There were 'numerous specimens of the most valuable and curious productions peculiar to China, Botany Bay, and the Cape of Good Hope' and a Linnaean garden with 'every herbaceous plant ... regu-larly classed' and labelled.[5] The arboretum proper consisted of an 'extensive and thriving plantation of forest trees and flowering shrubs, as well exotic as native' and extended for 'a considerable way' on banks opposite the new gardens whilst George Agar-Ellis considered 'the wood – the botanical garden – the conservatories' and other parts of the gardens to be 'all beautiful'.[6] Loudon made various visits and used Whiteknights extensively in the preparation of the *Arboretum Britannicum*, citing it eighty-five times.[7] But botanical significance was of secondary concern except where it enhanced rarity and demonstrated value. Collecting furnished improved mansions with works of art, antiquities and books just as it drove the parallel adornment of pleasure gardens with a plethora of prize arboricultural beauties. Almost as famous as Whiteknights was the Marquis's extravagant expenditure on items such as the only perfect copy of Boccaccio's *Decameron* in existence complete with an ornamented miniature picture for £2,260, then the highest price ever paid for a single book.[8] The lavish expenditure on books, antiquities and rare trees and shrubs was paid for with loans raised on the value of the estate, but the Duke ran into severe financial difficulties and was forced to sell the estate and many possessions. However, the Hoflands and many other creditors remained unpaid. As Loudon tactfully put it 'in consequence of a similar mode of proceeding in his transactions generally ... he soon found himself involved in debt and lawsuits' which 'greatly crippled his exertions'. However, on moving to Blenheim after being forced to sell the Whiteknights estate, he continued to acquire trees and shrubs such as many of his favourite magnolias.[9]

The Duke of Marlborough's mania for collecting was paralleled by the activities of William Spencer, sixth Duke of Devonshire (1790–1858), who employed Joseph Paxton to transform his estate at Chatsworth in Derbyshire. Paxton's experiences working on the Chatsworth gardens, especially the creation of the arboretum, pinetum and giant hothouses, and the Duke's patronage directly inspired the Great Exhibition of 1851 and the subsequent formation of the Crystal Palace at Sydenham. Employment at Chatsworth and the Duke's patronage also enabled Paxton to rise to pre-eminence in Victorian horticulture and landscape gardening and to promote a significantly different, more elitist vision of arboriculture from that of Loudon. Encouraged by Paxton, the Duke of Devonshire developed the Chatsworth gardens after turning away from national politics following his efforts to ensure the passage of the Great Reform Act of 1832 which he regarded as a final settlement rather than a prelude to more fundamental reforms. After undertaking improvements on the house the Duke began to care for his plants 'in earnest'. The 'old greenhouse' was converted into a stove, the garden greenhouse constructed, 'the arboretum invented and formed', and construction of the orchid houses begun.[10] Pearson has emphasized how 'for a Duke who had imported marble by the ton from Italy, it seemed quite natural to collect the rarest plants on the same princely scale'. Chatsworth with its majestic art galleries, massive library, theatre and, through Paxton, 'above all the exotic richness of its gardens' became the Duke's 'own private pleasure-dome, where he could steadily create the idyllic life he wanted' encouraging visitors to marvel through the guidebook that he composed for their benefit.[11]

Like the Duke of Marlborough, and the Duke of Devonshire, the Holfords were equally industrious and adept at procuring the most recently discovered trees and shrubs just as they acquired numerous books, antiquities and artworks. Before transforming Westonbirt house and gardens, Robert Staynor Holford (1808–92) spent much of the considerable wealth acquired through shares in the New River Company in constructing Dorchester House in London and amassing a large art collection. Holford valued the Westonbirt trees and shrubs and returned Loudon's *Arboretum Britannicum* questionnaire with details of them in 1834.[12] Although he had begun collecting trees in 1829, Robert Holford does not seem to have begun to form a systematic arboretum until the 1850s and 1860s, and much of the arboretum still visible today was the work of his son George. The older manor house was replaced, although this was also removed in 1863 to make way for a much larger mansion by the architect Lewis Vulliamy, which was completed in 1871. At the same time, the estate lands around the house were extensively remodelled. The land near the house was raised several feet and levelled, forming terraces. The village including rectory, farm and cottages, was imperiously moved to the south west, behind a bank, so that it could be concealed from the house; only the church had to remain, and it was screened from view by beech trees and a high wall. The gardens surrounding the house were formed into Italian-style terraces with balustraded steps, fountains and an 'Italian garden', whilst the park towards the lake which covered the site of the old village to the south west was landscaped with lawns, large Pulhamite rockeries, a fernery and clumps of trees. The park to the north of the house which became the arboretum was extended and the main road between Tetbury and Bristol diverted away, thus the pleasure grounds and park offered complete privacy for the family whilst reinforcing their domination of local society. An intensive programme of planting was conducted by the Holfords between the 1860s and 1920s and after acquisition by the Forestry Commission it was accorded the status of national arboretum.[13]

Elvaston Castle and William Barron

Whilst Joseph Paxton was the most successful Victorian landscape gardener, his contemporary William Barron played almost as important a role in forming the image of the heroic landscape gardener. Like Paxton, Barron trained on private estates and a botanical garden, utilizing aristocratic patronage to develop a career as a professional landscape gardener. Barron was employed at the Edinburgh Botanical Gardens and at Syon House, Middlesex, for the Duke of Northumberland. Barron's Scottish background, general education and experience was an important factor in his success, and like Loudon's career, reflected the quality of Scottish education and attention to improvements in horticulture and agriculture that characterized the Caledonian enlightenment.[14] Barron employed these ideas and techniques creating multiple urban parks, cemeteries and green spaces and formed one of the most successful Victorian landscape gardening and nursery companies. The landscape gardening practices developed and employed at Elvaston, which Barron regarded as 'one huge pinetum, artistically treated', especially the hugely ambitious industrial-scale tree transplanting, propagating and grafting, transformed a largely featureless site into one of the most celebrated gardens in Europe and North America. Hundreds of trees, including very large and mature specimens, were moved across Derbyshire and adjacent counties creating an arboricultural equivalent of Isambard Kingdom Brunel and Robert Stephenson's civil engineering feats. Urban parks, gardens, cemeteries and other green spaces were rapidly transformed into apparently mature creations, a process that would previously have required decades of growth and management. The achievements at Elvaston and Barron's *British Winter Garden* (1852) promoted the use of evergreens in public and private spaces and Barron & Company undertook hundreds of national and international commissions. This helped to drive the new fashion in British, European and American gardens and secure Barron's status as one of the leading British arboriculturists. Consulted by Parliament, Barron also helped to establish forestry and arboriculture as professions in their own right in Victorian society.

Over two decades from 1830 at Elvaston, William Barron drained, landscaped, planted and turned the flat, muddy, marshy and unpromising grounds into a Gothic fantasy park to serve as counterpart to the redeveloped Gothic mansion. By 1850, the park featured thousands of trees, a large lake surrounded by elaborate mounds, trenches and ridges, rocks shaped into fantastic shapes and more formal geometric gardens dominated by masses of topiary, also moulded into wonderful shapes, towards the house. The gardens were created for Charles Stanhope, fourth Earl Harrington, and his wife Maria whom he had married in 1831. Close to the Prince of Wales, as Viscount 'Beau' Petersham until 1829, the earl became famous as a Regency dandy with a fondness for extravagant and exaggerated fashions, romantic gestures and the theatre. However, he also had a significant military career, rising to the rank of Colonel by 1814. Maria Foote was acclaimed for her beauty and talent as an actress, playing at Covent Garden and touring Britain in various productions until 1831. The couple were the source of much society gossip. As well as being an important retreat from the gossip of society, the transformation of Elvaston offered the opportunity for the lovers to create a private theatrical fantasy world, a new stage upon which to perform.[15] Loudon and his wife Jane visited in 1839 when work was still under way and were highly impressed with the formal gardens, the house and Barron's arboriculture, which they described in a report in the *Gardener's Magazine*; Loudon suggested that they should be opened to the public when they had reached maturity.[16]

As at Whiteknights, the transformation of the Elvaston gardens succeeded other estate improvements. Under the third Earl, major changes to the castle were undertaken by James Wyatt and Robert Walker, although parts of the earlier manor house were retained. The fourth Earl had the east wing rebuilt and interiors refurbished in the Gothic style under Lewis Cottingham the leading archaeological Gothic revival architect. The entrance hall became the 'hall of the fair star' and the castle was decorated with symbols of medieval chivalrous love and ideals with lances and swords, suits of armour, gold, black and scarlet paint and with mottoes and shields. These themes were continued under Barron's direction in the formal architectural gardens south of the Castle based upon enormously long, closely clipped evergreen hedges transformed into fabulous shapes, also intended to celebrate the ideals of knightly chivalry. The Alhambra garden featured a Moorish temple with a statue of the Earl and Countess, Charles kneeling at Maria's feet. The Mon Plaisir garden, also known as the garden of the fair star, was the most celebrated and featured elaborate topiary and statues inspired by seventeenth-century designs. The centre consisted of a monkey puzzle tree, the tallest and probably then most expensive at Elvaston, surrounded by an eight-pointed star-shaped bed planted with golden holly with yews clipped in architectural shapes to form bowers for statues. Surrounding the star were eight bowers of arbor vitae panelled with *Cydonia japonica* which held semi-circular seats. The garden was enclosed by walls of yew and within these was a tunnel of arbor vitae which offered shade and views of other parts of the garden through cut windows. Finally, the adjacent Italian Garden, based on Tuscan designs, was also enclosed by tall yew hedges inside of which were twenty tall marble statues, according to Loudon it was 'richly furnished with vases, statues (any of which are in grotesque forms), richly gilt, basins, fountains and other works of art'.[17]

Almost equally celebrated at Elvaston were the avenues, lake, elaborate rock work and pinetum expressly designed to frame particular pictures from different parts of the garden and especially from the windows of the castle. The completely artificial yet naturalistic and picturesque lake was formed by 1830 and surrounded, especially on the north side opposite the castle, by equally artificial rock work featuring caves, grottoes and mounds. The rocks were moulded into fantastic shapes including small caves and arches and covered with Alpine plants, whilst some of the mounds created from the earth excavated when digging the lake were 50 ft tall. Barron created and augmented various avenues, the most spectacular being the grand east wing avenue which followed a line already developed from the castle extended to provide uninterrupted views of ten miles distance towards the Gotham Hills in south Nottinghamshire by removing intervening trees, thus serving to appropriate surrounding lands into the estate. At either side of smooth turf were circular flower beds and a line of Irish yews following a wavy pattern around. The trees were planted in straight lines, three rows deep, using twelve different species with particularly striking '*Picea nobi-lis*' (*Abies procera*), and inspired the development of evergreen plantations and avenues on other estates.[18]

After the death of the fourth Earl, the fifth Earl opened Elvaston gardens to the public in 1852 which created a sensation. However, at a cost of three shillings for admittance they remained the preserve of elite visitors.[19] According to the *Gardener's and Farmer's Journal* in 1852, Elvaston was without exception

> first for its great accumulation of rare evergreen trees and shrubs; and secondly, for the effect which are by means of these trees produced, that is, in the grouping and mixing trees according to their habits of growth, their various heights, but, above all, their colours, that in this way effects are produced such as are nowhere else to be seen.

It was remarkable that yew trees had been collected 'from all parts of the country', some of which were 'several hundred years old' and consisted of 'a mere outer shell, the whole of the centre being decayed and gone', yet these old yews were now 'thriving and growing with as much vigor as if they had been raised from the seed bed but ten years ago'. Likewise, Glendinning lavished praise on Barron's achievement. Looking across the lake from an elevated position amongst 'enormous columnar rocks' where a seat had been placed, he found that

> the view here almost baffles description; an immense expanse of water, swarming with water fowl, is exposed, as well as the bold sinuosities of the margin, with an inconceivable amount of artificial rock naturally disposed and planted with thousands of the most valuable plants that could be obtained.[20]

The Elvaston gardens and pinetum demonstrated the possibilities of varied colour and shape in gardens throughout the year and helped to make evergreens and topiary the height of fashion in Victorian Britain and beyond in Europe and North America. Barron undertook numerous private commissions between the 1850s and 1880s, many of which featured extensive arboricultural work. At Sennowe Hall, for instance, in Norfolk between 1858 and 1865, he managed the estate in Chancery for the trustees of Morse Boycott and, employing a Scottish woodsman and labourers, formed an entrance drive similar to Elvaston, with rows of cedars and Douglas firs.[21] One of the gardens inspired by Elvaston was Biddulph Grange, Staffordshire, where a large number of evergreens were planted in a pinetum amongst the spectacular rock work, many on mounds, golden yew hedges were formed, and a multilayered avenue formed by the owner James Bateman and the landscape gardener and painter Edward Cooke during the 1840s and 1850s.[22] Barron's arboricultural expertise was recognized by the select committee on forestry established in 1885 in order to consider whether to establish a British school for foresters who questioned him as their first witness in 1887. Emphasizing his experience of managing the estates at Gawsworth, Elvaston and Sennowe, Barron argued that too much 'rubbish' was being planted around the country and that if the right kinds of trees were selected and planted in the correct manner then British forestry could compete commercially with foreign imports to supply timber for important industries such as the railways. He recommended to the committee that a British forest school be established to educate a generation of foresters in botany and practical arboriculture who could then teach the next generation. He provided a list of recommended trees for planting around Britain, although not until they were at least 4 or 5 ft tall; especially favourite coniferous varieties included the Corsican pine, Austrian pine, Douglas fir and Oregon fir recommending that permanent forests needed to be divided by intermediate spaces for hardwood, game and in order to facilitate access to the timber to be harvested.[23]

Arboretums, Forestry and Estate Management

Arboretums provided an important economic resource, particularly when tied to agricultural production and forestry as part of the management of a productive estate. Stimulating local and national plant trading networks, they also sometimes launched the careers of nurserymen. Arboretums were distinguished from ordinary plantations, which tended to have only a few varieties of trees driven by economic and practical rather than botanical considerations, but distinctions between the two were blurred. Until the period of agricultural decline towards the end of the century, and especially the commercial decline of estates during the 1920s and 1930s, manpower requirements necessitated systems for training many workers in agriculture, forestry, horticulture and arboriculture. The relationship between orchards and kitchen gardens, forestry and the development of an arboretum is evident at Chatsworth, Elvaston, Holme Lacy and Eastnor. Despite their importance in the development of estate arboretums, gardeners, woodsmen and other staff tend to be ignored in secondary works which generally present parks and plantations as if they were the outcomes of the will of aristocrats or consulting landscape gardeners. This is also reflected in artistic depictions of nineteenth-century estate parks, gardens and arboretum which, just as villages were moved to create private spaces and reinforce authority, generally do not depict estate workers labouring. Yet, as Loudon recognized, gardeners had to acquire considerable knowledge and skill in dealing with the configuration and planting of novel species as well as the management of those below them, including detailed knowledge of soils, plant physiology and botany.

Most estate gardeners continued to be regarded as servants (Figure 7.1). However, gardening became increasingly recognized as a profession in its own right
requiring specialist forms of knowledge, systems of training and recognized
qualifications.[24]

Figure 7.1: Detail of *Berberis gagnepaini* with faceless gardener from A. B. Jackson,
Catalogue of the Trees and Shrubs at Westonbirt (London: Oxford University Press,
1927), plate 15. Reproduced from the authors' personal copy

Although much published correspondence is available in the Victorian gardening press, the survival of primary sources directly relating to gardening staff is
rare. Two valuable surviving exceptions relate to Westonbirt and allow comparison between notebooks kept by estate gardeners and landowners. A notebook
of Robert Staynor Holford lists trees acquired and reflects the Holford's fascination with measurements and memorializing the sources and individual histories
of different kinds of trees, such as the seven oaks planted in 1890 that had been

provided by 'Mr Estcourt raised from acorns given him by Lord Manvers from Sherwood Forest'.[25] Many had important personal associations such as the '*Acer Colchicum rubrum*' planted in the 'east side of clump at the end of the new terrace' on Holford's eightieth birthday, 10 March 1888. Compiled between 1880 and 1891, when Holford was in his 70s and early 80s, the notebook includes a mass of diverse arboricultural and estate information. Emphasis is placed on the starting and finishing dates of the building of the mansion, farmhouses, and lodges, the making of roads and rides (and their precise measurements) and the digging of the lake, but most information concerns dates of tree planting and plantation formation. There is much less detailed horticultural and silvicultural information than in Jonah Neale's notebook, but there is an attempt to list the principal arboricultural events.[26]

Neale was head gardener at Westonbirt between the 1830s and 1865. Apparently written for his own benefit, Neale's notebook records details of planting and other information and demonstrates his evolving understanding of trees, shrubs and plants. Rather than planting specimens and undertaking work merely at the direct instigation of the Holfords, the notebook suggests that gardeners were allowed considerable scope to work on their own initiative. Structured around an alphabetical list of species and miscellaneous information, the notebook was partly intended to further Neale's education and records tree species, dates of planting, rates of growth, and the movement of trees within the gardens and estate plantations. The statistical information suggests attention to arboricultural veracity but was not compiled consistently or systematically. Attention is given to the geographical origin of new specimens and when they were first brought to Britain, which appears to have been derived from originating nurseries, gardening publications and probably information relayed by Holford. For example, Neale notes that the *Araucaria imbricata* 'derives its name from the Araucanes a people of Chilli in South America' and had been introduced to Britain in 1795. The first one was planted at Westonbirt in 1840 or 1841 when 3 ft and cost as much as £20. By 19 May 1857 it had grown to 14 ft 9 in and was 1 ft ¾ in in circumference, 2 ft from the bottom, and by 5 September 1864, 23 years after planting, it was 21 ft 8 in high. A meteorological diary recorded lowest temperatures, first frosts and most serious frosts, including information concerning events before Neale's time. Early in his career Neale noted, for example, that there are 'five great oceans upon the globe', that the Atlantic lies 'to the south west of England' and that it is necessary to traverse this ocean to travel to the Americas although the geographical information pertaining to the origins of trees and shrubs is more sophisticated in later entries.[27]

Enthusiasm for novel exotic trees is evident, such as the *Cryptomeria japonica* or 'Japan cedar', first introduced from China in 1842 and Japan in 1861. Neale explains that it was found in China 'on the island Heushan and southern moun-

tainous districts of Japan'. Holford had purchased a plant in 1845 that was six
inches high for three guineas which Neale potted in April 1846 and kept in the
greenhouse. The tree did so well that in July 1847 Neale planted it in a hole 6 ft
in diameter and 3 ft deep which was partially refilled with soil and half with red
sandy loam, breaking the pot it was in rather than turning it out of the pot. By
19 May 1857 this tree had reached a height of 17 ft 6 in. Experiments, trials and
attempts to acclimatize new trees are also recorded, comparisons being made
between the same varieties growing externally and within greenhouses. One
of the longest entries for the '*Wellingtonia Gigantea*' reflects Victorian fascination with the tree after its discovery and early importation to Britain during the
1850s. Neale notes how Wellingtonias were 'introduced by Mr Lobb from of the
Hills in California South America [*sic*]' and that 'they grow 300 feet high' so that
the oldest would have begun life 'in the days of the Judges', been 'quite a youth
in the time of David' and have been 'about 1200 years old' at 'our blessed Lords
Incarnation'. Four Wellingtonias were purchased from James Veitch & Company
at Chelsea for £6. Two one foot high examples of these were planted very promi-
nently on either side of the gate opposite the new lodges on 8 March 1856 and
cuttings were rapidly taken from the lower branches because of the value of these
trees. The Wellingtonias took well and by 1860 the first planted had reached
heights of between 7 and 9 ft with girths of up to one foot five inches.[28]

Estate Arboriculture and Nurseries

Closely observed by visiting gardeners, estate nurseries were crucial to arboriculture as places where seedlings and cuttings were prepared in pots or sheltered and
isolated planting beds for permanent planting. Comprised of both glass houses
and open – but usually walled – beds, they can be regarded as labour-intensive
plant laboratories where experiments were undertaken to elucidate optimum
conditions for the growth of trees and shrubs, such as soil composition, light
and situation. At Eastnor in 1878, Thomas Baines noted that 'quantities of the
glossy leaved *Viburnum Sieboldii*' were being grown along with a collection
of different ash trees in the nursery, which was being expanded. A broad walk
through the nursery passed through grounds planted with younger trees and
shrubs whilst in one position were being tested 'the hardiness of the later introductions of Japanese trees and shrubs about which there is uncertainty as to their
standing here'.[29] The development of estate arboretums was also encouraged by
the symbiotic relationship with commercial nursery companies. This is evident
in two ways. Firstly, nurserymen marketed, labelled and supplied many of the
trees and shrubs that were planted on British estates, sometimes having special
contracts. Nursery companies served as foci for the importation of new varieties
from around the world and even if not actively involved in plant-hunting expe-

ditions, often financed them and were usually the first to be informed of new discoveries. Gardeners were also trained in nurseries and pursued their careers in country estates and vice versa, whilst nursery companies recommended staff to estate managers and landowners. At Elvaston, the nursery developed by William Barron during his time as head gardener from the 1850s and 1860s subsequently formed the basis of his commercial nursery business which moved to a forty-acre site at Borrowash near Derby in 1865 after his resignation. Barron claimed that his nursery had purchased 'every new plant introduced into this country' during the previous two decades. Like other major Victorian nursery companies, his success partly depended upon the relationship maintained with other landscape gardeners, journalists, collectors and botanical writers who, in turn, required the cooperation of nurserymen and their network of contacts and agents. Such a relationship is evident in the role of nurserymen such as Loddiges in Loudon's work, but it is also clear from the assistance provided by Barron to George Gordon in the production of *Gordon's Pinetum*.

The Barron Company specialized in Coniferae but offered all manner of other trees, shrubs and plants deciduous, ornamental and flowering trees and shrubs, American plants such as rhododendrons, roses, stove and greenhouse plants and fruit trees and had international agents from whom consignments of seeds were received. Editions of the printed catalogues claimed that their stock of Coniferae had 'become celebrated both at home and abroad as being perhaps the best in the trade'. Between 1868 and 1874 the firm had 'constantly exhibited at all the principal shows in the United Kingdom' and had 'invariably obtained first honours' whilst international commissions were obtained. Several young men from Germany, France and the Netherlands were employed, whilst Barron's son was educated under Prince Herman Puckler-Muskau (1785–1871) and Eduard Petzold at Muskau in Silesia and in the Netherlands. This cooperation with two of the leading European landscape gardeners and arboriculturists continued as Barron's grandson attended a government school in Potsdam whilst Petzold's son worked at the Borrowash nursery for three years. Petzold was director of gardening at Muskau between 1852 and 1881, the author of many textbooks on landscape gardening, horticulture and arboriculture, including *Arboretum Muskaviense* (Gotha, 1864) and the designer of numerous European public parks.[30] The Elvaston experience allowed Barron to respond to criticisms made by William Gilpin and others that evergreens had insufficient variety of colour for parks and gardens by promoting new varieties. The most famous varieties developed at Elvaston and associated with the nursery included the *Taxus elvastonensis aurea* (*Taxus baccata elvastonesis*) or golden Elvaston yew, a type that was 'a bright orange colour, and unlike all other golden or silver yews,' 'not variegated but a self colour ... by far the most brilliant of any in winter'. This was described in *Gordon's Pinetum* as 'by far the most brilliant of any of the golden varieties in the winter time'.[31]

There were, however, limitations to the kind of collections that nurseries could acquire. Although some companies, such as Loddiges of Hackney, developed major complexes of gardens, glass houses and arboretums, as display for pleasure was not the primary purpose, these specimens were sold off, cut back or removed once they had grown to a certain size, thus destroying rather than preserving their collections. Just after Loudon had had sketches taken of all their plants, Loddiges had most of their timber trees removed in 1832 and 1833, because they had reached a stage of development where they were crowding out other plants. Stools and young plants remained for the propagation of the next generation of plants, but Loudon called this an 'incalculable loss' of 'a collection of specimens such as could be found assembled together nowhere else in the world' as it had prevented him from viewing many trees when in flower.[32] Specimens could be nurtured on estates over years whereas commercial nurseries needed to regularly replenish their stock to attract sales.

The interdependency of nursery companies, large Victorian estates and their arboretums is also evident at Westonbirt and Eastnor. At Westonbirt, Robert Holford and Jonah Neale's notebooks suggest that the estate was a major customer of nursery companies who seem to have supplied many of the most celebrated and unusual trees and shrubs, although the names of these are usually not given. A *Cedrus deodara* planted in the pleasure ground towards the village for instance, one of the earliest specimen trees obtained for Westonbirt by Holford in 1837, was bought for [£5.5.00] from Knights of Chelsea, who had bought it from the sale of the old Horticultural Society collection for making cuttings. Knights also supplied other plants and trees. Page's nursery in Southampton supplied other specimens including a large cedar planted in 1836 although James Veitch of King's Road, Chelsea took over as major suppliers of specimens during the 1850s. In 1854 Neale records that a *Picea amabilis* was purchased from Veitch for £3 and planted at the end of the pleasure ground whilst specimens of *Picea bracteata*, 'Pinus macrocarpa' and *Pinus muricata* or 'edgariana' were bought from the same company at the same price.[33]

The larger nursery companies and estates each provided training grounds for gardeners who passed between them during their careers, whilst landowners and agents often sought recommendations for prospective gardeners from nursery companies. Veitch & Son's Royal Exotic Nursery at Chelsea played a key role in the appointment of gardeners at Eastnor. In 1854, William Coleman had been taken on by the nursery with whom he maintained a close relationship. In May 1854, Coleman was sent by Veitch to act as gardener to Lord Cloncurry at Lyons, County Kildare in Ireland and having returned to England in 1859 was provided by Veitch with 'the choice of two first class situations', one of which was the position of head gardener at Eastnor.[34] His successor George Mullins also worked at Veitch and Sons before becoming a foreman at Ashgrove, Sevenoaks, from where he returned to Veitchs's nurseries before finding another temporary appointment and becoming gardener for Lady Henry Somerset at Reigate in 1895 and taking over the Eastnor gardens in 1897.[35] Of course, Nursery companies also provided most of the seeds, trees and shrubs for estate arboretums and made recommendations about which of their stock would fare best in local conditions. Surviving notebooks of garden expenditure kept by Coleman reveal that purchases for Eastnor were made from a variety of nurseries including Richard Smith of Worcester and Barron at Elvaston, although between 1860 and 1871 when much of the arboretum was planted Veitch's was the main source.[36]

Design of Estate Arboretums

The different reasons for creating and using estate arboretums influenced their design and management, particularly the varied requirements of forestry, botany, rural sports, landscape gardening, private pleasure and public amusement. Local geology, topography, climate, hydration and other geographical and bio-geographical factors also shaped the design and management of planting schemes. As we have seen, there was also an important tension between the varied demands of collecting and botanical significance. Some nineteenth-century estate arboretum designs were intended to be systematic and representational, striving to include examples of as many species and varieties as possible; others were intentionally picturesque with an emphasis upon exotics. Attempts were made to combine systematic with aesthetic approaches by forming geographically-representational arboretums whilst in the later nineteenth and early twentieth centuries there was a reaction against the imposition of exotic collections.

When Paxton arrived at Chatsworth the gardens were still dominated by the formal work of George London and Henry Wise set in a very extensive park designed by Lancelot 'Capability' Brown. One of Paxton's earliest large-scale changes was to plant the pinetum and later arboretum on the slopes behind the palatial neo-classical mansion, the objective being to naturalize as many foreign tree and shrub species as possible (Figure 7.2). The arboretum was partly formed from the existing pinetum and required extensive ground work including the diversion of a stream two miles away from its original course on the East Moor to ensure that the site was well watered. This also provided an additional romantic and naturalistic feature amongst the trees. The system of planting was informed by both taxonomy and aesthetics, with specimens of the larger trees and shrubs planted on little hills along a path. Smaller shrubs such as azaleas and heathers were planted in groups around 'the towering crags and forests protecting them from the keen north and east winds, with convenient pools of water ... amongst these exotic beds'. Most existing timber trees were removed from the site and the pinetum was planted at the warmest part of the estate around the pond next to a grotto on the hillside. Loudon praised Paxton for taking advantage of the space available at Chatsworth to provide sufficient room between each specimen to allow each to grow into its natural shape. The 1,670 rising to 2,000 species were arranged across some forty acres following the chain of orders in Loudon's *Hortus Britannicus* in seventy-five groups on both sides of an ascending walk which made a circuit of the hill behind the house reaching its zenith near the start of the waterworks and concluded westwards down the hill towards the stove and back to the house.[37]

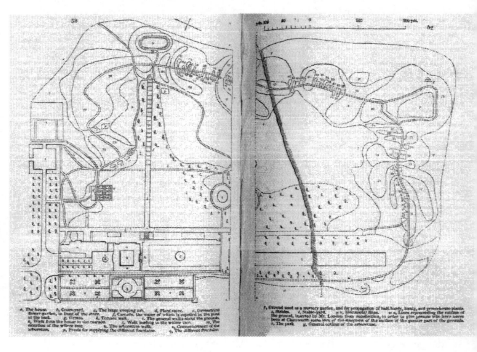

Figure 7.2: **Plan of Chatsworth Arboretum,** *Gardener's Magazine* (1834), p. 52.
Reproduced from the author's personal copy

Every planted specimen at Chatsworth was carefully labelled with scientific name, country and year of origin, height in native country, English name and date of planting, all painted onto steam-hardened oak 't'-shaped tallies. William Adam was 'surprised and delighted, on ascending the walk' along the 'mazy turns' up the cliff to 'find both sides adorned with trees and shrubs from every climate ... a noble and truly patriotic undertaking', whilst Paxton and Loudon recommended that arboretums on this model be established in nine out of every ten British country seats. As such, Paxton was keen to emphasize that this horticultural improvement had been achieved without costing the Duke sixpence as the produce of the removed plantations had financed the entire scheme.[38] The Chatsworth arboretum served as an important model and stimulus for the development of other estate arboretums. This was because Chatsworth was visited by thousands and described in Paxton's publications, but also because of the major Victorian landscape gardeners who trained under him including Edward Kemp (1817–91) and Edward Milner (1819–84). Paxton's pinetum and arboretums at Chatsworth were arranged botanically along curving paths, with views from the slopes behind the Duke's palace providing most of the picturesque beauty. At the Biddulph Grange, Bicton, Elvaston and Westonbirt arboretums the focus was more on the beauty and picturesque effect of individual or grouped specimens laid out without botanical sequence. Carriage rides and viewing bays were incorporated around the Westonbirt arboretum providing avenues so that exotic trees could be enjoyed by the Holford family and their friends from horse or carriage.[39] At Eastnor there was also a carriage ride three miles long from the lodge at British Camp in the Malverns to the Castle 'flanked by evergreen as well as deciduous trees and shrubs', including yews, the wild service tree, *Arbutus* and many varieties of *Crataegus*.[40]

The formal Elvaston pinetum occupied a level sixteen-acre grassed field, although Barron regarded the whole gardens as 'properly speaking ... one vast pinetum, artistically treated'. The formal pinetum was planted exclusively with evergreens from 1835 around what had formerly been the central drive, the eastern side with *Pinus* specimens and the west with *Abies* and *Piceas*. Stock was obtained economically using Barron's evolving methods of transplantation, from nurseries and through propagation of cuttings, partly because, as Barron emphasized, the Earl had never been a member of the Horticultural Society and could not easily obtain trees first hand. As it was often only possible to obtain single or small numbers of rare trees, propagation and grafting provided an opportunity to enhance the collection without large expenditure. At the centre of both sections were straight turf avenues planted with Irish yews at the front, backed by golden yews opposite the openings. The next line was *Araucaria imbricata* and then behind these two rows of deodars which were grafted onto the cedars of Lebanon. The northern and southern ends of the avenues were planted with *Cedrus deodara* and *Cedrus libani* and at the other end different species of Taxaceae and Cupressaceae; Loudon was astonished at the quantity of *Thuyas*, red cedars, white cedars, hemlock spruces and variegated and common yew. The plan of the pinetum at Elvaston combined attention to symmetry, space and geometry with aesthetics and taxonomy. There is no mention of labelling which may have been regarded as unnecessary in a private garden. For Glendinning, 'if any artificial assemblage of trees can reach the sublime in gardening' the Elvaston pinetum was such an example and it was 'difficult, nay impossible to convey any adequate idea of the impression' produced by such a 'noble plantation'. The pinetum contained almost every variety of coniferous plant 'known or obtainable in European gardens', some of the 'most rare and valuable kinds', in such a quantity and size never before established.[41]

The topography at Eastnor was more varied than at Elvaston. According to Alexander Cramb, head gardener at Tortworth Court, the pinetum was extensive and 'of considerable altitude' with hillside planting that favoured the growth of many different types of coniferae. The pinetum did not follow a 'geographical distribution of genera and species' which Cramb thought sensible given that 'from the inequality of growth and other causes, the general effect would be greatly reduced'. Whilst the formation of pinetums was being considered or had been decided upon, geographical arrangements were 'almost certain to take possession of the mind'. Yet although this might appear to be 'the correct line to follow', Cramb considered that they could be a 'grievous mistake'. The sixty-two-acre pinetum contained a large variety of trees and shrubs set on a hillside above the lake and castellated mansion.[42] The formation of the Eastnor arboretum was aided by the relatively mild climate and reasonable level of rainfall, although the alkaline quality of much of the soil determined by the limestone underneath discouraged the planting of Victorian favourites such as rhododendrons. The grounds also contained 'one of the finest examples in the country of *Picea bracteata*' whilst further on were 'all the best established varieties of Holly, fine vigorous trees, forming in themselves by contrast of form, habit and colour, one of the most interesting features that a garden can possess'. Baines noticed a 'dense bush of the Chinese privet (*Ligustrum lucidum*)', a 'healthy vigorous example of the Californian Silver Fir (*Picea lasiocarpa*)', many Wellingtonias, some already 35 ft high by 1878, Mexican pines (*Pinus montezumae*) and the Japanese green-leaved holly-like *Osmanthus ilicifolius*. There was also a 'fine specimen of the Himalayan *Picea webbiana*' which seemed to do well, and 'several varieties of evergreen Oaks' which had been raised from acorns collected by Earl Somers in Chaldea.[43]

Despite the differences in the nature of landscape and terrain, there are marked similarities between the Eastnor and Westonbirt arboretums. At the former, Coleman shaped the planting in ornamental clumps with single specimen trees which 'rounded off the angles, and blended in a very pleasing way the margin of the water and the accompaniments of the park.' He 'studiously avoided uniformity of outline' and 'wisely brought out the rugged features of the surrounding country'. Some very large-scale planting of newly 'discovered' trees was undertaken. These included over two hundred Wellingtonias which were growing well by the 1870s.[44] According to the detailed analysis of Thomas Baines (1823–95), Eastnor held a 'prominent position ... amongst the palatial homes of England'. In addition to the 'massive grandeur' of the castle and 'splendid scenery of the surrounding district', the extensive and thickly wooded grounds were 'famous for the number and size of evergreen trees and shrubs that they contained. For A. Barker, who visited somewhat later in 1888, the park scenery was 'of the greatest beauty and grandeur', but the wealth of conifers, planted as single specimens or in groups was 'almost unique in this country'. Of the many trees and shrubs that Baines noticed were many yews growing freely and the Chinese evergreen shrub *Photinia serrulata*, which was doing 'remarkably well'. Amongst the many evergreens were 'dispersed beautiful thriving specimens of Cedar of Lebanon' which stood 'in bold relief to the others,' there was also 'a beautiful example of the Japanese Cryptomeria japonica in wondrous fine condition' and nearby a 'fine Mount Atlas Cedar' (*Cedrus atlantica Glauca*) 50 ft in height which 'like many others here to be met with' had been raised from cones gathered by Earl Somers on Mount Atlas around 1859. The Earl had brought to bear his passion and knowledge of tree collecting which he had undertaken since childhood and his seventy acres were adorned with 'innumerable splendid trees and shrubs'. Newer introductions of taxaceous and coniferous trees were 'intermixed with fine examples of deciduous subjects'. Most of these were quite old but remained 'vigorous' and had attained great size such as 'some stately examples of Ash'. Baines considered that 'the effect ... collectively produced' by this arboricultural assemblage was 'infinitely superior' to that resulting from collections of evergreens alone which was 'too often' the form of the 'modern pinetum'. As at Westonbirt, the general scheme was attained through combining diversity with grouping together of varieties. Some trees and shrubs were planted singly or in groups of two or three, whilst from the 'undulating character' of the grounds the use of the same trees and shrubs in places had the 'effect of the difference in position and association with others' presenting both novelty and views of interest to informed observers.[45]

The Westonbirt arboretum developed from picturesque planting of exotics and the creation of plantations for various other purposes including hunting and shooting and was not originally conceived as an arboretum. The layout of the gardens is generally attributed to William Sawry Gilpin, although Holford's agent Edmund Rich seems to have been involved also. Aspects of Gilpin's *Landscape Gardening*, a copy of which is in Westbonbirt library, seem to have been followed although the example of Elvaston and the involvement of Barron in the design of one of the avenues also seem to be significant. The positioning of exotic specimens, use of exotic evergreens and formation of the avenues in the arboretum and Silkwood are reminiscent of Elvaston. Probably a more important factor in the development of the arboretum was the exchange of ideas and rivalry between Holford and other landowners such as Earl Somers. Broadly, as Symes has emphasized, there was 'an overall concept of house, gardens, park and arboretum' each forming 'part of a single grand design', although the arboretum took decades to form. The arboretum was linked to the house and gardens primarily through the use of avenues and rides which provide framed views of the house from different places, whilst the original entrance to the arboretum lay opposite the lodge gates on the Bath Road.[46] However, as at Elvaston, the arboretum was regarded as a private affair for the consumption of family and friends and it was not until the end of his life that George Holford encouraged the publication of a comprehensive account of the collection. This contrasted with the Duke of Devonshire, the Duke of Bedford and the fifth Earl of Harrington, who were eager to open their arboretums and pinetums to the public and publicize the collections.

The order that Westonbirt trees and shrubs were purchased is apparent in A. Bruce Jackson's catalogue published in 1927 and surviving documents, especially Robert Holford's and Jonah Neales's notebooks.[47] Cedars are amongst the most striking trees in the gardens and some were the first of their kind to be planted in England, such as the deodar cedar and Atlas cedar planted on the western lawn in 1837 and 1841 respectively, and the incense cedar and Japanese cedar acquired during the 1840s and 1850s. Other trees that were amongst the first to be planted in Britain were Wellingtonias and Californian redwoods, varieties of the Lawson cypress, monkey puzzles planted between 1840 and 1857 and the golden yews favoured at Elvaston.[48] The scheme of planting, design of the arboretum and use of exotics was not random but configured according to picturesque principles. Trees and shrubs disguised the bounds of the natural-looking lake whilst artificial rocks were 'stratified so as to appear geologically authentic'. In the pleasure gardens and park around the house some special trees were planted on their own and others were arranged in groups with isolated varieties including four types of cedar. The groups seem to have followed a basic pattern with tall columnar trees such as Lawson Cypresses, incense cedars or Wellingtonias occupying the centre of each. Smaller bushy trees were planted around these including the Hiba (*Thujopsis dolabrata*) from Japan, the juniper and green and golden Dovaston yews. These groups were also usually tied together with types of variegated holly including silver- and golden-leaved varieties.[49]

The success of large tree collections on estates in Herefordshire, Gloucestershire, Devonshire and other parts of the south west reflected relatively favourable climatic and geological conditions for the planting of exotics. At Westonbirt, it was found that many American trees and shrubs did not do so well as Japanese varieties, hence the scheme of planting came to feature many of the latter. In contrast, arboretums and tree collections in other parts of Britain faced greater difficulties. When Loudon questioned the design of Paxton's arboretum at Chatsworth, Paxton retorted that the particular conditions of the Derbyshire Peak, including the harsh winters, meant that some varieties did not thrive. At low-lying Elvaston on the other hand, the problems stemmed from peaty acidic soil prone to flooding and winter frosts which precluded many staple British trees as well as exotics and necessitated the ambitious programme of drainage engineering before major planting could begin.

Large estates offered opportunities for innovatively designed arboretums unavailable in smaller villa, urban or public gardens. The experienced landscape gardener Charles McIntosh complained about the tyranny of linear arboretums and the 'dotting' of trees in defiance of local conditions, arguing that it was only on large country estates using carefully positioned clumps of trees that successful arboretums could be realized. In McIntosh's elitist view, great estate collections could only be successfully presented using the spaces and resources of large parks, whereas they could easily look ridiculous in smaller villas and public gardens.[50] One approach that he favoured to counter the rigidity of systematic collections was the use of geographical layouts and he included a detailed description and plan of the pinetum established by head gardener John Spencer from 1848 at Bowood, Wiltshire, for the Marquis of Lansdowne. According to McIntosh and Spencer, as aesthetically pleasing and locally sensitive creations, geographically arranged pinetums and arboretums were botanically superior to the kind of linear arboretums long advocated by Loudon because they could provide more than single examples of specimens and better demonstrate the changing international vegetable characters. Spencer argued that the systems 'generally adopted' of planting coniferae arranged 'according to their height and habit' as at Dropmore and Chatsworth, or botanically according to 'the several divisions and subdivisions' established by botanists were neither fully satisfactory. Although the former plan was more aesthetically successful, both schemes were 'liable to confuse parties' wishing to examine conifers nationally which was 'far the most important feature in studying their general character' and now had to wade through numerous trees and shrubs from all over the globe at once.[51]

The Bowood pinetum occupied land that had previously been an orchard and nursery-ground surrounded by 'an irregular belt of forest trees' and 'undergrowth of evergreens'. The six-acre site was enclosed by a wire fence and surrounded by lawn. At various points along the belt on the side next to this were 'some fine cedars of Lebanon, red cedars, pinasters, and other ornamental trees'. McIntosh noted that it had not yet been decided whether to leave closely mown grass in-between the trees and shrubs or to contrast the exotics with 'native heaths, ferns, and very low-growing shrubs, in imitation of wild scenery', which he favoured. The plan adopted by Spencer allowed 'the whole of the different species at present introduced' to be planted so as to 'attain their ultimate size and character' whilst demonstrating international plant geographies. One large division consisted of species indigenous to Europe and Asia and included the conifers of northern and central Europe, Spain, Italy, Greece and Asia Minor which were succeeded by 'Syrian and Caucasian species' then 'Himalayan and varieties from Central Asia, the division ending with the cryptomaria and other coniferae of

Northern China and Japan'. The relative unimportance of African species in
Spencer's eyes meant they were presented as just a small group, whilst Australasian conifera, 'though magnificent and highly important in an economical point
of view' were 'too tender' to bear the British climate. The second great division
was devoted to North and South America passing from Canada through Mexico, Chile and Peru, represented in the latter case by the aracauria. A deliberate
contrast was provided along the central axis with the hardy coniferae of the 'Old
World' occupying one side of the central walk and those from America the other.
Additionally Spencer proposed that one of the central groups could be devoted
to 'the Himalayan species' and another 'to the principal Mexican species'. Others
representing 'the more characteristic species of the Old World would be positioned in 'immediate contrast with large taxodiums, *Pinus lambertiana,* and the
smaller, though more beautiful long-leaved kinds from Mexico and California'.
These, with the 'new Cupressus macrocarpa which equals the cedars of Asia in
size' were intended to demonstrate the difference between coniferae of each
hemisphere. Two, three or more 'duplicates' of each specimen were planted to
'produce immediate effect', although these could be removed as they grew taller,
especially where large varieties were concerned.[52]

Tree Transplanting

According to Loudon, transplanted trees stood as testaments to the 'employment of skill and expense' by owners and park promoters. Landscape gardening
was 'greatly inferior in beauty to the imitate creation of a painter from the
same groundwork and materials' and 'no comparison between the powers of
landscape-painting and those of landscape gardening [could] be instituted, that
[would] not evince the superior powers of the former art'. This was primarily
because wood provided the great source of beauty in every landscape and the
source of this beauty depended on 'accidental circumstances' in the progress of
trees from planting until maturity, which could not 'be said practically to be
under the control of the gardener'. However high the aim of the gardener, however much the 'natural effects of time' were studied and 'however correctly we
may imitate them', at the end of all work, 'any wood of art will always be far inferior to a wood of nature under the same circumstances'. For Loudon, landscape
gardening was therefore limited to 'picturesque beauty' and the production of a
harmonious and agreeable 'assemblage of objects'. Large-scale tree transplanting,
however, offered greater opportunity for landscape gardeners to aspire to the
higher beauty of the artist in their response to nature by simulating and appropriating the 'natural effects of time'.[53]

Although tree transplanting was not new, it was undertaken on a much greater
scale using novel techniques and equipment during the 1820s and 1830s, which

had a major impact on Victorian arboriculture and horticulture.[54] Sir Henry Steuart used a technique in which trees were placed into holes with bare roots whilst others, including Loudon, recommended that balls of earth should be allowed to remain around the roots in some circumstances. Only some botanical institutions or large Victorian estates tended to have the wealth, resources and patience required for large-scale tree transplanting which came to be regarded as a heroic Victorian endeavour and reinforced the status of landscape gardeners such as Paxton and Barron as conquerors of unruly nature. For instance, in April 1830 Paxton oversaw the transportation of a large weeping ash tree from a nursery company in Derby to Chatsworth on an improved machine loaned by the Strutt family based upon Steuart's principles. The tree weighed almost eight tons with the ball of earth attached and took eighteen hours to cover the twenty-eight mile distance between Derby and Chatsworth. The tree was so large that 'the gates and wall at the entrance to Chatsworth Park were ... obliged to be taken down, and the branches of some of the trees in the park lopped off', and it took a week to complete the operation. Paxton was able to exploit the knowledge of tree transplanting gained and subsequently moved many large trees, including palms from Walton-on-Thames to the Great Conservatory at Chatsworth drawn by eleven horses and the transportation of the giant palms from the Loddiges collection to Sydenham in 1854 for the Crystal Palace Company.[55] Paxton and Barron were able to achieve almost immediate effects with mature and established trees in what would previously have taken decades or centuries, providing instant satisfaction and much wider possible configurations for owners and gardeners. The technology they employed and the scale and success of tree transplanting work was unprecedented, providing an instant arboriculture for private gardens and public places commensurate with the ambitions of aristocracy and middle class in industrial society. The expanding railway network was used to transport trees, which also carried parts of Barron's machines and his specially trained workforce – invariably sold as a package – around the country (Figure 7.3). Barron became adept at publicizing his transplanting methods and the movement of large and old trees through the streets was much photographed and reported in newspapers such as the transplanting of the Buckland yew in 1880, a tree celebrated by Loudon with a documented age of 800 years.[56]

Figure 7.3: **Removal of large tree in Royal Botanic Society's garden, Regent's Park using Barron's machine watched by J. C. Sowerby, secretary of the committee and others,** *Illustrated London News* **(1855). Reproduced from the authors' personal copy**

The first of Barron's tree transplanting machines was devised in February 1831 and in November a 43 ft-tall and 48 foot-diameter Cedar of Lebanon was moved into the gardens at Elvaston which had grown by the 1870s from having a trunk of 2 ft in circumference to one of 10 ft. Another 72 ft high was moved more than two miles in an upright position and yews from six to eight hundred years old, oaks and larches from 40 to 50 ft high and large spruce and silver firs even in the middle of summer, were, it was claimed, moved 'without losing a leaf'. The method worked by preparing the tree for two or three seasons by digging around to cut the spreading roots and by filling this with fresh soil. The transplanter was then assembled around the tree and the exposed ball of roots and soil was wrapped and tied before being winched up to the transporter.[57] By the 1870s, machines had been constructed for the Duke of Portland, the Duke of Manchester, others of the gentry and nobility, and the Royal Botanic Gardens at Kew. The latter was purchased in 1866 and became known as 'the Devil' by the gardening staff who one winter moved sixty trees from between two and seven tons. Commissions were obtained from throughout Britain and abroad and machines were supplied to aristocracy, gentry and other private clients, institutions and urban government authorities. By the early twentieth century machines had been supplied to international clients including the City of Freiburg, the Grand Duchy of Baden, the International Exhibition at Kingston, Jamaica, the Indian army and the Maharajah of Bharatphur. Others also adopted Barron's methods with considerable success including the Scottish landscape gardener William Tillery at Welbeck Abbey, Nottinghamshire, for the Duke of Portland.[58] Testimonials appeared in editions of Barron's catalogues illustrating the geographical and social spread of the market.[59]

Of course, such statements were deliberately selected by Barron to reflect well on his company and methods and we seldom hear if anger or hostility resulted from old trees being moved from one location to adorn private parks. Yet partly through Barron's efforts, transplanting became common in Victorian parks and gardens. At Westonbirt, for instance, Robert Holford's notebook describes how two fifty-eight and fifty-nine foot elm trees were transplanted from 'the Elm Grove opposite the Hare and Hounds' and 'planted in the new South Pleasure Ground' in front of family members and friends, whilst in May 1887 a *Cupressus lawsoniana* 13 ft 7 in high was 'moved from Arboretum to South Pleasure Ground'.[60] Although this form of collecting, as Barron emphasized, could be used to move great and beautiful trees to more suitable sites or to prevent them from being spoiled by other trees, tree transplanting was an aggressive and imperious act akin to moving villages to make way for gardens that involved denuding other places sometimes of very old and loved specimens. At Elvaston trees were taken from places where they had stood for centuries for the private pleasure of one couple. Barron provides very little information concerning original locations and seems to have been unconcerned with the effects of denuding places of well-loved trees. In this respect there are parallels between tree transplanting and the exploitation of British and colonial labour and natural resources by Victorian industry and the Duke of Devonshire's travels in Italy to search for sculptures and antiquities. Tellingly, criticism concerned the question of aesthetic artificiality inherent in Barron's schemes rather than issues of possible exploitation. Barron acknowledged that there was some hostility towards the unnaturalness of Elvaston, especially the large-scale transplanting, clipping and shaping of trees and shrubs, the architectural qualities of the design and the arrangement of rocks. However, he thought that these demonstrated how well coniferous trees would 'yield to art, at the discretion of the operator' and, so long as the rules of nature were 'properly attended to', would be of service to mankind, like the breeding of edible crops and fruit trees. How could it be reasonable to 'object to art being applied to alter the forms and habits of ornamental trees in the garden or landscape to suit styles of artistic gardening'?[61]

Impact of Estates on National and International Botany

Through their employment at the influential botanical gardens of Kew and Edinburgh and on great country estates, the careers of Barron and Paxton demonstrate the interconnections between country estates, nurseries and national botany. They were able to utilize this experience and the connections with academic botany and international botanical networks to obtain knowledge, specimens and recognition. The sixth Duke of Devonshire and Paxton took a keen interest in global plant collecting, the Duke offering patronage and financial support for international plant hunting expeditions, striving to purchase at the earliest opportunity novel varieties and incorporating new finds such as orchids in the Chatsworth gardens and arboretum. Employment at the Horticultural Society gardens at Chiswick in the 1820s provided Paxton with much experience of managing exotics and facilitated his introduction to the Duke of Devonshire, who leased the grounds to the Society and had special access. Paxton was gardener in the Chiswick arboretum when the first exotic conifer seeds from the Americas were being sent to Britain by David Douglas, James Macrae and other explorers. By the later 1830s, based at Chatsworth, Paxton had gained considerable national standing as a horticulturist and gardening writer in his own right as his gardening magazines and monograph on the dahlia demonstrate. He also assisted John Lindley in reporting critically upon the state of the royal gardens following the accession of Queen Victoria, including the Royal Botanic Gardens at Kew which received particular condemnation for their failure to advance botanical knowledge and supply fruit and flowers. The Duke, who in 1838 became president of the Horticultural Society, subscribed to various expeditions. The botanist John Gibson was sent to the forests of Assam under the auspices of Dr Nathaniel Wallich, head of the Calcutta Botanical Gardens, to find rare species and especially new orchids for the Chatsworth collections, and he obtained a striking flowering *Amherstia nobilis*. The Duke gave his name to a banana plant, the dwarf *Musa cavendishii*, which had originated in China and been obtained by Paxton for the Chatsworth hothouses.[62]

Another expedition supported by Paxton and the Duke was a three-year expedition to the north-west coast of North America in 1838 under the auspices of the Hudson Bay Company, which was undertaken by two of the staff from Chatsworth, Robert Wallace and Peter Banks. Paxton organized support for the expedition which he obtained from gentry, aristocracy and leading nurseries, seeking advice from William Hooker in Glasgow and John Lindley in London and planning a route in co-ordination with the Company. The expedition ended in disaster after Wallace and Banks drowned on the Columbia River when their boat struck a rock before they could begin collecting seeds and plants. However, this tragedy did not dent the enthusiasm of Paxton and the Duke for collecting as the completion and planting of the grand conservatory or 'Great Stove' at Chatsworth demonstrates. Designed by Decimus Burton and Paxton to house the Duke's large collection of tropical and subtropical plants, flowers and trees, the great conservatory was the largest hothouse constructed in the world when completed and planted in 1840 and was the model for Paxton's Great Exhibition and Sydenham Crystal Palace buildings.[63]

The impact of global and colonial plant exploration upon estate arboretums is also evident at Eastnor Castle. Until the 1850s planting had been dominated by North and South American varieties of trees and shrubs, especially examples from California and Mexico. However, as a result of the destruction caused by the unusually harsh winter of 1860–61 these were replaced by many specimens from Japan as the country was explored by western plant hunters and novel varieties reached Britain. Casualties of the harsh winter at Eastnor included many *Pinus insignis*, of which 150 were destroyed, and also *Cupressus macrocarpa*, of which 130 died from 8 to 40 ft tall, although those planted at a higher elevation fared better. These were replaced with many Japanese specimens collected by John Gould Veitch and Robert Fortune which seemed to fare well during the British winters and even, as William Coleman noted, the spring frosts to which Eastnor was susceptible.[64] The acquisition of novel trees and shrubs for Eastnor was facilitated by the Somers-Cocks family connections with the colonial service. The third Earl's cousin (who was also grandfather of the sixth Earl) Arthur Herbert Cocks (1819–81), served in the Indian colonial civil service from 1837 rising to Deputy commissioner of the Punjab in 1850 and retiring to England in 1863 where he served as a Worcestershire JP. The third Earl's wife, the strongminded countess Virginia Somers-Cocks (née Pattle) (1827–1910) and a major influence on the development of the Eastnor estate, was daughter of James Pattle (1775–1845), an official in the Bengal civil service, and his wife Adeline de l'Etang (1793–1845). Imbued with the colonial ethos, Virginia was brought up in Calcutta, the site of one of the leading colonial botanical gardens. Her father, who entered the civil service in 1790, became senior member of the Board of Revenue and oldest individual in the administration and through her surviving sisters in Calcutta, Virginia retained strong Indian connections.[65]

Systematic estate collections and horticulture also had an important relationship with botanical publications. Gardeners, woodsmen and landowners provided considerable descriptive and statistical information concerning tree collections, including numerous measurements and in turn received valuable public attention and horticultural and scientific recognition (Figure 7.4). The *Horticultural Register* (1831–5), *Paxton's Magazine of Botany and Register of Flowering Plants* (1834–49), *Gardener's Chronicle* and *Pocket Botanical Dictionary* (1840), *Flower Garden* (1850) and *Calendar of Gardening Operations* were inspired and informed by Paxton's experiences at Chatsworth as well as Loudon's activities. Paxton described the experience of forming the Chatsworth arboretum in an article in the *Gardener's Magazine* whilst Barron's achievements at Elvaston were detailed in his *British Winter Garden*. Estate owners and gardeners were generally keen to see accounts of their work in print and to assist writers by providing tours and additional information such as the data provided to Loudon for the *Gardener's Magazine* and *Arboretum Britannicum*. Some publications were very closely associated with estate arboretums, notably the catalogues of systematic collections sometimes published at great expense by landowners. These included Hofland's extravagant celebration of Whiteknights, James Forbes' description of the Woburn pinetum for the Duke of Bedford and Clinton-Baker's *Illustrations of Conifers* (1909) based on the Bayfordbury pinetum. Jackson's *Catalogue of Trees and Shrubs at Westonbirt* published by Oxford University Press in 1927 aimed to be a thorough record and included multiple photographs of trees and shrubs. These works magnified the importance of estate tree collections whilst contributing to botany by providing detailed descriptions and illustrations of individual specimens and some accounts of techniques used for tree planting, management and care.[66]

	August 11, 1892			March 24, 1900		
	Height.	Girth at 3 feet.	Spread of Branches.	Height.	Girth at 3 feet.	Spread of Branches.
Wellingtonia, planted 1857 (a)	66.6	9.3	26	74.11	10.7	28
Picea sitchensis, planted 1845 (b)	91.9	9.7	45	105.10	11.3	...
Pinus monticola, planted 1850 (c)	67	5.6	18	79.2	6.2	22
Araucaria imbricata, planted 1847 (d)	42.6	4	9	51	4.8	9.8
Abies Pinsapo, planted 1847	34.3	6.6	...	42.6	7.10	...
„ magnifica, planted 1867	31.9	2.7	9	43.3	3.8	11
Pseudotsuga Douglasi, planted 1847 (e)	86.6	8.10	24	97.4	9.10	27
Abies grandis, planted 1852	64.2	4.8	22.6	79.10	6.10	35.8
Tsuga albertiana, planted 1860	56	5.5	32	72.1	6.4	40
Abies nobilis, planted 1847	75.4	6.1	...	92.8	6.6	...
„ nordmanniana, planted 1854	52.6	4	...	74	4.9	...
Tsuga brokeriana, planted 1862 (f)	30	4	15	39.6	4	15.4
Cedrus Deodara, planted 1842 (g)	51.3	6.8	26	61.2	7.4	36
„ Libani (h)	65.10	11.8	...	67	12.5	...
Cryptomeria japonica, planted 1852	36.3	4.2	26	41.7	4.3	...
Libocedrus decurrens (i)	34.8	3.6	10	38	4.5	14
Thuya gigantea, planted 1862	46	3.6	21	57	3.7	...
Cupressus lawsoniana, planted 1859 (j)	...	...	...	48.7	4.2	...
Spanish Chestnut	...	...	...	...	17.10	...
„ (k)	...	...	...	...	19.2	...
Silver Fir (l)	...	...	...	...	11.3	...
Cupressus lawsoniana erecta viridis	...	...	...	90	2.8	7
Picea ajanensis, planted 1885	...	...	...	25	1.7	10.6
Abies brachyphylla, planted 1885 (m)	...	...	...	14	...	9.7
„ Veitchi, planted 1885 (n)	...	...	...	20.9	1.4	10.2
„ amabilis, planted 1885	...	...	...	14.11	...	9.5
„ concolor violacea, planted 1885	...	...	...	20.1	1.5	14.9
English Yew (o)	...	...	...	30	14.3	...
English Yew	...	...	...	...	10.8	...
Tsuga albertiana (at Roman Bridge) (p)	...	...	...	75	4.3	...
Picea orientalis, planted 1832	30	2.7	...	49	3	...
„ morinda, planted 1857	47	...	...	...	...	...
Pinus Jeffreyi	...	...	...	57	4.9	...

(a) At the ground this tree measures 16.9; cones freely. (b) There are six others about the same size, and all are growing freely. (c) Most of these have lately got a fungoid disease, viz., peridendrum. (d) Many of these lost branches, and some were killed by frost in 1864-5. They cone freely, and young ones are growing from seed. (e) A great many others about the same size, and all perfectly healthy. (f) A beautiful tree quite distinct from the others; long, drooping branches. (g) About sixty trees growing in the grounds averaging 90 cubic feet. (h) Age unknown, but probably not less than 150 years. (i) Probably thirty-five to forty years of age. (j) Two trees, recently taken out, measured 19 cubic feet and 14 cubic feet. (k) At ground this tree measures 29 feet. (l) Inclined to go back. (m) Will become a handsome tree. Coned last year. Some fertile. (n) Very apt to lose its leader either by birds or wind. Coned last year. (o) Very old; possibly 900 years. Many others of the same age and size. (p) Quite a different form from the others, the lower branches being quite table-form.

Figure 7.4: Growth rates of conifers at Murthly Castle, Perthshire originally compiled for the Scottish Arboricultural Society, E. T. Cooke, *Trees and Shrubs for English Gardens*, 2nd edn (London: Country Life, 1908), p. 128. Reproduced from the authors' personal copy

Conclusions

Estate arboretums were stimulated by emulation and rivalry just as the mid-Georgian Whig oligarchs had competed to modernize and improve their houses and estates as symbols of magnificence, power, good taste and refinement. The publication of estate arboretum catalogues and the movement of visitors and gardening staff ensured the exchange of information as well as specimens either through cooperation or subterfuge. Rivalry between the Duke of Devonshire and the Duke of Bedford, for instance, is manifest in the private notebooks of Cavendish and Paxton. When the Duke of Bedford and his gardener James Forbes designed an arboretum and other horticultural improvements at Woburn which were celebrated in a series of special publications, Paxton privately imputed them to jealousy and made criticisms in his *Horticultural Register*.[67] A more friendly form of rivalry between landowners is evident in tree collecting networks such as that involving Earl Somers, Robert Stainer Holford, Lord Delamere at Vale Royal, Sir Philip Egerton at Oulton Park in Cheshire and the Earl of Ducie at Tortworth Court, which helped to foster various significant estate arboretums. Together they exchanged ideas, seeds, cones and novel specimens and supported plant hunting expeditions, whilst they all tended to reject systematic and geographical groupings of some earlier arboretums, preferring planting for picturesque effect instead. Through such means estate arboretums significantly impacted on British and international arboriculture and horticulture. On the other hand they also had a more oppressive aspect, demonstrating the continuing power and influence of the British aristocracy and landed gentry. Most were established as private ventures primarily for personal aggrandizement, the pleasure of family and friends, and to a lesser extent scientific and economic reasons. The emphasis on well-managed, labour-intensive and well-resourced horticulture for private leisure and economic exploitation often produced different kinds of arboretums from those in botanical gardens, nurseries and public parks. Demanding and sometimes dangerous colonialist plant hunting expeditions cut across delicate global environmental differences and unbalanced British habitats. Major schemes of tree transplanting robbed the surrounding countryside of treasured specimens to adorn the interior of fenced parks. The removal of villages, banks and screens ensured privacy and controlled access to what remained in many cases private domains and, temporarily, helped to maintain social status. This was the other side of the British estate arboretum ignored in celebratory catalogues.

8 PUBLIC URBAN ARBORETUMS

Introduction

The Derby Arboretum and Loudon's publications encouraged the planting of greater varieties of trees and shrubs and the provision of arboretums in public parks and gardens and the formation of public arboretums. Public arboretums were partly inspired by the discourse of rational recreation as expressed by the Select Committee on Public Walks of 1833 and partly intended to ameliorate the problems of urban industrial expansion and to provide a taste of rural pleasures in urban settings.[1] Like Ruskin's intentions for architecture, arboretums were an attempt to restore the 'loss of fellowship with nature', to bring a piece of the countryside into towns and provide an opportunity to experience the foreign and exotic. As the *Walsall Observer* put it after the formation of a local arboretum, the 'tendency of the age is towards the creation of an artificiality ... incompatible with natural and real life'; the park would help 'restore that balance' which 'busy commercial and industrial life has rendered artificial and unnatural'. Arboretums were places for urban refreshment where, as Charles Kinglsey observed of galleries, town dwellers could escape the grim smoky 'city-world of stone and iron' by taking country walks and wandering 'beneath mountain peaks, blushing sunsets ... broad woodlands ... green meadows ... overhanging rocks [and] rushing brooks'.[2] Rational recreational objectives help to explain similarities between arboretums as 'living museums' and other nineteenth-century civic institutions such as galleries, museums, libraries and mechanics' institutions. Pleasure was to be combined with education and science for individual improvement and the public good. As visitors walked amongst interesting plants and systematic collections, reading labels and perusing guides, so they would learn about the natural world and botany.[3]

There were, however, many disagreements concerning public parks in the early and mid-nineteenth century and little political consensus concerning how they might be best funded, designed and managed for rational recreation, reflecting broader political disputes concerning the scope of local government. As we shall see, Loudon's argument that arboretums were ideal for public parks was challenged by Paxton and others who regarded them as more appropriate for botanical and horticultural institutions and elite country estates which had the expertise and resources to manage them properly. The cost and skills necessary to maintain public arboretums was another difficulty heightened by contradictions within discourses of rational recreation which tried to reconcile popular education with spatial control and regulated, conditional access for the middle classes. Whilst there was generally a trend towards free public parks, there was much local variation and many institutions continued to rely on entrance charges and subscriptions to support upkeep, planting and facilities. Even where parks or plantations designated as arboretums were established these were often very different from Loudon's gardenesque conception and notion of a systematic collection. The processes of establishing public parks and arboretums varied considerably across the British Isles and they were never purely the outcome of the will of landscape gardeners, borough engineers, nurserymen or other designers. Analysis of local factors is required to explain the relative success, character and development of public parks and arboretums shaped by donors, governing committees, subscribers, local government, the police and public opinion and varied local geographical, economic, social and political circumstances. To appreciate the causal interplay of this rich skein of factors, and the contingency of urban arboretums, it is necessary to look behind the façade of civic celebratory rhetoric manifest in town directories, guides and newspapers, to re-situate these institutions as contested, contingent and changing spaces. This underscores the difficulties inherent in 'restoring' sites back to their 'original' state using landscape gardening plans never fully enacted as originally intended.

The Foundation of Public Arboretums

Whilst Loudon's publications and the example of the Derby Arboretum helped to promote his vision of public arboretums and there was a discernable public Victorian arboretum movement, other examples took many forms. Great attention was paid to the design and management of the Derby Arboretum by park promoters from elsewhere who sent observers to examine the design and requested detailed information concerning management. Local rivalry also spurred the Derby committee to action as when a supporter of the construction of a crystal palace at the arboretum in honour of Paxton, noted that 'the new gardens at Nottingham are to be opened shortly with great splendour'. The 'Men of Derby' should 'keep up [their] arboretum', and 'suffer not so prominent and bright a star to be banished from [their] horizon.'[4] The fact that the Derby Arboretum was provided by a wealthy donor prevented widespread emulation of the plan.[5] At Leicester there were demands that a similar arboretum be created for public education and 'the general good', whilst at Nottingham experience of the Derby Arboretum and local civic rivalry helped to generate a campaign for a similar institution.[6] At Lincoln, Walsall, Worcester, Lowestoft, Ipswich and other towns there were also campaigns for public arboretums between the 1850s and the 1870s although major systematic formal collections were usually not established and the creation of parks greatly depended on local factors. An arboretum was established at Lincoln in 1872 after the larger, more industrialized centres of Derby, Nottingham and Leicester had acquired public parks. As the deliberations of the Nottingham committee show, there was pressure to employ well-known landscape gardeners although in practice most parks were designed by local nurserymen, borough engineers, gardeners and sometimes architects rather than major figures such as Paxton, Barron or Edward Milner. Milner was a protégé of Paxton and his assistant on various projects including Princes Park, Liverpool (1842), Buxton Pavilion Gardens, Moor Park in Preston, the People's Park in Halifax and the Crystal Palace Gardens at Sydenham.[7]

The Nottingham Arboretum opened in 1852 was regarded as the jewel in the crown of radical and extensive enclosure measures that allowed housing to be built on the common land surrounding the town, relieving the desperate over-crowding and ill health that marked life in the town prior to the 1840s. Although long opposed by many on the corporation and most freeholders, the enclosure campaign and public walks were supported by mixed political group of rational recreationists including the Sherwood Forest group of writers, radicals, liberals, landowners and some Tory opponents of the Liberal corporation.[8] At Worcester, a coterie of liberal businessmen and others campaigned for the creation of a public park resulting in the Worcester Pleasure Gardens designed by William Barron and managed by a company that relied upon subscriptions and income from special events on a site provided as a freehold by the Bishop of Worcester for £3,500. The gardens cost £5,000 obtained from share issues of ten pounds each with the corporation providing £1,000 on condition that free access was granted one day a week.[9] The risks of such ventures is underlined by the serious financial difficulties experienced by the company which had to sell off the land after a proposal for support from the rates with the backing of the Marquis of Lansdown was rejected by a local referendum.[10] Likewise, the Walsall Arboretum and Lake Company which opened subscription gardens in 1874 designed by Wolverhampton nurseryman Richard Lowe was in financial difficulties by the later 1870s, and the arboretum was eventually taken over by the council.

The problems of funding and access were minimized at Ipswich where after much debate during the 1840s the corporation decided to lease land from William Charles Fonnereau of Christchurch Park to form what became known as the Upper Arboretum. Adjoining land was leased by a committee and rented as the Lower Arboretum, a scheme that proved successful as the latter survived until 1922 when the two gardens were absorbed into Christchurch Park, which opened in 1891. Whilst this arrangement was complex and resulted in two smaller arboretums, it meant that risks to middle-class ratepayers were reduced and disputes concerning funding and access were small compared to those at Nottingham and Derby. Following the success of the Ipswich Museum which was acquired by a private committee who developed it and then handed over to the corporation afterwards, a similar arrangement was pursued with the Upper Arboretum, which was conveyed to the town with most of the landscaping done in 1853. Given that Rev. John Stevens Henslow, professor of botany at Cambridge University and rector of Hitcham in Suffolk was instrumental in the establishment of the Ipswich Museum, it is likely he encouraged the formation of the arboretum especially given his wider interests in popular scientific education and role in enlarging the Cambridge University Botanical Gardens and Arboretum.[11] The Upper Arboretum and a specially completed house were rented out to a nurseryman, William Brame Jeffreys, on condition that he completed 'the whole of the pleasure ground ... free of all expense to the town', a scheme intended to provide a botanically interesting public collection at no direct cost to the corporation. The arrangement also seems to have been successful as Jeffreys was still living in the house in 1881 when the corporation undertook to carry out repairs after a dispute. The two arboretums were designed and maintained differently. Designed by William Pontey who had worked at the Derby Arboretum, many kinds of trees and shrubs were planted in the 'very picturesque and well laid out' Upper Arboretum which also had 'pleasant gravelled paths, flower beds and lawns' and was free to the public at all times. Only open to subscribers and the paying public for sixpence, the Lower Arboretum was considered 'very beautiful' and 'formed in a quiet undulating hill and valley', with a lake in the lower ground, although it was not an arboretum in any formal sense. Visitors to the Upper Arboretum were initially confined to gravel paths, but from around 1860 they could wander more freely. With its admission charges the Lower Arboretum was regarded as 'of course, more exclusive', aided by its proximity to Christchurch Park from which it was separated by an 'almost invisible wire' and hosted exhibitions of the Ipswich Horticultural Society attended by the 'middle and upper classes of the neighbourhood'.[12]

Despite public support for the establishment of arboretums securing land often presented difficulties. The elaborate civic ceremonies and symbolism evident at the opening ceremonies of public arboretums and parks often disguised considerable disagreement even within the elite concerning the design and management. At the opening ceremony of Edward Milner's Lincoln Arboretum, for instance, the Bishop invoked a pious, elaborate and sustained metaphor of the park as manifestation of the Garden of Eden. He proclaimed that being 'engrafted in Christ', visitors might be like 'trees planted by the waterside' bringing 'forth their fruit in due season, in all piety, charity, and holiness of living' and reminded of the transitory nature of corporeal existence. However, when the corporation had initially approached the church as the largest landowner in the city for permission to use some of their land for an arboretum in 1868 this had been refused. The corporation committee then 'deeply regretted' the actions of the dean and chapter in refusing to enfranchise the lands required for the arboretum angrily noting that they had 'not been asked to make any sacrifice of income to promote the health, comfort and prosperity of the citizens'. As the arboretum project had received the sanction of 'the local authorities' and the 'Lords of Her Majesty's Treasury', they looked upon this as 'further proof of the impolicy of leaving the management of large masses of national property in the hands of ecclesiastics' who were too incompetent to deal with estates of which they were but the trustees.[13]

Design and Development of Public Arboretums

As we have seen, the development of the gardenesque by Loudon and his promotion of the Candollean natural system in the *Arboretum Britannicum* and Derby Arboretum were intended to facilitate the combination of systematic botanical planting with aesthetics. Of all the Victorian public urban arboretums, that at Nottingham appears to have followed most closely the spirit of Loudon's recommendations (Figure 8.1). The design was by Samuel Curtis (1779–1860), former proprietor of the *Botanical Magazine* and designer of Victoria Park in London, who had close connections with the county having worked as the agent to the Duke of Newcastle. From 1808 Curtis had planted a major arboretum of British, American and European trees at Glazenwood, Essex and was working on an even more ambitious garden at La Chaire, Jersey containing probably the largest number of species growing outdoors in any British garden because of the temperate climate.[14] The geology and topography of the Nottingham site was very different from that at Derby resulting in a different kind of arboretum. The land was very hilly and sandy owing to the sandstone underneath which meant that saplings did not thrive well without much care and attention. At seventeen acres, the Nottingham Arboretum was larger than the one at Derby yet scarcely appeared 'so extensive' from the fact that all the boundaries were visible. Some regarded this as an advantage because it provided a 'charming view of the whole ground' while at Derby there was 'no landscape view whatever'.[15] Lodges in the Tudor Gothic style for refreshment, relaxation and the garden staff similar to those at Derby were constructed although an ambitious and 'very beautiful' design by Paxton for enclosure of the corridors of the refreshment house in ironwork and glass were rejected in 1853 because of the projected cost.[16] Two types of rows of cedars were planted along the central path, Lebanon and deodar, which were alternated with fast growing limes to provide early shade. Trees and shrubs were selected as hardy enough to survive the local climate. Specimens were arranged according to their botanical families 'in the order pursued at the Derby Arboretum', with porcelain labels attached, again, as at Derby, specifying the Latin and English names, date of importation and height.[17] The number of specimens at 1,010 was around the same as at Derby, each being arranged according to the 'natural method' which was defined as grouping or classifying plants together 'according to the greatest number of points in which they resemble each other'. This was chosen so as to assist the memory and the judgement on the basis that bringing specimens together of like properties would illustrate the principle that external properties were accurate indictors of similar 'internal properties'. Thus visitors could easily compare plants from one order or family with another, especially when in flower, and 'by pursuing a diligent course of study in this direction' inspired by the arboretum, 'an excellent practical acquaintance may be obtained on a subject that is little understood by those who reside in our towns'. In contrast to the dead collections of a museum, this 'access to a living collection of specimens' would also help gentlemen to select the best trees and shrubs for the ornamentation of their grounds which would save expense and trouble.

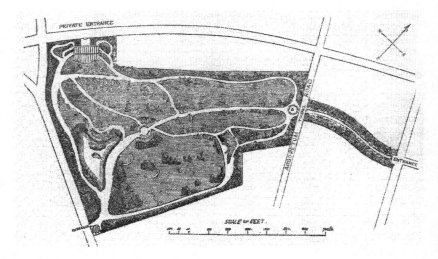

Figure 8.1: Plan of the Nottingham Arboretum and Arboretum Approach, *Illustrated London News*, 15 May 1852. Reproduced from the authors' personal copy

A campaign to transform the Royal Victoria Park in Bath into an arboretum during the 1850s also demonstrates the impact of Loudon's ideas and the Derby and Nottingham institutions whilst underscoring the complexity of relationships between paper arboretums and planted manifestations. The most prominent supporter of the venture was the registrar and surgeon, Frederick Hanham, who compiled a catalogue of trees and shrubs in the park. Hanham's work was, however, much more than a mere catalogue. It was really a manifesto for public arboretums as rational recreational institutions and a 'measure of his hearty wish to co-operate' in forming an arboretum, to increase the 'value and attractions of the park' by 'combining science with recreation and pleasure'.[18] A public arboretum – one of those institutions 'becoming so numerous and fashionable' – would help to preserve rare and valuable trees, stimulating interest by adding science to picturesque beauty leading to admiration of the works of God and 'an exalted life of nature', whilst facilitating 'mental training', youthful observation and discrimination.[19] He emphasized that the Derby Arboretum was 'visited annually by a large number of persons, strangers to Derby, on account of, and solely for' its 'correct and scientific ... collection of trees and shrubs'. There was no reason why, in 'scientific and practical terms', just as the Derby park provided an authoritative northern collection, so a Bath public arboretum would 'in addition to being a place for recreation and pleasure', become a similar regional collection of living specimens for reference. In the face of the shortening season and the 'serious diminution' of annual subscriptions, an arboretum with special emphasis on evergreens would enhance Bath's position as a desirable winter resort and promenade as evidenced by the 'vigour and health and longer season of trees and shrubs in the vicinity'. The 'mild, sheltered and temperate' climate, which he claimed was evidenced by his experience as registrar and medical officer, would attract new visitors and residents.[20]

Hanham went to great lengths to devise a 'correct and scientific nomenclature' for the plants with labels and a catalogue, although the information concerning the description and location of specimens was highly subjective and localized. The maple-leaf plane (*Platanus acerifolia*), for instance, was described as being situated on the upper lawn and between the obelisk and pond. The book also included details of specimens that he considered necessary for the committee to purchase in the future.[21] Tellingly, Hanham looked to the Derby Arboretum and 'the celebrated and extensive pinetum' at Elvaston Castle for authoritative comparison before Kew, Chiswick, Regent's Park and other national collections. However, the work was difficult and the process of comparing took some years, particularly concerning the Ulmus, Crataegus and some of the Coniferae, which, he consoled himself, were 'generally considered difficult by botanists' to identify because of their 'protean character'. The disappearance and introduction of new trees and shrubs provided further problems. Although the committee tried to implement his plan by buying new specimens that were 'wanting to complete certain classes', including the auctioned contents of a local nursery, he emphasized that these were included as 'desiderata to be restored or otherwise, an undertaking of this kind, it must be obvious, at the period of its publication, would always be incorrect' and the 'great deal of time and labour expended in vain'.[22] The apparent botanical comprehensiveness and objectivity of Hanham's published arboretum was thus achieved by incorporating various walks and common lands never originally included within the park boundary, including trees and shrubs that had disappeared and others that might never appear. His work concerned an arboretum that only partially existed supplemented by an integrated 'virtual' idealized arboretum of specimens that might never be replaced and others probably never planted.

Some Victorian parks and public arboretums were less like the Derby Arboretum and some were arboretums in name only, the term really being employed as an alternative name for an ordinary public park with no necessary pretensions towards systematic planting. On the other hand elaborate planting schemes within public parks were not always named arboretums and Barron, for instance, formed numerous and varied plantations at Peel Park, Macclesfield (1852), Locke Park, Barnsley (1877), Abbey Park, Leicester (1882) and other places. Another of Barron's creations, the Worcester Pleasure Gardens, quickly became the Worcester Arboretum although his design featured broderie parterres surrounded by serpentine walks with terraces, promenades, flower beds, a bowling green and other sporting features enclosed in elaborate iron gates and ornamental palisades.[23] The thirteen-acre Lincoln Arboretum site was laid out at a cost of over £8,000 by Edward Milner who was able to exploit the dramatic qualities of the location to provide a series of terraces commanding extensive views of the surrounding countryside. Much of the plan was formal with a central walk or 'grand promenade' along a terrace running east to west lined with a double row of lime trees and at the centre was an iron and glass building with climbing plants and benches for refreshments. Below the terrace further down the slope were gardens adorned with clumps of trees and shrubs, statues and park furniture. Water played an important role in the design and there were two fountains at either end of the grand terrace, two drinking fountains and a lake situated towards the south eastern corner crossed by two bridges.[24] Land on the steep north west corner was donated in 1894 for an extension and laid out by Edward Milner's son Henry Ernest (1845–1906) using a mixture of old trees transplanted from the original arboretum and new specimens brought from local nurseries. Despite the picturesque qualities of the Lincoln design, like the Derby and Nottingham arboretums, the 'various shrubs and plants from foreign lands' were all 'carefully labelled with their proper names' to provide 'an instructive means of recreation'.[25]

Public Arboretums and the Urban Environment

Alongside their cultural and botanical significance, public arboretums and arboriculture had an important impact on urban development. Often situated towards the periphery of towns where common lands were historically located, open spaces remained and land was relatively cheap, public parks and arboretums fostered surrounding developments of middle-class villas. Sometimes these were part of the original plan and in other cases they were constructed because of their proximity to the parks. At Worcester middle-class houses for 'respectable inhabitants' on the 'London plan of ... small suburban villas' with ornamental gardens were being constructed on Sansome Fields before the arboretum was developed.[26] The establishment of the Ipswich arboretums coincided with a rapid expansion in the urban economy represented – and encouraged – by the growth of the town as a port. In 1849 one guide confidently commented that in the tide of prosperity 'Ipswich will rise to importance and greatness, and become ... THE EMPORIUM OF THE EASTERN COUNTIES'. In addition to new public institutions such as the mechanic's institutes and museums and the creation of public walks and cemeteries, this new found prosperity was represented by the formation of suburbs with large villas and gardens on the higher ground to the north away from the industry and harbour. Villas on Henley Road in Ipswich were inhabited by 'merchants and professional men' because it had 'the advantage of facing the public Arboretum and the park beyond' which preserved the front prospect 'for generations to come'. The establishment of an arboretum was supposed to ensure the character of the area offering prospects of growing trees that ordinary public parks might not do, and was closely tied to the relocation of the Ipswich Grammar School on the opposite side of the road.[27]

At Derby, middle-class villas were constructed around what became the chief entrance to the arboretum from the railway station in an area designated Arboretum Square. At the centre of the square was an elegant ornamental lamp whilst the whole was overlooked by the grandest arboretum lodge of glass and brick surmounted with a statue of Joseph Strutt constructed in around 1850. Other sides of the park were also surrounded by fairly large houses. However, the rapid development and proximity of heavy engineering and manufacturing industries and urban expansion fostered by railway development changed the socio-economic character of the area and working-class terraced housing came to predominate. At Lincoln space was set aside in the arboretum plans at the north of the Monk's Leys for the construction of large villas along the crest of the hill level with the cathedral which commanded extensive views of the surrounding countryside whilst terraced houses were constructed at the bottom of the hill reflecting the enduring spatial division in the city between upper and lower class residential districts. This aspect of the plan, however, attracted criticism as it entailed the sale of public land. Advocating repossession, the *Stamford Mercury* regretted that 'a portion of the ground, and that the best part, was sold for building sites' which 'not only curtailed the available space' but 'inflicted on the Arboretum the presence already of two ugly buildings'. Industrial and residential development created problems for gardeners and park committees especially at Walsall, Derby and Nottingham, where smoke damage destroyed rarer species and forced them to plant smoke-resistant but common species such as limes and London planes.

The Nottingham Arboretum was the most closely integrated into a major urban improvement scheme which was one of the most ambitious in mid-Victorian Britain. The Nottingham Enclosure Act of 1845, which enclosed 1,069 acres, gave the council statutory authority to develop and maintain an arboretum and to landscape a total of approximately 130 acres of walks and grounds, including an area known as 'the Forest'. Construction standards were defined, including the width of larger roads, piped water supply and number of rooms. Enclosure commissioners oversaw the sale and development of land, while a corporation enclosure committee developed the recreational land including the arboretum.[28] An elaborate and extensive series of sports grounds, parks, gardens, cemeteries and tree-lined walks were created connecting the Forest recreation ground with the arboretum and St Anne's Hill. Names such as 'Robin Hood's Chase' resonated with the landscape, history and myths of Nottinghamshire encouraged by the work of the 'Sherwood Forest' group of writers; at the opening ceremony the mayor, William Felkin, claimed that the outlaw who wandered the surrounding hills would have enjoyed the arboretum.[29]

Tree Planting and Arboretums in Cemeteries

Increasing attention was also given to the provision of burial grounds in urban areas during the nineteenth century. This provided an opportunity for additional tree planting and sometimes arboretum provision. British writers and naturalists from Thomas Gray and Gilbert White to William Wordsworth had long celebrated the melancholy beauty and richness of churchyards, including their trees. Urbanization, industrialization and population expansion had a detrimental impact upon the state of many older burial grounds, particularly in urban areas. They also helped to change perceptions of death and cemeteries and to idealize rural burial grounds as sanctuaries of repose and contemplation. It is well known that the poor state of ancient London churchyards became a source of scandal, rapid population growth and deaths having long outstripped medieval provisions. Established customs, practices and ecclesiastical laws discouraged changes. Horrified descriptions of multiple burials and casually discarded human remains shocked society and the improvement of burial grounds and foundation of new municipal cemeteries became an important objective for reformers such as Edwin Chadwick. The campaign for new burial grounds was part of the Victorian municipal revolution which saw the development of sewerage and water supply systems, hospitals, the police and other measures intended to regulate, rationalize and improve urban environments. Medical, statistical and scientific evidence was marshalled against old practices and in support of new burial grounds, tree planting and the provision of garden and arboretum cemeteries being advocated to facilitate healthy decomposition and dissipate noxious emanations. As established symbols of death, sacredness, venerable antiquity, vitality and renewal, trees also embodied and imported the idealized tranquillity of rural burial grounds, inducing uplifting moral and religious feelings and tempering the stark modernity of industrialized commercial and municipal cemeteries.[30]

The design of these new urban cemeteries became a subject of intense debate involving social reformers, clergy, and politicians. Architects, municipal engineers and landscape gardeners proffered rival schemes. Growing requirements for urban cemeteries induced gardeners, nurserymen, architects, borough engineers and landscape gardeners to take advantage of these demands by offering rival designs and services in a competitive marketplace, staking claims for professional territory. The role of architects in designing structures such as chapels, roads, gates and monuments meant that they often had some imput into overall cemetery designs. Still feeling the impact of the Evangelical revival, many within the Anglican church feared that new private or municipal burial grounds might weaken its control over cultural and social practices, while effacing old divisions between consecrated and nonconformist burials was also opposed. In functional terms death was reinterpreted primarily as a problem of public health and the safe disposal of bodies away from urban populations to prevent the spread of disease. The design and planting of garden and arboretum cemeteries and the use of burial grounds as public walks was much debated. Particular kinds of tree came to be favoured for burial grounds for cultural, aesthetic, and scientific reasons, and Loudon in particular promoted the idea of arboretum cemeteries as sites of public rational recreation, as well as religious institutions. Arguments over the design and planting of commercial and municipal cemeteries demonstrate continuing tensions between post-Enlightenment rationality and religion and between Anglican Toryism and Liberal progressivism during the nineteenth century.[31]

In an attempt to allay fears of the secularization of death, between 1800 and 1850 attempts were made to apply the principles and practices of British landscape gardening to the design of burial grounds. This was encouraged by the use of cemeteries as public walks in towns such as Liverpool. Replanting and redesign were encouraged by moral and religious precepts of natural theology and perceived as an ideal opportunity for spiritual and moral fulfilment engendered through enjoyment and contemplation of nature. Encouraged by Loudon, Collison, Barron and others, the garden cemetery movement fostered much more ambitious tree planting within cemeteries, sometimes including the development of arboretums and pinetums. Landscape gardeners such as Loudon and Barron produced a series of influential cemetery designs with major tree collections paying careful attention to the introduction of new types, the measured use of deciduous species and the configuration of planting, paths, buildings and other features (Figure 8.2). Planting provision was intended to promote moral, spiritual and natural theological contemplation and rational recreation and to help allay fears of the secularization of the burial process, particularly given the cultural and spiritual associations of venerable trees such as yews. Coniferous types were favoured for aesthetic and scientific reasons and because they provided shelter all year round, whereas the leaves of deciduous trees allegedly retarded the process of decomposition and harboured noxious gases which caused disease.

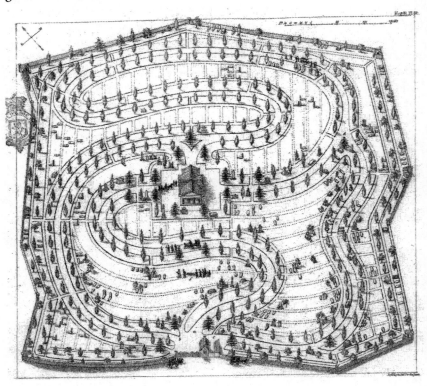

Figure 8.2: Design for a cemetery on hilly ground from J. C. Loudon, *On the Laying out, Planting and Managing of Cemeteries and on the Improvement of Churchyards* (London, 1843), figure 41. Reproduced from the authors' personal copy

As public walks, it was maintained that cemeteries could fulfil important rational recreational as well as a spiritual functions by fostering botanical education. Encouraged by Loudon in the *Gardener's Magazine, Arboretum Britannicum* and *Cemetery Interment,* burial grounds such as Abney Park in Stoke Newington incorporated major systematic tree collections. Just as Loudon strove to promote natural history as a form of secular wonder, so his post-Enlightenment rationalization of burial ground design and planting provoked opposition. This was not allayed by public health fears and the supposed rational recreational and educational benefits of arboretum cemeteries. This may explain why although Barron's Belper Cemetery was effectively a pinetum, the trees and shrubs were configured aesthetically rather that systematically as at Abney Park. As in other public arboretums and botanical gardens and parks, there were tensions in burial grounds between the desire for public access, the need to maintain income by encouraging numerous interments and tasteful monuments and efforts to support systematic collections. Garden and arboretum cemeteries provided a means by which the severely and starkly functional problems posed by disposal of the dead could be masked and alleviated. Just as the Victorian neo-Gothic hankered after a mythic medieval golden age which preserved the spirit of religious craftsmanship, so these tree-lined gardens of repose and memory were intended to lessen the impact of industrialization and provide comfort in the face of urban mortality.[32]

Critiques of Loudon's Arboretums

Whilst the *Arboretum Britannicum* was well received in Victorian society and became a standard work particularly in abbreviated form as the *Encyclopaedia of Trees and Shrubs,* Loudon's post-Enlightenment vision of systematic collections in public parks and arboretums attracted some criticism. As we have seen, criticisms of Loudon's Derby Arboretum design were made by the American landscape gardener Andrew Jackson Downing, who disliked the isolation of specimens and the nature of the mounds, and compared it unfavourably with Robert Marnock's Regent's Park botanic garden.[33] Although he did not directly criticize the Derby Arboretum, in his *Magazine of Botany* Paxton argued that systematic arboretums were inappropriate for public parks without other interesting landscape features. He questioned the fashion to 'collect *all* the known species of particular genera or of every genus containing hardy ligneous plants' and plant them in 'beds or masses over a greater or less extent of surface according to their systematical affinity'. Such schemes were not always 'fit for general adoption' and became 'absolutely disgusting' when too often or improperly repeated. Paxton objected to arboretums 'except in places where there are unusual facilities for making them pleasing', maintaining that they could only be of subsidiary interest and never the primary object of a park. Arboretums existed to present an aggregated view of all the trees and shrubs that could be cultivated in British gardens 'so that their peculiar or relative natural beauties or singularities may be at once discovered' and 'to afford the means of investigating the affinities and value' of their classes. As such they were a kind of botanical laboratory. Landscape gardening, on the other hand, for Paxton was primarily an aesthetic and practical activity. He did not doubt that it was 'advisable and useful to familiarize the public with every variety of tree and to ascertain decidedly its true character' and that therefore 'a national or suburban arboretum must be extremely serviceable in many respects', but considered that it was 'immediately obvious' many of these schemes fell short in terms of the degree of recreation offered, and 'influence on the minds and morals of visitors.' These complaints reflect the fact that relations between Paxton and Loudon had at times been strained following the latter's criticisms of Chatsworth and the *Magazine of Botany* before they experienced a rapprochement. In fact, elsewhere in his *Magazine of Botany* and *Horticultural Register* Paxton advocated ideas very similar to Loudon's including the need for upmost diversity in estate arboriculture, an elaborate geographically representational national scientific garden and multiple urban subscription botanical gardens.[34]

Paxton argued that pleasing arboretums required an 'exceedingly irregular surface' with varied direction and 'sides composed of old plantations of trees ... in apparent disorder' with 'only one principal walk' such as his own at Chatsworth. Prepared 'hillocks and artificial plots' such as those employed by Loudon at Derby ought to be avoided whilst large trees that could be left standing and every 'practicable deviation from order and system, and every opening that admits a view of the distant park or country', including streams, were to be 'assiduously sought, in order to relieve the necessary wearisomeness of artificial classification'. All attempts at forming an arboretum 'in a garden of limited extent, or flat surface' were 'sure to result in a displeasing and almost unbearable dullness' if botanical arrangements were followed. It was 'only when the natural character of a plot' was 'in itself beautiful' and other gardening arts were employed and that 'the other divisions of the pleasure-grounds are immeasurably more extensive, that an arboretum is at all tolerable'. Schemes associating 'plants in a landscape solely on account of their generic alliance' regardless of whether they augmented 'each other's beauty' or which introduced a quantity of species that had 'neither interest nor ornament to recommend them' unless they had any other purpose, ought to be 'decidedly repudiated'.[35]

Other landscape gardeners criticized aspects of Loudon's arboretum designs. Charles H. J. Smith (1810–95), an experienced Scottish landscape gardener, enthusiastically adopted and developed many of Loudon's ideas concerning arboretums in his influential *Landscape Gardening* (1853), devoting a chapter each to 'The Arboretum' and 'The Pinetum'. Smith agreed that arboretums were best laid out 'on the principle of natural orders' which exhibited 'most conspicuously those more external and prominent characters which are apt to strike the eye of even a casual observer. He agreed with Loudon's fundamental proposition that arboretums provided 'a happy union of scientific and popular elements' and were 'extremely interesting to the botanist ... presenting to the outward eye, in material and living presence' various forms in which 'more refined relations' entered into 'abstract and recondite arrangements'. They were also of great general interest to students of nature demonstrating the global importance of trees collectively and individually. Like Loudon he believed that arboretums had only recently been introduced as 'decorative accompaniments to country residences and public gardens' encouraged by the London Horticultural Society and Kew garden collections. Smith asserted that 'now no moderate sized country residence or public park and garden can be considered complete' without an arboretum.[36]

Smith believed that a natural arrangement based upon Lindley's *Vegetable Kingdom* rather than the Candollean scheme followed in the 'great' *Arboretum Britannicum* was more effective. This was primarily because Smith had found arranging arboretums according to Candolle's 'three great divisions' to be 'not a little unwieldy and unmanageable', whereas Lindley's system allowed 'a more unfettered distribution of the materials' and 'more abundant opportunities' to elicit picturesque effects. Far from application of the Candollean system solving the problem of arboretum arrangements as Loudon had hoped, taxonomic disagreements continued to make the introduction of systematic arrangements into gardening problematic. On botanical and aesthetic grounds, Smith was less enthusiastic about 'linear arrangements' than Loudon had been, emphasizing the fact that arboretums could only ever be approximations or representations of the natural world, 'of a very fragmentary character'. Taking his cue from Lindley's analysis of the position of each natural order in the *Vegetable Kingdom*, he argued that this was because such distributions of trees and shrubs were formed on 'superficial space' of two dimensions. Smith's solution to the problem of demonstrating the 'perfect co-ordination of affinities' in arboretums and pinetums was to aspire to a more three-dimensional presentation akin to 'the distribution of stars in the firmament'.[37]

The importance that arboretums had come to occupy in mid-Victorian landscape gardening is evident from the fact that Smith included a detailed thirteen page 'synopsis of the natural orders and genera of plants' intended to inform the arrangement of systematic collections in a treatise on parks and pleasure grounds. The synopsis, which Smith admitted was unusual within gardening books, was included in a section on the 'scientific treatment of the arboretum'. However, returning to the Linnaean analogy between taxonomy and geographical maps, Smith regarded this foray by gardeners into the world of botanical systematics so important that he considered an omission of discussing arboretum planting to be akin to discussing 'the topography of a country without a map', and he tried to reassure his readers that though the terminology might appear 'formidable', it was easier to the untrained eye than it might appear. The practical need for the synopsis and the botanical demands being placed upon gardeners by the fashion for arboretums is evident in Smith's work. To those who thought the taxonomic discussion unnecessary he responded that he would 'have considered a much slighter sketch a boon the first time we were called to lay out an arboretum.' Smith's list of 'alliances, orders and genera' was based on Lindley's *Vegetable Kingdom*, although for detailed information regarding each species, he employed Loudon's *Arboretum Britannicum* and *Encyclopaedia of Trees and Shrubs*. However, the use of Loudon and Lindley's books was not without problems and Smith tried to correlate references to Lindley's scheme with the *Encyclopaedia of Trees and Shrubs* classifications.[38]

Smith reckoned that there were approximately 1,500 species and a 1,000 'botanical varieties of hardy trees and shrubs' available from most nurseries. However, despite his relatively conservative species delineations, Loudon included hundreds more. On the other hand, most of Lindley's 'alliances and orders' omitted ligneous and hardy plants. Whilst Smith expected most to favour the planting of each genus on its own, he suggested that these could be combined into orders, then marshalled together as 'allied races or alliances', thus harmonizing botanical theory with the aesthetic observations of onlookers. Faced with the realities of the British climate, Smith was more conscious than Loudon of the impact of severe weather upon over-ambitious collections. Climate, soil characteristics and other factors delimited the content of planted Lindleyan orders, causing 'immense disruption of affinities' and providing an arboretum arrangement that could be 'at best only fragmentary'. While linear arrangements along walks with straight or curved lines assuming spirals or other shapes made botanical sense and had been 'adopted in some nurseries for the sake of mercantile convenience', they sacrified beauty to utility. Following Lindley's *Vegetable Kingdom*, which placed orders in the centre, two orders with the closest affinity to the left and right and two orders with weaker affinities over and under at right angles, Smith suggested that stars were a much more natural shape for arboretum arrangements. Although this 'stellar' arrangement could lead to 'great complication of figure' and possible confusion, he thought it 'ought to be adopted on the ground'. Taking inspiration from the established analogy of botanical systems with maps, he argued that orders in the natural system could be demonstrated just in the way coloured English maps showed how individual counties related to each other coterminously and 'bounded by frontier lines of varying form and length.' It was possible to harmonize the picturesque with this vision by using methods such as interspersing with evergreens, incorporating lawns in-between clumps of trees, placing trees and shrubs together of varying heights, creating 'an arboretum within the woods'. Arboretums could be planted for decorative effect, especially around villas, and there were various ways in which 'botanic interest [could] be modified and its scientific rigor softened' without sacrificing exactness.[39]

Finally, Smith augmented his general analysis of arboretums with analysis of more specialist forms including quercetums, aceretums, salicetums and pinetums, devoting a chapter to the latter as 'unquestionably the most important of all the special collections of trees'. Smith reckoned that there were in cultivation about twenty hardy conifer genera with almost 280 distinct species and varieties, most of which were evergreens. Again he provided a 'synopsis of the arrangement of the pinetum' but this time based upon Austrian botanist Stephan Ladislaus Endlicher's *Synopsis Coniferarum* which corresponded with Lindley's *Gymnogens* class but differed in sequential order from that presented in the *Vegetable Kingdom*. Smith admitted he could not 'profess to adjudicate' on the respective 'scientific merits' of the two systems encapsulating the difficulties presented to gardeners and landowners in choosing which system to represent in their arboretums that he was trying to assuage. He thought that Lindley's scheme might be 'better adapted to express the external relations of the order' within a general system but admitted that the reader might 'adopt either, as it suits his convenience', which was hardly calculated to solve the problem. Like Loudon, Smith accepted that pinetums might tend to be sombre owing to their dark coloured foliage but he argued that they could be relieved by careful attention to topography, local climate and through using different sized and coloured varieties.[40]

It is unlikely that Loudon would have objected to most of Smith's arboretum ideas which clearly owed much to his own work. Other landscape gardeners were, however, much less satisfied with Loudon's vision. Whilst Smith had made some alterations to Loudon's arboretum inspired by Lindley's latest work and expressed doubts about more linear schemes, he shared Loudon's belief idea that arboretums could combine botany with beauty and were appropriate for many parks and gardens. Other influential mid-Victorian landscape gardeners were much less enamoured and in his *Book of the Garden* (1853), Charles McIntosh (1794–1864) included a strong critique of the Loudon arboretum arguing that more obviously systematic arrangements tended to be ugly and that it was difficult to combine beauty and botany except through informative geographical arrangements. First of all, and without naming Loudon, McIntosh attacked the tendency to incorporate exotics in shrubberies. It was a 'fatal mistake' to 'desire to possess collections of trees and shrubs more valued for their rarity' than for landscaping effects. It was 'vain to expect pictorial beauty by congregating together genera and species' arranged upon principles of systematic order which could only be done in the context of the 'arboretum and fruticetum' in nurseries or botanical gardens. McIntosh agreed with Smith that arboretums were a 'comparatively modern designation' for planted collections of as many hardy genera, species and varieties of trees within a given place arranged 'according to a botanical classification'. Ignoring the *Arboretum Britannicum*, he attributed their popularity to Loddiges' Hackney Arboretum and the example of other British nursery arboretums such as that of Peter Lawson & Son in Scotland. However, though this 'mode of arrangement' with single individual representatives of each group 'might satisfy' botanists, it was 'exceedingly ill-adapted to culture' and to the 'production of pictorial effect' in other kinds of gardens. Furthermore, single specimens 'from many accidental causes' such as soil differences often represented 'exceedingly ill the true character of its kind' providing a 'very erroneous' notion of its 'habit, size and character'.[41]

McIntosh was particularly scathing towards Loudon's notion that arboretums might be suitable for public gardens, suggesting that attempts to do so at Kensington Gardens and 'other public arboretums' had little beauty. Despite the fact that 'space, diversity of situation, climate and nation's purse' had been available and the collection was now well established, 'few strangers passing through' recognized 'the existence of an arboretum' except for the labels 'painted white and lettered in black' that 'offensively arrested' the eye 'at every step'. Although these provided scientific and English name, native country and date of introduction as Loudon had suggested, they were 'of such portentous dimensions as to have no proportion to many of the specimens'. This 'excellent opportunity' of 'replanting a noble public garden' according to either the 'ancient' or 'modern' style, had been 'thrown away', when if it had been undertaken upon 'correct principles', could have been 'one of the finest arboretums in the world.' It was only in 'private parks' and due to the 'good taste of private individuals' that any 'good exemplification' of the arboretum was to be found. For McIntosh, these were merely continuations of the 'taste for ... ornamental trees' that had 'long existed in this country' as exemplified by the 'magnificent grounds' of Holkham, Sion, Studley, Woburn and Painshill. In these places trees and shrubs were planted in clumps and in 'general situations adapted to them, such as weeping willows and deciduous cypresses by the lake at Sion and cedars of Lebanon on the 'elevated knolls' of Claremont and Painsill without evidence for 'stiff formality' or 'rigid systematic arrangement'. He attributed the 'revival of a taste' for exotic ornamental trees to Loddiges, the London Horticultural Society and the 'noble encouragement' provided by the Duke of Bedford. Most damning of all 'in planting what has in general been denominated arboretums', too much weight had 'been laid on the term' and concept without much attention to anything beyond 'the number of species and varieties of trees it contains'. This mindless collecting impulse had resulted in pieces of ground set apart, clustered with trees, often 'destroying the amenity of the whole place' and having 'no connection or harmony with the surrounding parts'. Small arboretums were even worse because they presented specimens 'crowded together', requiring half to be removed before they could 'develop to their natural characteristics', or causing 'universal destruction to the whole'. Hence, far from being desirable for spaces of all sizes as Loudon maintained, arboretums 'upon a limited scale' were never 'ornamental or useful'.[42]

McIntosh could not deny the arboretum fashion had encouraged 'the introduction of many trees previously unknown even to botanists', but contended that it was not in gardens or shrubberies that the 'relative merits' of these trees could be 'fairly tested', especially where just one example of each was planted. He suggested that the best solution for those wanting systematic collections was to select one natural order and try to make their collection 'as complete as possible' as situation and soil allowed. McIntosh recommended that collections of conifers or pines would be ideal, providing a 'much better idea' of the 'true characters and value' of each kind than if planted 'in systematic order without regard to soil, situation, or association with surrounding objects.' He eagerly proposed pinetums as a beautiful form of arrangement particularly where a geographical plan was adopted in place of Linnean botanical presentation. He cited the examples of Dropmore in England and George Patton's pinetum at Carnies, Perthshire and Humphrey Graham's pinetum at Belstane, West Lothian as being some of the best Scottish examples. Systematic planting was as ill-suited to pinetums as it was to arboretums as 'however convenient' it was for 'catalogues of plants or in the arrangement of dried specimens in a herbarium', it was inappropriate for most gardens. One important solution (as Loudon had actually suggested) was geographical planting which had been 'hitherto ... almost disregarded' apart from the geographical arrangement of Messrs Vander Mallen at Brussels. McIntosh noted that John Spencer, head gardener at Bowood, had designed a complete pinetum for the Marquis of Lansdowne featuring mixed groups of trees and shrubs by countries and regions, which was more botanically useful and aesthetically pleasing than more linear systematic arrangements. He provided a plan and full description in his *Book of the Garden*.[43]

McIntosh's ideal arboretum was conceived 'upon a grand scale', 'judiciously extended' over a large park, 'grouping each natural family by themselves, and multiplying individuals' positioned in groups according to the aesthetics of landscape gardening and practical circumstances such as fitness of soil and shelter. Only 'the whole estate' and certainly not 'parks of ordinary dimensions' nor 'dressed grounds around the mansion' offered a wide enough field for successful arboretums. Hence, exotic oaks could be positioned in 'close connection' with native varieties or with beech, ash, elm or other kinds, which was far superior to 'merely sticking in solitary specimens'. Not only was McIntosh opposing linear arboretums with solitary representatives, he also retreated from gardenesque emphasis upon the individuality of trees which made up for the limited aesthetic opportunities presented by systematic collections. For McIntosh, with the exception of commercial nursery and botanical garden arboretums, landscape gardening considerations always triumphed over the botanical display of individual specimens. Arboretums of the 'highest character' placed sizeable groups of trees in harmony with existing larger trees forming a 'natural undergrowth' essential for 'pictorial planting'. Similarly, grouping or arranging natural orders along 'the margins of pre-existing plantations' and by the sides of drives allowed collections to 'blend with the woods already formed' and to be seen with 'advantage whilst riding or driving through them'. Surrounding trees could be cut away or thinned as necessary to allow space and protection for the 'fullest development' of exotics. Where space was more limited, arboretums could be formed along the boundary lines of plantations or carried 'in a graceful, circuitous manner through the park' as 'so well exemplified in the grounds at Bicton'. Another method was to 'dedicate a space on each side of the approach to the mansion of a mile or two in length' and to group trees and shrubs so as to allow views through as at Preston Hall in Scotland, the 'first and only example' in the country of 'how such an arboretum ... can be effectually arranged'.[44]

The 'perfect' arboretum formed at Bicton, Devonshire, by head gardener Robert Glendinning, John Rolle, first baron Rolle (1756–1842) and his wife Lady Louisa Barbara Rolle (with advice from Loudon), provided a 'spacious grass drive' through the 'extensive assemblage' of trees, allowing each to be 'distinctly seen' but following 'strict attention ... to scientific arrangement'. McIntosh emphasized that at Bicton there were 'no dug borders crowded with coarse and commonplace annuals and flowering plants' such as those that 'disfigure' Kensington Gardens. Neither were there large garishly coloured labels to offend the eye. Even the wire fence that protected the trees from the 'ravages of cattle and hares' was laudable being 'so constructed and arranged as to be scarcely visible – securing protection without the appearance of a boundary line'. In fact, it was essential that a successful great park arboretum should look as little like a systematic collection as possible, so that only discerning observers could recognize it for 'anything like collection, far less systematic arrangement, should be disregarded.' McIntosh objected to the aesthetic appearance of linear arboretums and also Loudon's profound belief that they were potentially suitable for urban parks, gardens or even streets. McIntosh's vision was of an elite arboretum with carefully positioned exotics amongst the lawns backed by plantations. Excluding the public, surrounded by wire fence and carefully managed smooth lawns, free from obtrusive and vulgar labels or flower beds, traversed by grass rides for horses, carriages or private strolls, this was realized in parks such as Elvaston and Westonbirt, the preserve of aristocrats, gentry, large landowners and their servants.[45]

In his edition of Charles McIntosh's *New Modern Horticulturist* (1872), James Anderson made similar criticisms of the use of arboretums in gardens to those of Paxton, whilst recommending the *Arboretum Britannicum* as 'to this day the most valuable work of its kind' for advice concerning ornamental planting.[46] Anderson argued that numerous foreign exotic species should continue to be assiduously introduced into British gardens and his book carried lengthy reports from various British arboretums and pinetums. In a special section on arboretums he defined them as a 'division of the pleasure ground dedicated to the cultivation of a collection of useful and ornamental trees and shrubs', whilst emphasizing that the term tended to refer to 'a selection of the most interesting trees and shrubs' rather than 'a complete collection arranged in strictly botanical order'. While Anderson contended that arboretums could be planted 'in any convenient part of the pleasure grounds if large', if the ground was small, he recommended that trees 'should be blended through the lawn, shrubbery, and flower garden' and if planted with taste would 'add much to the beauty of the whole'. Trees 'beautiful as well as rare' were welcome in flower gardens, shrubberies and general plantations. However, there was 'no very great desire to see arboretums in all their formality of style' in many parks and gardens. According to Anderson they could 'only be tolerated' as 'nursery quarters for new and rare trees, designed to be uplifted when of sufficient size.' He regarded arboretums as therefore 'more things of the past than of the present or the future' and while being 'very interesting in a botanical point of view', collections of rare and exotic trees were only valuable to landscape gardeners or villa owners if they could be aesthetically rather than botanically arranged. Anderson enthusiastically promoted the use of trees devoting a large part of his textbook to their description and advocacy and admitted that 'there is a grandeur and an interest which strikes the most superficial observer' in observing them, particularly new types of ancient trees which turned 'many places now no longer noted for horticultural science' into sources of fascination.[47]

The difficulties of reconciling aesthetics and botany in the light of such criticisms are evident in the disputes between William Nesfield, William Hooker and the governing committee of the Royal Botanic Gardens, Kew, as they strove to develop an ambitious pre-eminent national and imperial arboretum from the 1840s. This arboretum greatly extended a collection of trees and shrubs that had been established in the royal gardens at Kew and laid out by William Aiton, which was augmented during the 1770s and afterwards with many specimens obtained from the colonies. However, by the 1830s, the arboretum was regarded as too small compared to those of Loddiges and the Horticultural Society and there was pressure for reform. A committee appointed to enquire into the state of the royal parks and gardens, ironically as a cost-cutting exercise which included Paxton and Lindley concluded that it was 'not very extensive' and the plants 'too much crowded'. They concluded with support from horticulturists such as Loudon that changes were required and these were instigated after the garden passed from control of the Crown to the Commissioners of the Woods and Forests and Hooker was appointed director in 1841. Some forty-five acres was acquired from the pleasure grounds and Nesfield was commissioned to design what was described as a national arboretum that would be worthy, as Hooker remarked in his annual report to the commission, 'of Great Britain and serviceable to its extensive possessions and foreign relations'.[48]

Originally a watercolourist, Nesfield initially considered himself to be constrained by Hooker's specification that a taxonomic display of trees and shrubs be paramount, and the latter objected when the former produced an ornamental and geometric design (Figure 8.3). It was only after Robert Glendinning the nurseryman and former head gardener at Bicton, Devonshire, had submitted a rival plan and there were disagreements amongst the Commissioners of Woods and Forests that Nesfield's scheme was finally accepted. Using skills developed as a painter, Nesfield tried to combine taxonomy with aesthetics in his arboretum, taking account of changing shapes, sizes and colours in the positioning of specimens. Some thousand trees and shrubs including a pinetum were placed in irregular clumps with spaces between offering views towards ornaments, buildings and the pleasure gardens, the plan involving some felling, thinning and repositioning of established specimens. The taxonomic groupings of over 2,000 species were intended to facilitate botanical education and were designed to offer cohesion, space and visual stimulus partly centring upon a system of walks, Decimus Burton's new Palm House and the formal geometric gardens that surrounded it. The aspiration towards national eminence of the Royal Botanic Gardens enhanced by the radical re-design of the arboretum and gardens, however, increased the popularity of the gardens and challenged their botanical pre-eminence by threatening to create a public park rather than what was primarily a botanical institution.[49]

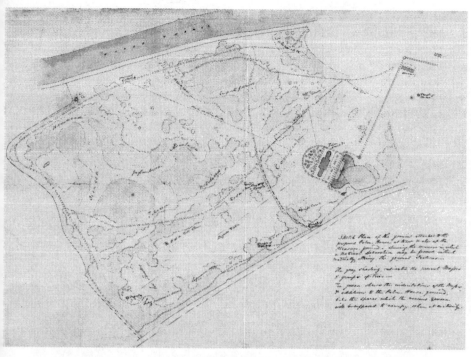

**Figure 8.3: W. Nesfield, sketch plan for the 'natural arboretum' at Kew.
Reproduced by permission of the Royal Botanical Gardens Library, Kew**

In the late Victorian and Edwardian periods growing hostility towards the introduction of alien exotic species increased criticisms of the widespread use of arboretums in gardening and landscape gardening. Leading the charge after 1870 was William Robinson for whom 'native' trees held a special place in the wild garden. Robinson complained that 'the passion for the exotic is so universal' that many of the finest native varieties were never planted, whilst money was 'thrown away like chaff for worthless exotic trees like the Wellingtonia, on which tree alone fortunes have been wasted'. The costly 'regulation pinetum' prevalent on many estates, was 'not by any means the best way' of growing trees and the isolation and 'dotting about of specimens' was 'very far from artistic'. Robinson favoured blending flower gardens with lawns and trees, framing views with their naturalistic forms (Figure 8.4).[50] Similarly, in her treatment of tree planting, especially the woodland garden at Munstead, Gertrude Jekyll, the distinguished British garden designer, offered the strongest contrast with formal arboretums and the kind of botanical schemes they represented and took Robinson's ideas to their logical conclusion. About half of Munstead was devoted to the wood- land garden but the principle she employed to determine planting schemes was

to 'appeal to the little trees themselves and see what they had to say about it'.
Jekyll tried to leave existing trees in position where possible and to emphasize
interesting combinations and patterns of species that were already presented.
Rather than introduce exotics, she combined silver birch with holly and left the
sections of oak, beech, Scots pine and Spanish chestnut as, according to her, 'the
preponderance of one kind of tree at a time [gives] a feeling of repose and dignity'. Five mown grass paths were formed from the lawn near the house into the
wood whilst colour was provided by rhododendrons and other flowering bushes.
There was a fern walk through birches and oaks, a rock walk and various carefully cultivated patches of flowers in glades.[51]

Figure 8.4: Garden framed in trees showing natural forms at Ketton Cottage near
Stamford in Lincolnshire from W. Robinson, *English Flower Garden*, 8th edn (London:
John Murray, 1900), p. 50. Reproduced from the authors' personal copy

Conclusion

Public urban arboretums and arboriculture played a significant role in the development of Victorian public parks. Promoted by Loudon, public arboretums
combined the character of botanical gardens with plantations, landscape gardens
and public leisure parks. As 'living museums' they were celebrated as botanical
counterparts to other rational recreational institutions offering an immediate
and practical experience of botany and horticulture unobtainable from dried
botanical collections.[52] They were promoted by paternalistic reformers and

public health campaigners such as Edwin Chadwick as a means of improving the health, morals and demeanour of the working classes.[53] The popularity of public arboretums and parks was fostered by the railway, newspapers and the burgeoning popular press. However, the suitability of systematic labelled taxonomic collections for public parks was questioned on aesthetic grounds by Paxton, Anderson and others who did not fully share Loudon's post-Enlightenment vision of popular botanical education. Parks and arboretums also had an important impact on wider urban development encouraging the construction of middle-class suburban villas in their vicinity and the development of other walks and gardens. Encouraged by the publicity surrounding British parks and visits from American landscape gardeners such as Andrew Jackson Downing, Charles Mason Hovey and Frederic Law Olmsted, the model of the public arboretum was adopted in North America and played a major role in the development of US public parks and urban planning.[54]

9 THE TRANSFORMATION OF VICTORIAN PUBLIC ARBORETUMS

Introduction

Public urban arboretums did not generally fulfil the original objectives of rational recreationists. Popularity, cost of maintenance and public clamour for access facilitated their decline as botanical educational institutions and they became increasingly indistinguishable from other urban leisure parks to an extent unanticipated by Loudon. This was reflected in demands for wider access, the abolition of privileged private subscriptions and the desire for municipal funding and control. The process was hastened by the success of special festivals, the demand for sports and recreational facilities and the provision of ponds, aviaries, lakes and other ornamental features. Public parks and arboretums became important centres for civic and patriotic display and communal feeling with processions and ceremonies, military events and the provision of imperial monuments and trophies. As we have seen, arboretums can be represented as idealized, objectivized collections akin to printed arboretums in arboricultural treatises, like laboratories. However, the reality was that they were fragile creations requiring constant maintenance and many utilized glass houses to lessen the impact of climatic and seasonal changes. Systematic botanical tree collections in urban areas presented special problems of maintenance and preservation, usually without the workforce and resources available on landed estates. The effort required to maintain significant public urban arboretum collections is evident from the well-documented difficulties that curators and staff had in accommodating the demands of public access, pollution and the encroachment of industrial and residential building.

The problems of securing suitable land, legal ambiguity of local authority management, difficulties of raising money, political disputes and other factors meant that a large number of public gardens and arboretums were either never realized as planned or underwent serious changes in order to remain afloat. These usually involved a reduction in the importance of systematic labelled botanically significant collections and a move towards public leisure parks just when the sciences were becoming more professionalized and institutionalized. Furthermore, although the interpretation of public urban arboretums is inevitably shaped by elite biases inherent in the sources and the tendency to emphasize the importance of civic and national pride and social cohesion, there is evidence of many contested responses within local communities. Arboretums were intensely policed and managed places intended for rational recreation and approved leisure activities reinforced by the provision of symbolic monuments, furniture and trees and shrubs, but there is evidence of serious divisions within elites and significant challenges to their management and interpretation of arboretum spaces.

Public Arboretums as Parks

Lodges, sculpture and park seats were always, of course, included from the beginning in public arboretums, but the proliferation of ponds, lakes, fountains, aviaries, skittle and bowling greens, pagodas, guns and other features illustrates their decline as botanical educational institutions. Many parks came to depend upon income obtained from fetes and special events which relied on the proliferation of special features other than more esoteric landscape and arboricultural attractions. At the Walsall Arboretum, for example, the boating lakes dominated the park and festivals such as the illuminations were extremely popular annual occasions, whilst at the Lincoln and Ipswich arboretums additional lodges and fountains were added. At Derby the immensely popular anniversary festivals, originally intended to honour the birthday of Joseph Strutt, quickly dominated the social calendar and park finances. Tens of thousands of visitors brought in by special trains came to enjoy events that included dancing, circus acts, marching bands, balloon launches and theatre. The committee came to rely upon income from these festivals as subscription numbers declined, and laid on ever more elaborate and spectacular events. In 1843, 7,000 attended and the entertainment included bands, dancing, refreshments and a balloon launch. By 1848 30,000 attended, attendance being encouraged by early shop and factory closures. During the 1860s and 1870s, astonishingly over 30,000 people regularly crowded into the fairly small Arboretum bringing in receipts of between £600 and £800 when annual subscriptions were only providing about £250. Where private companies such as Walsall Arboretum and Lake Company and the Worcester Pleasure Gardens Company managed public arboretums the pressure for special events to supplement landscape gardening and planted attractions was even greater.[1]

Public parks and arboretums became centres of civic pride, patriotism and display, but this tended to put pressure on systematic tree collections. Lavish opening ceremonies reported in local and national newspapers celebrated the public functions of parks even where they were not funded or managed by corporations. Such ceremonies were also motivated by civic rivalry and emulation and intended to present an image of social unity and coherence invested in – and manifested by – these public places. Most shops, businesses and schools were closed for the day to encourage attendance. At the opening of the Nottingham Arboretum in 1852, the sheriff, town clerk, magistrates, members of the corporation, enclosure commissioners and inhabitants processed through the town into the park supported by bands and witnessed by an estimated crowd of 25,000. Elaborate ceremonies accompanied the opening of the Lincoln Arboretum in 1872 which included processions, speeches from civic dignitaries, a blessing from the bishop and dancing fairies with wands. At Lowestoft the opening of the Arboretum Hill Gardens in 1874 by the improvement committee was taken to signify the town's new wealth and status as both port and resort, and its attraction to the gentry and middle classes. The civic nature of the Worcester Pleasure Gardens once acquired from the private company was underscored by the corporation's payment of a fifth of the cost on condition that some gratuitous public access and the involvement of the mayor, sheriff and councillors in the opening ceremonies in July 1859. After these the *Worcester Journal* contended that the Arboretum and arrival of the railway demonstrated that the city had become a regional centre.[2]

The role of local politicians in the foundation and management of Victorian parks and arboretums is reflected by the number of plaques and inscriptions that survive listing their names in these institutions. At the Nottingham Arboretum members of the original enclosure committee were immortalized on a plaque on one of the entrance lodges, while a bust of Samuel Morley (1809–86), a wealthy textile manufacturer and social reformer who served as MP for Nottingham, remains to this day the first feature greeting visitors at the entrance. Other inscribed sculpture, objects and park furniture presented by local bodies included the drinking fountains near the Ipswich Upper Arboretum entrances and in the Nottingham Arboretum. At Lowestoft, a bridge linking the Arboretum Hill Gardens with Gunton Cliff donated by Alderman William Youngman, first mayor of the new corporation, was opened in 1887 to celebrate the diamond jubilee. At Lincoln, the Dean presented a statue to the Arboretum whilst a bandstand provided by the corporation was embellished with a commemorative plaque carrying the names of the mayor and park committee. Easily the most striking piece of sculpture in the arboretum was the stone lion presented by Councillor F. J. Clark, chemist and druggist of High-Bridge, standing on a pedestal immediately in front of the central flight of steps up to the terrace. As well as being an obvious symbol of British imperialism, the lion was interpreted by the Bishop as an emblem of civic pride with the three lions of Lincoln emblazoned on the town crest. It was also perceived half-jokingly as a symbol of inter-urban rivalry given that as scripture had predicted the lion would lie with the lamb, so trains would carry the 'lambs' of Nottingham to lie down with the 'Lion' of Lincoln in the Arboretum.[3] Military and naval campaigns were also commemorated in Victorian parks and arboretums. Russian guns captured during the Crimean War were acquired by the Derby, Worcester and Nottingham arboretums after elaborate civic ceremonies. At Derby, cannon were placed in front of the 'crystal palace' on the extension. At Nottingham a bell tower was constructed with a medley of Indian and Chinese features to house a sacred bell looted from a Chinese temple during one of the opium wars which was celebrated in an inscription around four sides of the supporting mound. At the four corners were placed two Russian army guns captured during the Crimean War and two replicas, great attention being given to the positioning of these trophies even to the extent of consulting serving gunnery officers.[4]

Trees and shrubs donated to parks and arboretums by wealthy patrons, subscribers and civic dignitaries and often commemorated with celebratory plaques also became symbols of civic and patriotic pride as well as private munificence. At Nottingham members of the corporation and their families planted an avenue of limes, while from the 1890s at the Upper Arboretum, Ipswich, an avenue was set aside for trees and shrubs planted by successive mayors each of which received a plaque. Other trees and shrubs were planted by donors and subscribers to celebrate national events. An oak planted in the Ipswich Upper Arboretum to commemorate the marriage of the Prince of Wales to Princess Alice of Denmark on 10 March 1863 was marked with an inscribed plaque while a silver cedar planted near the west entrance on 23 April 1864 commemorated the 300th anniversary of the birth of Shakespeare.

Pressure for Public Access

The tension between rational recreational objectives and the requirements of retaining botanically significant collections in public parks and arboretums was most obviously manifest in the vigorous and sometimes passionate debates concerning the terms of access. Like other mid-Victorian parks, most public arboretums experienced funding difficulties because of their relatively ambiguous status as gardens funded by local taxation and managed by local government. Even where parks were donated by private individuals, controversy was not avoided. It was maintained that charging for admission provided the funds necessary to maintain the botanical qualities of arboretums and preserve them as ordered places. However, this seemed to contradict rational recreational imperatives which were unachievable if most of the public could not gain admittance. There was only limited public access to the Derby Arboretum until 1881 despite Joseph Strutt's intention that it be a gift to the town. Serious disputes concerning the funding of the Nottingham Arboretum were exacerbated by political divisions which affected the quality and variety of trees and shrubs displayed and therefore the botanical status of the institution. Although the Tory *Nottingham Journal* led the campaign for free access as a means of challenging the local Liberal hegemony during the 1850s, other Tories and some Liberals feared opening the park to all classes would threaten the distinctive characteristics of the Arboretum. Most Liberals on the corporation and Tory allies feared that the borough rate would have to increase to pay for additional police required to prevent 'injurious and disorderly acts' involving 'young persons', and that the Arboretum would become 'a mere place of recreation' instead of 'an ornamental pleasure garden'.[5]

Divisions within governing elites are evident concerning all aspects of urban arboretum design and management but particularly concerning questions of funding and access. The original regulations of the Nottingham Arboretum stipulated that it was to be open *gratis* on Sunday, Monday and Tuesday, except for special fete days when a charge would be levied, while special rates were offered for family membership. Smoking was prohibited in refreshment rooms and grounds along with spirits, liquor, wine and dogs, whilst on Sunday refreshments were banned and the Arboretum opened at 12.30 p.m., the rules being enforced by superintendents, groundsmen and policemen.[6] At Walsall, under the auspices of the Arboretum and Lake Company, the arboretum was originally only open to paying subscribers admitted through an entrance lodge. At Worcester, privileges were also available according to the scale of donations given with £25 allowing lifetime free admission for the donor and family, £10 giving the same privilege except for fete days, whilst amounts between £5 and £10 obtained the same on payment of an additional 10s. 6d. per annum and £1 1s. allowed equal privileges for donor and family on an annual basis. Access to the Derby Arboretum remained strictly controlled for forty years, with other conditions such as a ban on alcohol.[7] Yet it was often stated, for example during the opening ceremonies, that a primary objective of the committee was to encourage all classes, especially members of the 'industrial classes', to visit the park.[8] Strict access controls and admission charges served to exclude the working classes from entry for most of each week, suggesting that the rhetoric of rational education was not fully believed by most subscribers. It was expected that most members of the working class would be working throughout the week apart from Sundays. In fact the primary purpose of urban parks generally was more to support the Victorian family as an institution allowing men, women and children to promenade fashionably in directed and instructive leisure activity. Where access was unconditional, as in the Manchester public parks from 1846, strict regulations enforced by parks staff and police governed conduct.[9]

Limited access was believed to be a financial necessity for some Victorian parks, especially before direct corporation funding was legalized. The responsibility of corporations was sometimes unclear despite their involvement. The *Westminster Review* observed in 1841 how remarkable it was that the ratepayers of Derby could not, by law, contribute annual funds to the Arboretum.[10] Indeed the financial position remained uncertain, as subscriptions alone barely covered expenses, which was partly why charges continued to be made on five days a week until 1882, the public only being allowed in for free on Sundays, except, significantly, during the morning service, and on Wednesdays from dawn to dusk.[11] Under the terms of the original conditions, general family tickets, which included servants, cost 10s. 6d. per annum, tickets for a family and their visitor cost 1 guinea, while individual tickets were 7s. per year. Boarding school pupils were admitted for 2s. 6d. per annum, while single admittance for anyone was 6d., with children going in for 3d. On festival days the committee generally charged for all except subscribers.[12] But in 1854, 'in order to gratify the wishes of several subscribers, who feel a laudable desire to see the usefulness of the Arboretum extended as widely as possible and ... made more available to the working classes', the park was opened additionally on Saturday afternoons for free. This was only from 3 pm to dusk and during the months from April to June, and was only provisional, suggesting that there was much opposition from subscribers and it was feared that too many free days would reduce subscription numbers and cause problems of order. The experiment was abandoned because 'scarcely any of the class in whose benefit it was intended ... availed themselves of the privilege' whereas 'others of a higher class, for whom the change was not designed, and who have the means of becoming subscribers' had 'reaped the benefit'.[13]

The fact that the Derby Arboretum came to be largely funded by the popular annual festivals, yet remained an exclusive institution for five days a week, meant that the majority were funding the pleasure of an elite. The extent to which the festivals came to dominate could not have been predicted by Strutt, and little was done to adjust the entrance terms according to changing social conditions and public demand. So for most of the year, in terms of institutional organization, the Arboretum remained wedded to a model of public institutions of the 1830s and 1840s, such as the subscription libraries and museums. Some people tried to gain admission without paying which suggests resentment at the charges. The *Derby Mercury* reported of fete days that many people 'obtained admission ... by falsely representing themselves as friends or servants of a subscriber or as being members of his family'. However improbable it appeared to the committee, 'that persons considered respectable should endeavour by falsehood to obtain entrance to the grounds in order to save the trifling sum of 6d', the fact was 'placed beyond dispute'.[14] Another area of dispute was the family ticket, which specified that all members of a family and servants were entitled to be admitted, but because there was only one ticket this sometimes created confusion.

By the end of the 1870s it was becoming increasingly difficult to justify the lack of free access to a public facility, donated to the town, for five days each week, though the issue divided on party political lines. In 1867 Michael Thomas Bass (1799–1884) the brewing magnate and Liberal MP for Derby had donated a recreation ground adjoining the Derwent, which allowed completely free access to all.[15] The Parliamentary Reform Act of that year had given about one third of all adult males in Britain the right to vote, but more to the point (since all women and two thirds of men were still excluded), the principle of government support for social institutions, so that they could be free at the point of demand, had been conceded.[16] The council's scruples of 'economy' and Tory opposition were overcome in 1881, when it was finally agreed to take control of the Arboretum from the rates and allow unlimited gratuitous access. At a large civic ceremony in November 1882, the grounds were declared officially open to the general public.[17] However, this park had become very different from Loudon's original conception of a 'living museum'.

Arboretums as Contested Spaces

Political disagreements surrounding public parks, arboretums and their tree collections did not only concern access but other aspects of usage and they remained highly contested places as the quarrels concerning a campaign to erect a large statue of the Chartist Fergus O'Connor (1794–1855) in the Nottingham Arboretum reveal. O'Connor had been elected MP for Nottingham in 1847 but as he did not have a very strong connection with the town and was disliked by some Liberals as well as Tories the plan generated petitions and counter-petitions, letters to the newspapers and a council debate. The statue was only finally allowed after a special committee with serious disregard for the powers of subjective interpretation ordained that it must include no political gestures or controversial inscriptions threatening the 'neutrality' of the park and could only be regarded as a work of art rather than a political statement.[18]

There is considerable evidence that middle-class authority and control of space in public Victorian parks and arboretums was challenged. More fundamental challenges to elite governance of public arboretums was provided by behaviour perceived as illicit but which reveals an intention by other social groups to appropriate, mould and experience the space in different ways. The contested nature of spaces within public arboretums is evident from the types of behaviour that predominantly male middle-class management regarded as constituting a threat to order and authority. The apparent purposelessness

of working-class and youthful appropriation and manipulation of the arboretum space and their lack of reverence towards trees and shrubs threatened the
rational recreational foundations and botanical collections. As trees and shrubs
grew they became collaborators in these freedoms providing darkness and cover
whilst mounds and hills such as those at Ipswich, Derby, Lincoln and Nottingham, chosen or developed for their picturesque qualities, provided inviting
slopes for play. Contesting for arboretum spaces was particularly marked during
festivals and special events when large numbers of adults and children of different social classes mixed together.

Hostility towards working-class behaviour is evident in the language used to
report the crowds who attended opening ceremonies of the Nottingham Arboretum. The Tory *Nottingham Journal* complained that many 'knowing dodgers,
up to a thing or two' and the 'unwashed' who helped themselves to refreshments
without paying had come to take part, 'knocking crockery, eatables, and drinka-
bles into one vast irretrievable chaos'. One member of the committee reported
that the Arboretum had been 'transformed into a play ground' with people
roaming about 'at their own sweet will, regardless of shrubs or flowers, or the
preservation of the peace' and some youths were arrested for quarrelling over a
ball. All of this was reported in the *Journal* under the outraged headline 'DIS-
GRACEFUL ABUSE OF THE ARBORETUM' while the mayor complained
of the destruction and misconduct on 'the property of the public' which ought
'ever be held sacred'. The youths were 'annihilating that which, thirty years hence,
would be a large and beautiful tree, shedding a beauty on the scene and providing a comfortable shelter for the public'.[19] Reports of behaviour interpreted
as unacceptable included loud talking, factory workers kicking an Indian rub-
ber football around, attacks on structures and fences and damage to trees and
shrubs. By 1857 half of the 500 botanical labels in the Nottingham Arboretum
had been destroyed and youths had carved 'hieroglyphics' into trees which also
hid amorous activities, the committee complaining that 'the grossest indecency
was sometimes witnessed in the walks'. In response, balls were confiscated by
park officials, policemen were stationed in the Arboretum and notices were displayed threatening punishments and offering rewards including one on which
were to be painted the names of those convicted 'of doing damage to any of the
recreation grounds'.[20]

Although interpreted by the authorities as rowdiness or mindless acts of
vandalism, such activities represent challenges to middle-class expectations of
rational behaviour and decorum, and working-class appropriation of arboretum
spaces for pleasure and enjoyment. Carving names and other symbols onto trees
might be interpreted as a celebration of their symbolic status and longevity and
as territorial markers equal to systems of botanical labelling while damage to the
tallies suggests lack of sympathy for middle-class values. Public arboretums at

Derby, Nottingham, Lowestoft, Lincoln and Ipswich were supposed to replace – and often entailed the appropriation and radical transformation of – common lands, and these activities also suggest lingering dissatisfaction at the loss of these spaces and the character of their replacement. At Ipswich the Upper Arboretum appropriated a 'very extensive hilly meadow' called Bolton which formed a 'vast playground for the children of the town, a cricket ground' and a 'pleasant and healthy place of resort' where 'thousands would congregate in parties'. So despite the public arboretum, 'very few of the old inhabitants' did not 'regret the loss of the green hills and plains of Bolton' without its 'gravelled paths', shrubs and flowers, 'lodge-keepers closing the gates at fixed hours' and 'any of the restrictions and refinements of carefully-managed pleasure grounds'.[21]

Difficulties of Maintaining Systematic Collections

Besides political disagreements and financial problems there were other reasons why public arboretums tended to decline as botanically significant institutions. The lack of expertise of gardening staff in managing exotic specimens was another factor, although it is difficult to know how decisive that this was. Additionally, committees were under pressure to provide striking, but not necessarily botanically interesting displays. Once landscape gardeners such as Loudon, Milner, Paxton or Barron had completed the work of overseeing the laying out and planting of public gardens, they usually had little to do with subsequent management and did not provide the same degree of continuous attention given to aristocratic employers. Although they usually provided recommendations for park management, the problems faced by gardeners, governing committees and local government meant that these were not always followed consequently placing additional pressures on collections. Loudon's very detailed instructions on how the Derby Arboretum was to be managed were intended to ensure that the quality of the collections were maintained in a relatively small space. However, some of the work required was highly interventionist and probably unrealistic, such as the requirement that some older trees 'be removed in the course of a few years' and that 'whenever any one of these ... reaches a greater height than 40 or 50 feet' they should be taken away. If this was not done, Loudon claimed that the 'rapid-growing large trees' would 'soon overtop the slow-growing smaller ones and the shrubs, and ultimately destroy all the finer kinds'.[22] This policy was never implemented and Loudon's prediction proved accurate as a large green foliage canopy allowed little light onto the ground during the summer and contributed to the demise of delicate trees and shrubs. This mirrored the state of collections in other public arboretums where larger trees were usually allowed to grow without restriction to the detriment of most smaller and delicate varieties. Rapid expansion of industrial towns fostered demands for residential and commercial land and while most parks and arboretums had originally been situated on urban peripheries as at Lincoln and Derby, they became surrounded with industrial development which hastened the destruction of specimens and spoiled views through smoke pollution.

Labelling was usually regarded as essential in denoting the status of systematic public tree collections and part of the definition of an arboretum. It was also, of course, considered as prerequisite for botanical education and therefore as essential for fulfilling rational recreational objectives although there is evidence that the standard of labelling was exaggerated and idealized in guide books. The quality and maintenance of labelling provides a good indication of the determination of gardeners and governing committees to maintain botanically significant collections in public arboretums. The system of tallies identifying plants with Loudon's catalogue in the Derby Arboretum remained incomplete for the first five years. This was the basis for one of the few complaints that were made, or at least published, in the local newspapers, which for the most part were universally supportive. In 1844, an 'Observer' complained about the:

> gross and unpardonable piece of inattention, ignorance, or perversity, exhibited ... the nomenclature of the trees and shrubs; and whatever the talented Directors can advance in favour of the practice of appending wrong names to the plants ... [this would] confuse the young naturalist by appending such names as "Pendula" to a nearly-erect growing species of Poplar, or Serratum to a smooth-edged Holly. Besides such names as Tenniflorum for Tenuiflorum; wholly-leaved for woolly-leaved, or strait-branched for straight-branched.[23]

And it was 'by no means an extravagant supposition' to suppose that 'numerous similar cases [were] to be met with'. If the errors were not corrected the 'Observer' promised to send his own list to the newspaper for publication 'as a means of facilitating the inhabitants of Derby to escape the acquirement of those errors they are now exposed to' in the pursuit of knowledge. Whilst too much store should not be set upon the testimony of one individual, who also complained about 'indelicate vases placed near one of the pavilions' which had a 'decidedly immoral tendency' causing 'pernicious merriment' amongst younger visitors, there were clearly practical difficulties in maintaining the standard of labelling required in Loudon's scheme.[24]

Many of the trees and shrubs in British arboretums were, of course, foreign and relatively new to the English climate, it would have been difficult and expensive to devise the best conditions for their care, a task which must have required much trial and error and experimentation. The British winters were a particular problem and in their first annual report of 1841, the Derby Arboretum committee reported that 'a few of the more tender plants were killed by the unusual severity of the last winter' but promised that their places would 'be supplied in the proper season'. The next year it was remarked that some of the plants (frustratingly the reports seldom specify particular types) had 'already attained a size and beauty almost beyond expectations'. The few that had perished 'either from the severity of the last winter, or from other causes' would, it was emphasized, 'in due time be replaced'. There remained at this time a commitment to Loudon's original plan and it was not until the 1850s that the wisdom of continued adherence to this scheme in defiance of practicality and circumstances began to be questioned. In 1854 the committee reported with 'mortification', the 'disastrous consequences of the last unusually severe winter' when not less than a hundred trees and shrubs had been killed by the 'hideous frost'. Yet many of these had hitherto 'been regarded as hardy pines' and had successfully 'withstood the influence of every winter season since the grounds were first planted'. Some, including 'several varieties of pines introduced into this country' recently 'from California and the Himalayas' and 'several sets of cedars, cypresses, junipers, magnolias, bays, heather', were 'very rare and expensive and to renew them all would 'require no small outlay'.[25] As a result, for the first time, the committee officially questioned the wisdom of following Loudon's scheme remarking that although it was 'no doubt desirable as far as possible to preserve specimens of every plant, the name of which is enclosed in the printed catalogue', it was still 'questionable how far it may be wise to replace those which may be subject to destruction whenever a winter of undesirable severity shall occur'.

Labelling seems to have been discontinued at the Derby Arboretum during the 1870s which therefore marks a significant stage in the 'decline' of the institution as a Loudonesque arboretum whilst labelling at the Nottingham Arboretum, never as complete as that at Derby, also seems to have ceased during the later nineteenth century. At other parks designated as arboretums such as those at Worcester and Walsall, there does not appear to have been much if any systematic labelling despite the fact that the original scheme had envisaged labelling specimens, which brought some criticism. The decline of public arboretums as botanical institutions did not go unchallenged by staff and public and in some later Victorian and Edwardian parks efforts were made to provide labelling inspired by the Derby Arboretum such as at the Belper River gardens.[26] At the Walsall Arboretum, Thomas Everton (1843–1919) the first superintendent employed by the corporation, tried to remedy the lack of labels as part of a general initiative to protect established trees and plant new ones. Everton planted many trees in the town on local parks, common lands and streets and policed the arboretum putting forks through balls that strayed on shrubs and walking sticks through bicycle wheels. He also established representative groups of conifers in a section of park he designated the pinetum, went against fashion to introduce unusual plants to common ornamental bedding and returned to the original plans by labelling specimens.[27]

Public Access to the Royal Botanic Garden Arboretum, Edinburgh

The arboretum in the Royal Botanic Garden, Edinburgh, also illustrates the difficulties caused by political disagreements and attempts to reconcile systematic scientific collections with open public access in Victorian urban arboretums. By the 1870s it was widely recognized that the Edinburgh gardens were too small to accommodate the collections, its numerous commitments and the ambitions of the staff. In 1878 for instance, many lectures and demonstrations were held that were attended by 412 medical, pharmaceutical and botanical students requiring 47,280 plant specimens despite the fact that the government allowance was inadequate. In 1874 to 1875 the idea of a new arboretum on adjoining land was proposed as a much-needed extension by Professor John Hutton Balfour (1808–84) to 'materially promote the study of arboriculture', 'extend the capabilities of the garden as a school of science' and rapidly increase the number of students of botany and forestry with an eye to colonial service. It was also supposed to help prevent smoke damage from possible buildings nearby and protect adjoining mature trees which offered shelter to the rest of the garden. However, during the 1870s government of the botanic garden was in the hands of the First Commissioner of Works funded by the Treasury but decisions also required the consent of the Lord Provost and Town Council of Edinburgh. Balfour was professor of medicine and botany at Edinburgh University by virtue of the Town Council and also Regius keeper of the Botanic Garden with the title of Regius professor of Botany under the Crown. The Treasury replied to Balfour's proposals by arguing that it was a local Edinburgh matter which ought to be funded from local rates rather than a national or imperial concern, the government only supplying annual expenditure. The City therefore agreed to purchase the land and transfer it to the government for a scientific arboretum and place of public recreation. However this arrangement proved unworkable.[28]

On the appointment of Alexander Dickson (1836–87) as Regius professor and keeper in 1880, he was not given clear authority over the new arboretum which adjoined the botanical garden but was walled off. John Sadler (1837–82) was appointed curator at the botanic garden in 1879, but also separately curator of the Arboretum, and began planting thousands of trees. Dickson requested additional funds and formal powers to control the arboretum from the government, but because the status of the arboretum as scientific collection and public park remained unclear and it was inaccessible on Sundays (backed by the powerful Sabbath Alliance), the matter was difficult to resolve. Dickson feared that an arboretum in an openly accessible park would allow plant thefts and therefore pursued a policy of 'aloofness from the general public' according to his successor Isaac Bayley Balfour (1853–1922), even keeping the 'best plants' in 'retired and locked places' for a few special visitors whilst his staff supposedly discouraged information on the basis that folk would 'soon be as wise as ourselves'. This resulted in two systematic collections adjoining each other but separated by a long wall and governed by different rules. Dickson kept the gate between the two locked, forcing visitors to walk three-quarters of a mile around to pass between them until he received the money and authority to manage both and station a gatekeeper in-between. After the Treasury requested mediation from Edinburgh University it was suggested that a sheltered gatekeeper be provided by the Board of Works and the Treasury under the authority of the Regius keeper, and that the arboretum would be governed by the same opening hours and regulations as the botanic garden. Only in 1882, after further disagreements involving the First Commissioner of Works, the Treasury and Dickson, was the manned entrance between the two finally opened, but the partial separation reinforced by the high stone wall remained.[29]

When an attempt was made to solve the problem of dual government by the Treasury and Comissioner of Works by placing the botanic garden under the authority of Edinburgh University in the Universities (Scotland) bill of 1889, Balfour, Dickson's successor objected partly because the issue of the arboretum was still unresolved administratively and financially, as it would have remained under the Crown whilst the botanic garden was run by the University. Interestingly, he also objected because he considered it would overstretch University funds and threaten the scientific status of the botanic garden and plans for a Scottish school of forestry whilst curtailing public access and enjoyment, arguing that the garden and arboretum should be joined and managed with the royal parks and pleasure grounds so they could be laid out as one. Backed by the University, Edinburgh Council, the Trades Council and most of the public this is indeed what happened, and with the Universities (Scotland) Act of 1889, the arboretum and botanic garden were placed under the First Commissioner of Works and Public Buildings by the Treasury. Despite continued opposition from the Sabbath Alliance, the new arrangement was so popular that over 27,000 people visited on the Sundays of April 1889 to take advantage of what *The Scotsman* described as the 'humanising, innocent, and elevating pleasure of strolling through a Botanic Garden'.[30]

Combining the gardens and arboretum, increased popularity and practical exigencies impacted upon the changing design and management of the systematic collections. Balfour instituted Saturday evening lectures in elementary botany and horticulture so the general public could combine recreation with scientific education. The gardens were integrated and reorganized with the assistance of new curator Robert Lindsay (1846–1913) and general foreman Adam Dewar Richardson (1857–1930), an important figure in the Scottish Arboricultural Society, who later served as curator and embarked upon a career as a landscape gardener. Striking and instructive specimens from each genus were selected to inform the public, a new park keeper was appointed, paths were widened to accommodate large crowds and the high stone wall separating arboretum and botanic garden was removed. It was originally intended that all the species of each genus would be placed as close together as possible in the arboretum, however through time as various trees and shrubs were used for shelter and practicalities, public demands and aesthetics took priority over taxonomy. The accommodation of multiple new specimens from across the globe, especially from Sino-Himalaya and North America, also resulted in wider dispersal, thus the modern Royal Edinburgh Botanic Garden arboretum largely combines a late-Victorian scheme with subsequent redesigns and plant discoveries.[31]

Contradictory Legacies of Industry: Endowment and Smoke Damage

Rapid industrial growth furnished much of the wealth for Victorian urban transformation evidenced in the growth of civic pride and the power and authority enjoyed by town halls in the period. But it also caused difficulties for urban arboretums. The formation of the great Victorian parks, libraries and other public institutions was one of the most obvious manifestations of this civic pride. Through endowments and the wealth of the burgeoning middle class some of this money was, as we have seen, invested in public arboretums and in the transformation of planting within other kinds of public park. However, the economic prosperity that funded the acquisition of novel varieties also contained the seeds of their destruction and facilitated the transformation of public urban arboretums into ordinary leisure parks. Smoke damage had a major impact upon arboretum collections although some steps were taken to counter its effects.

One of the greatest threats to urban arboretums was the increasing smoke caused by domestic chimneys and Victorian industry, especially after the 1830s when many manufactories moved decisively to adopt steam power which added to the black clouds belching from heavy industry furnaces. The changing position is evident from the fact that Loudon did not include analyses of smoke damage in his *Encyclopaedia of Gardening*, *Encyclopaedia of Agriculture* or *Arboretum Britannicum* despite incorporating the latest ideas from meteorology, agricultural chemistry and cognate fields. Later horticultural and arboricultural works such as the *Illustrated Dictionary of Gardening* edited by George Nicholson, curator of the Royal Botanic Gardens at Kew provided detailed analysis of smoke effects informed by recent science, whilst still downplaying the damage that was caused. From the 1830s, national and local government health reports also provided more detailed analysis fostering greater awareness of the detrimental impact of industry in large manufacturing towns such as Manchester, Liverpool and Nottingham.

According to Nicholson's *Illustrated Dictionary*, most of the smoke came from soot in large industrial towns consisting of carbon with small quantities of other components formed in the combustion of fuel. All were 'familiar with the black coating that settles on everything from an atmosphere polluted with smoke' which clogged the stomata of leaves and twigs and blocked the light necessary for growth. However the danger was not as 'frequent as might be imagined', a lot of the apparent soot being 'far oftener composed of fungi' whilst 'carbonic acid (the most abundant product of combustion)' in the atmosphere was 'necessary to green plants' and 'a most important food'. Furthermore, most soot could be removed by syringing or washing. The greatest danger to plants in industrial areas came from poisonous by-products of combustion, especially sulphurous acid gas resulting from the almost inevitable presence of sulphur as an impurity in coal. Experiments by German agricultural chemist Julius Stockhardt (1809–86) and others had demonstrated that where the gas was present in the atmosphere at 1 part in 1,000,000 and especially 1 part in 40,000 traces, it could be detected in leaves, in the latter case causing severe browning and withering. Whilst presenting a detailed description of the symptoms the *Illustrated Dictionary* commented that the experiments were usually conducted upon plants in confined air under bell glasses. However, it ought to be remembered that 'in the open air' plants were 'seldom exposed to the continuous action of the gas', and the danger was therefore 'considerably less' than the amount in the air might indicate. Even so conifers were known to suffer given that they renewed their leaves less frequently than deciduous trees, leaves becoming dull green and then brown-green at the tip, although the effect was very similar to that induced by frost and it required chemical analysis to demonstrate that sulphurous gas had been responsible. Deciduous trees fared the best as they renewed their leaves, whilst they were 'injured less by exposure to the gas' during darkness and winter. Hydrochloric acid gas resulting from the emission of chemical works was one of the most injurious gases, and it was admitted that 'the only thorough remedy' was the 'stoppage of the emission of the gas'.[32]

In response to smoke damage, some horticulturists and landscape gardeners made determined efforts to recommend resistant varieties or tried to breed others that were able to do so. Nicholson's *Illustrated Dictionary* offered a selection of trees and shrubs that were 'best calculated to withstand the smoke and chemical impurities of atmosphere' that 'abounded in most large manufacturing towns'. It was claimed that trees and shrubs that produced leaves late in the year, such as elms, willows, poplars and laburnums, were 'the best suited to the purpose' because they did not suffer to the same degree from fires in winter and early spring. Within the list, trees and shrubs that were better adapted for towns in the midlands and south of England were also distinguished from those more suited to Scotland and the north. In general terms, it was emphasized that most coniferous varieties were unable to survive 'the effects of the atmosphere of a densely populated town'. *Pinus sylvestris* and similar varieties, *Platanus occidentalis* and *Populus tremula* were regarded as suitable for planting in northern towns, whilst *Pinus orientalis* was better for midland and southern centres.[33]

The impact of smoke damage on trees in the Derby Arboretum is well documented, the growth of heavy industry in the adjacent Litchurch area resulting in a thick pallor of smoke from multiple chimneys. In 1842, government commissioners were told by local physician William Baker that 'such is the state of the air' within Derby gardens that 'none but deciduous shrubs can be kept alive; evergreens become never-greens, and a miserable existence of three or four years is the usual span of their lives'.[34] The problem was emphasized with greater frequency by the management committee and became acute as domestic and industrial smoke grew with the expansion of heavy industry in the Litchurch area encouraged by the Midland Railway locomotive works. Reference is made to smoke damage in the arboretum in the 1861 report, which states that 'from the growth of population in the neighbourhood and the consequent increase of smoke, many of the more delicate class of specimen evergreens are gradually dying out and the Committee fear it will be impossible to retain them'.[35] These fears were reiterated in reports of 1862 and 1863 when 'the health of the more delicate shrubs' was thought to be in danger. The committee's response in procuring 'a few specimens of the coniferous tribe of plants lately imported into this country' demonstrates how Loudon's scheme was being adapted.

Despite the allusions to smoke damage it was never suggested even by Baker that the amount of smoke be reduced, despite extensive evidence for its deleterious impact on the health of people and plants. In 1882 the *Derby Mercury* commented that 'the committee had no control' over the causes of the death of the plants.[36] Yet, it is salient that some of the wealthiest local industrialists whose income derived from the furnaces and manufactories causing the smoke were the most loyal benefactors of the arboretum. Thomas Wright, for instance, was owner of the large Britannia Foundry ironworks in Duke Street and a member of the arboretum committee who donated ornamental vases and gave other bequests.[37] James Haywood, owner of the extensive Phoenix Foundry on Nottingham Road was another member of the committee who served as Mayor of Derby from 1849 to 1850 and contributed a circular cast metal seat to be placed near the east lodge around a large tree in 1847. Similarly, Andrew Handyside, another prominent iron-founder who took over the Duke Street works, supplied the arboretum fountain and subscribed for the crystal palace fund in 1857.[38] Gradual destruction of the arboretum was regarded as a small price to pay for industrial progress and wealth creation, or perhaps the causal relationship between botanical decline and industrial pollution was not widely recognized or discussed by the committee. After all, most of the arboretum committee members and benefactors – whether industrialists, professionals or investors – were largely dependent on industrial prosperity.[39]

By the 1870s these changes seem to have been largely accepted as the Derby Arboretum was joined by other parks in the town. In 1875, the *Derby Mercury* noted of the intention to keep up 'the original condition of a botanic garden as devised by Mr. Loudon' that 'various circumstances have combined to interfere with the carrying out of this plan in its integrity'. It cited the 'growth of the town in the direction of Litchurch', which had 'rendered the atmosphere unfit for the preservation of those rare specimens of trees and shrubs, of which the catalogue is now all that remains to show what the arboretum once was in the days of its glory'.[40] The 'supply of plants and flowers' had been 'modified and arranged to suit the altered circumstances of the case' so that the arboretum, 'still presents the aspect of a well-kept town garden, enriched and embellished with noble trees'. Here was candid recognition that though the arboretum was no longer unique and in fact unsuitable for Loudon's plan it could still successfully occupy the lowlier position of a 'well kept town garden'. Yet in 1878 one subscriber complained about the state of the arboretum paths which he had never seen 'in a much worse condition than they have been for some time past'. Walking on them was like 'walking on peas, barefoot'. In consequence, 'what few visitors there are prefer to walk on the grass, and no wonder', frequently getting into trouble with the curator. This subscriber painted a damning picture of a park in which no young people ever came and where there were 'few visitors', but it is notable that these complaints concerned lack of neatness rather than the disappearance of the systematic collection.[41] In 1882 the *Mercury* thought the arboretum noteworthy for its 'unadorned barrenness' from the space left by the death of 'the trees and shrubs for which the Arboretum was originally celebrated'.[42]

The extent of the transformation of Loudon's arboretum into a Victorian leisure park was formally recognized in a report by the new curator, Thomas Husbands, in 1882 when the park was acquired by the council and opened freely to the public. He had been asked to detail the measures that were 'required to be done to place the arboretum in a satisfactory condition'. Husbands regretted that 'owing to the great increase of buildings round the Arboretum with their attendant smoke and other noxious fumes', Loudon's 'original idea of making the place a complete arboretum or collection of trees and shrubs' could not 'fully and exclusively be carried out'. Although it seems likely that poor management of the trees and shrubs contributed to their demise, as Husbands had been assistant curator of the Sheffield Botanic Garden he was probably sympathetic to Loudon's aim of maintaining a comprehensive arboretum if that would have been possible. Husbands recommended that the shrubberies be replanted with 'such trees and shrubs as are known best to stand the smoke of towns' such as *Aucuba japonica*, Austrian pine, guelder rose, holly, laburnums, limes, lilacs, planes and yews. In addition to repair of the walks, buildings, toilets and drainage, money was to be spent on hardy bulbs, perennials and hardy flowers. This was to allow as 'many of the flower beds as possible' to be filled with 'hardy flowering plants for the winter and spring' such as daisies, crocuses, wallflowers and polyanthuses. Land was set aside for the propagation of hardy perennials. These were practical and aesthetic decisions, largely dictated by force of circumstances. It was not simply about economy, indeed the curator recommended that more needed to be spent and extra staff hired to maintain the flower beds. The arboretum remained renowned locally for the beauty of its flower gardens and many other features such as an aviary and the bandstand concerts for at least another seventy years and gave much pleasure to many. But by the early 1880s the committee formally decided that it was no longer desirable or possible to keep the arboretum as an arboretum, which is significant as probably the last time that Loudon's design could have been easily restored.[43]

Conclusion

The attention to shrubberies and flowers at the Derby Arboretum reflected changes taking place in other parks and arboretums nationwide which became dominated by glorious and elaborate floral displays of carpet bedding and other effects. These combated pollution problems to some extent being replenished annually and could easily be replaced after harsh years, whilst striking effects could be achieved over large spaces for reduced cost. Although such changes do not constitute a 'decline' in terms of reduced popularity or pleasure, coupled with much greater provision for sports and games, they do underscore the change from the Loudonesque vision of botanical education and rational recreation towards ordinary pleasure grounds. They parallel the experiences in other parks, gardens, museums and rational recreational institutions where, whilst the sciences and natural history remained extremely popular nationally, diverse donations had to be accommodated to avoid alienating wealthy donors and increase subscription revenue. At the Birmingham Botanical Gardens, for example, by the middle of the century despite the resignation of most of the governing committee in protest, more popular colourful and floral displays were being provided for aesthetic reasons rather than botanical necessity or interest. At Derby a museum and library was formed combining scientific collections with antiquities, works of art and general literature. Similarly, as Alberti has shown so well, curators at the Manchester Museum and earlier institutions in the city were under constant pressure to provide classified and formally arranged scientific collections and accommodate bequests from donors whilst satisfying bifurcating audiences.[44]

CONCLUSION

He faced, across half an acre of lawn, what the previous owners had called their 'arbo-retum'. Ludovic thought of it merely as 'the trees'. Some were deciduous and had now been stripped bare by the east wind that blew from the sea, leaving the holm oaks, yews, and conifers in carefully contrived patterns, glaucous, golden and of a green so deep as to be almost black at that sunless noon.

Evelyn Waugh, *Unconditional Surrender*[1]

By the mid-twentieth century the term arboretum was relatively commonplace. It referred to a place where collections of trees were grown and displayed sys-tematically, sometimes planted according to botanical taxonomies, labelled and catalogued. For some, such as Evelyn Waugh, arboretums had become hack-neyed; the quotation above describes an arboretum in the garden of a 'large, requisitioned villa in a still desolate area of Essex', in 1943. Here the term is con-sciously pompous and affected, describing the remnant of a small tree collection in the garden of a modest house; but as we have seen, in the nineteenth century arboretums were perceived as innovative and exciting places.

This study has explored the development of arboretums in the period, argu-ing that there was a close relationship between botany and arboriculture and that the latter encouraged the adoption of the natural system in Britain. It has exam-ined the emergence of tree collections intended to provide botanical education including botanical gardens, private estates and public sites where collections were displayed. Although neglected in the history of science, arboriculture and arboretums should be recognized as one of the most important and enduring fusions of nineteenth-century science, botanical culture and landscape garden-ing. Like botanical gardens and other systematic planted collections, arboretums represented a powerful arboricultural version of the Edenic myth, the life-affirming narrative of rekindling Platonic order and perfection in an imperfect world, or a precious botanical ark preserving the essence of nature through the stormy vicissitudes of time. They have also been held to epitomize oppressive, objective, masculine, imperialistic, post-Enlightenment nineteenth-century sci-ence.[2] Arboretums were vital and contested sites for exploring and elucidating dense interrelationships between the sciences, aesthetics, landscape gardening, religion, politics and economics.

Loudon's protean post-Enlightenment arboretum model had a tremendous impact on British culture and society, particularly through the *Gardener's Magazine* and *Arboretum Britannicum*, encouraging the proliferation of arboretums and exotic tree collections. Trees and shrubs were pivotal to his notion of recognition by art which was central to his reformulation of picturesque landscape gardening as the 'gardenesque'. It was primarily through the agency of trees and shrubs that Loudon strove to unite British picturesque landscape gardening with botany and horticultural improvement. During the 1830s he fully developed and publicized his conceptions of the gardenesque and the arboretum, replanted his Bayswater villa as an arboretum, published the *Arboretum Britannicum* and created the Derby Arboretum. Annuals came and went, buildings were often inappropriate or impractical, especially for small gardens, the geometric style, unless very sensitively applied, too easily offended against the dignity of nature and obliterated the subtle suggestion of artistic representation. For Loudon, only exotic trees and shrubs, especially if systematically treated as they matured over the years, encapsulated the taste and botanic discernment of owners, the skill of gardeners, and the novelty and success of the modern style for new and future generations. Encouraged by friends such as John Lindley, Loudon also saw arboriculture as a means of disseminating scientific ideas and practices within gardening and beyond to new public audiences. He embraced the latest developments in botany including the Candollean natural system and analysis of plant physiology and structure, which French and British tree studies had played such a large role in developing.

Stimulated in large measure by Loudon's advocacy, different forms of arboretum appeared in a variety of contexts including botanical gardens, private and estate gardens and public parks, and had a major impact on the cultures of Victorian natural history, even if his progressive egalitarian objectives remained unfulfilled. The relationship between Loudon's inherently international *Arboretum Britannicum* and the Derby Arboretum epitomizes the complex symbiotic relationships between written texts and arboretum places throughout the period and as well as planted arboretums he helped to inspire everything from descriptive regional studies such as James Grigor's *Eastern Arboretum* of Norfolk trees to major later Victorian and Edwardian surveys such as Ravenscroft's *Pinetum Britannicum* (1863–84), Elwes and Henry's *Trees of Great Britain and Ireland* (1906–13) and Bean's *Trees and Shrubs* (1914).[3] Scientific tree collections also served as inspiration for arboricultural works such as those concerning Woburn, Westonbirt and Bayfordbury which fostered further planting.[4] Other studies intended as comprehensive national or international surveys drew their material primarily from one arboretum, which they also helped to promote, such as nursery company James Veitch & Son's *Manual of the Coniferae* (1881) and, on a larger scale, Charles Sprague Sargent's *Silva of North America* (1891–1902), much of the information being obtained from the Arnold Arboretum, Boston, which he directed.[5]

Loudon's vision was challenged and adapted in various ways by Joseph Paxton, Charles Smith, Charles McIntosh, William Andrews Nesfield, Edward Milner, William Robinson and other landscape gardeners. Differences between arboretum theories and designs and the material realities of planted systematic collections underscore the many collisions between the multiple contingencies of nineteenth-century science, culture and society. Stimulated by global tree encounters and international collaboration, the *Arboretum Britannicum* and living arboretums in private gardens, country estates, botanical gardens, public parks and cemeteries, the British arboretum affected global botany and landscape gardening. Enthusiasm for Japanese gardens, for instance, was more than matched by that for Japanese shrubs and trees, but the introduction and acceptance of novel trees and arboretums was often a complicated and prolonged process. Although British scientific ideas, institutions and landscape gardening exerted an important influence across the Atlantic and around the empire, this was a reciprocal relationship. North America provided a major source of exotic specimens planted on both sides of the Atlantic, hence from the outset American and imperial collectors had an important stake in the British Arboretum. American botanists utilized their knowledge of American natural history to inform British botany and travelled between the two nations, maintaining contacts with scientific communities.

As part of his enquiry into public parks for the New York commissioners, Frederic Law Olmsted toured Elvaston and Biddulph Grange in 1859 having been recommended to see them as exhibiting 'the art of landscape gardening in higher perfection than any other in England'. At this time Olmsted was engaged with the New York nurseryman Samuel Parsons (1819–1906) in selecting a 'valuable collection of trees and shrubs' from Britain to be shipped for Central Park in the spring. He was 'greatly delighted … especially at Elvaston' and commented to Sir William Hooker that here was the 'most interesting collection of evergreens, arranged in a striking and beautiful manner' and the finest such plantation in Europe.[6] Inspired by Elvaston, Chatsworth and other British examples, arboretums and pinetums were developed in North America and Australia. Henry Winthrop Sargent, Downing's friend and executor, developed a celebrated pinetum at Wodeneth by the Hudson after a European trip, for instance. Similarly, Horatio Hollis Hunnewell nurtured an informally arranged pinetum inspired by Elvaston on his estate near Boston from about 1860.[7] British models for parks and arboretums were adopted that exploited the spaces and natural resources of the growing American empire. American arboretums, however, had special qualities, symbolizing the transition from a frontier society that perceived external nature as a threat, barrier or economic resource to be exploited towards urban industrial concerns with rural preservation. North American exploration and frontier expansion stimulated American botany, fostering attempts to study and preserve native species. Rapid economic growth in the United States, increasing political and imperial power and urbanized population demands for leisure and education encouraged the formation of multiple national and urban parks. One result was the formation of more arboretums than any other country by the early twentieth century, many of which were associated with scientific and educational institutions such as Arnold Arboretum.[8]

The varied designs, management practices and planting configurations of arboretums as sites of knowledge production, re-interpretation and dissemination, reflected tensions, disagreements and developments within scientific communities and wider nineteenth-century society heightened by the importation of global exotics. These concerned the presentation of taxonomies and scientific education, the role and function of scientific institutions, the difficulties of accommodating taxonomies with aesthetics, different kinds of usage, consumption and appropriations of arboretum spaces and degrees of popular access and control. In some instances, as many public urban arboretums demonstrate, this resulted in the abandonment of scientific educational objectives and collisions between theoretical concerns and practical, material contingencies which, in turn, shaped ideas, as the taxonomic changes demonstrate. The concept of the arboretum was adopted across the globe and exported back to countries and colonies that had supplied many trees to Britain. Wrenched from original climatic and geological contexts, trees and shrubs grew, of course, in very different ways from those originally observed in situ, thriving in unexpected ways, or succumbing to unanticipated diseases or predators. In this respect, nineteenth-century arboriculture demonstrates the importance of material goods, objects and the multiple agencies of living entities in shaping the history of natural history and scientific cultures. Interpreting arboretums as living museums or tree laboratories can downplay the implications of the 'living' part, and the results of floral, fauna and human interaction, particularly apparent in 'declining' arboretums where resources were reduced. Yet the 'situatedness' of trees and the importance of non-human agencies in nineteenth-century arboriculture should not be exaggerated. Arboretums remained predominantly highly controlled and managed places reflecting the knowledge and experience of centuries of horticultural practices, as the industrial scale of Victorian and Edwardian tree transplanting demonstrates.

One of the most influential Victorian thinkers to react against soulless mechanical arboriculture was John Ruskin, who saw natural history and aesthetics as interconnected endeavours. For Ruskin, who was steeped in nature study and whose writing career was launched in Loudon's *Magazine of Natural History*, particularly in the powerful, eloquent and influential analysis of arboriculture in *Modern Painters* (1843–60), unlike the comparatively 'dead and cold' crystalline depths of the earth, trees communicated with humanity 'through a veil of strange intermediate being', breathing without voice, moving whilst remaining static and passing 'through life without consciousness to death without bitterness', wearing the 'beauty of youth without its passion' and declining to the 'weakness of age without its regret'. This 'mystery' of a subordinate 'intermediate being' akin to animals encapsulated 'most of the pleasures' and 'most of the lessons' required from the 'external world', 'all kinds of precious grace and teaching' being 'united' in this link between earth and humanity. Trees signified the divine preparation of the earth for human needs providing a soft carpet, 'a colourful fantasy of embroidery thereon' spreading foliage to shade from the sun, shelter from the rain and break the winter winds, and moisture retention for the springs and mosses. Formed with the perfect qualities of hardness, malleability and adaptability, wood provided materials for building, tools and instruments. Seeds and fruits were made 'beautiful and palatable, varied into infinitude of appeal to the fancy of man' as medicines, balms, incense, oils or resin. Trees encapsulated 'fragility or force, softness and strength in all degrees and aspects' with their 'unerring uprightenss' like temple pillars and 'mighty resistances of rigid arm and limb to the storms of ages'. Whilst roots cleaved 'the strength of rock' and bound 'the transcience of the sand', crowns basked 'in the sunshine of the desert' or hid by dripping rocks and streams with their 'foliage far tossing in entangles fields beneath every wave of ocean'. Trees clothed 'with variegated, everlasting films', the summits of 'trackless mountains' and ministered 'at cottage doors to every gentlest passion and simplest joy of humanity'. Yet whilst he challenged landscape gardening detachment, post-Enlightenment utilitarianism and the insensibility of regimented collections and questioned the march of industry, urbanization and decline of organized religion, Ruskin's work still derived much from the botanical didacticism and aesthetic insights of Loudon's arboriculture and the picturesque. For both men, as Ruskin maintained, trees provided an Edenic vision of wonderful and enduring animate nature and the 'means by which the earth becomes the companion of man – his friend and his teacher'.[9]

NOTES

The following abbreviation is used throughout the notes:

ODNB *Oxford Dictionary of National Biography* (Oxford: Oxford University Press, 2004).

Introduction

1. J. C. Loudon (ed.), *The Landscape Gardening and Landscape Architecture of the late Humphry Repton Esq.* (London: for the editor, 1841), pp. 525–32; J. C. Loudon, *Hints on the Formation of Gardens and Pleasure Grounds* (London: John Harding, 1812), pp. 31–2; J. C. Loudon, *Encyclopaedia of Gardening*, 5th edn (London: Longman, Rees, Orme, Brown & Green, 1830), pp. 795, 935–94; D. Jacques, *Georgian Gardens: The Reign of Nature* (London: Batsford, 1983), p. 196.
2. J. C. Loudon, *Arboretum et Fruticetum Britannicum*, 2nd edn, 8 vols (London: Longman, Browne, Green & Longmans, 1844); J. C. Loudon, *An Encyclopaedia of Trees and Shrubs: Being the 'Arboretum et Fruticetum Britannicum' abridged* (London: Longmans, 1842).
3. G. Sinclair, *Useful and Ornamental Planting* (London: Baldwin & Cradock, 1832), p. 129
4. J. C. Loudon, *Observations on the Formation and Management of Useful and Ornamental Plantations*, 2 vols (Edinburgh: A. Constable & Co., 1804), vol. 1, p. 20; W. Aiton, *Hortus Kewensis*, 3 vols (London: G. Nicol, 1789); J. Smith, *Records of the Royal Botanic Gardens, Kew* (London: for the author, 1880), pp. i–vi; Loudon, *Arboretum Britannicum* (1844), vol. 1, pp. 41–4, 127–9; S. Daniels, 'The Political Iconography of Woodland in Later Georgian England', in D. Cosgrove and S. Daniels (eds), *Iconography of Landscape* (Cambridge: Cambridge University Press, 1989), pp. 43–82; S. Daniels, *Fields of Vision: Landscape Imagery and National Identity in England and the United States* (Cambridge: Cambridge University Press, 1993); R. Desmond, *Kew: The History of the Royal Botanic Gardens* (London: Harvill Press and Royal Botanical Gardens, Kew, 1995), pp. 85–103; C. Watkins (ed.), *European Woods and Forests: Studies in Cultural History* (Wallingford: Cabi International, 1998); S. Schama, *Landscape and Memory* (London: Harper, 2004).
5. U. Price, *Essays on the Picturesque*, 3 vols (London: J. Mawman, 1810), vol. 1, pp. 265–7.
6. B. Latour and S. Woolgar, *Laboratory Life: The Construction of Scientific Facts*, 2nd edn (Princeton, NJ: Princeton University Press, 1986); B. Latour, *Science in Action: How to Follow Scientists and Engineers through Society* (Cambridge, MA: Harvard University

Press, 1987); N. Jardine, A. Secord, E. Spary (eds), *Cultures of Natural History* (Cambridge: Cambridge University Press, 1996); J. Golinski, *Making Natural Knowledge* (Cambridge: Cambridge University Press, 1998); B. Lightman (ed.), *Victorian Science in Context* (Chicago, IL: Chicago University Press, 1997); R. Drayton, *Nature's Government: Science, Imperial Britain and the 'Improvement' of the World* (New Haven, CT: Princeton University Press, 2000); C. Yanni, *Nature's Museums: Victorian Science and the Architecture of Display* (New York: Princeton Architectural Press, 2005); D. Livingstone, *Putting Science in its Place* (Chicago, IL: Chicago University Press, 2003).

7. G. Nicholson, *The Illustrated Dictionary of Gardening: A Practical and Scientific Encyclopaedia of Horticulture*, 4 vols (London: L. Upcott Gill, 1889), vol. 4, supplement, pp. 450–57; C. S. Sargent, A *Manual of the Trees of North America* (New York: Houghton Mifflin, 1905), preface.

8. Loudon, *Hints on the Formation of Gardens and Pleasure Grounds*, vol. 30, plate XVII.

9. C. H. J. Smith, *Landscape Gardening: or, Parks and Pleasure Grounds* (London: for the author, 1852), pp. 248–50, referred to in B. Elliott, *Victorian Gardens* (London: Batsford, 1986), p. 118.

10. D. D. Pontin, *Arboretum Suecicum* (Upsalla, 1759).

11. J. W. Adamson, *English Education, 1789–1902* (Cambridge: Cambridge University Press, 1930); D. E. Allen, *The Naturalist in Britain: A Social History* (London: Allen Lane, 1976); C. Russell, *Science and Social Change, 1700–1900* (London: Macmillan, 1983); I. Inkster and J. Morell (eds), *Metropolis and Provinces: Science in British Culture, 1780–1850* (London: Hutchinson, 1983); R. M. Young, *Darwin's Metaphor: Nature's Place in Victorian Culture* (Cambridge: Cambridge University Press, 1985); L. L. Merrill, *The Romance of Victorian Natural History* (London; New York: Oxford University Press, 1989); P. Bowler, *History of the Environmental Sciences* (London: Fontana, 1992); P. F. Stevens, *The Development of Biological Systematics: Antoine-Laurent de Jussieu, Nature and the Natural System* (New York: Columbia University Press, 1994); Jardine et al. (eds), *Cultures of Natural History*; A. B. Shteir, *Cultivating Women, Cultivating Science: Flora's Daughters and Botany in England, 1760–1860* (Baltimore, MD: Johns Hopkins University Press, 1996); Lightman (ed.), *Victorian Science in Context*; Drayton, *Nature's Government*; C. Merchant, *Reinventing Eden: The Fate of Nature in Western Culture* (London: Routledge, 2004); Yanni, *Nature's Museums*; M. Daunton (ed.), *The Organisation of Knowledge in Victorian Britain* (Oxford: Oxford University Press, 2005); J. Endersby, *Imperial Nature: Joseph Hooker and the Practices of Victorian Science* (Chicago, IL: Chicago University Press, 2008); S. Alberti, *Nature and Culture: Objects, Disciplines and the Manchester Museum* (Manchester: Manchester University Press, 2009).

1 British Tree Cultures in the Nineteenth Century

1. D. Hudson and K. W. Luckhurst, *The Royal Society of Arts, 1754–1954* (London: John Murray, 1954), pp. 86–9.

2. Price, *Essays on the Picturesque* (1810), vol. 1, pp. 259–63; A. Bermingham, 'System, Order and Abstraction: The Politics of English Landscape Drawing around 1795' in W. T. J. Mitchell (ed.), *Landscape and Power*, 2nd edn (Chicago, IL: Chicago University Press, 2002), pp. 77–101.

3. K. Thomas, *Man and the Natural World. Changing Attitudes* in *England 1500–1800* (London: Allen Lane, 1984); Daniels, 'The Political Iconography of Woodland', pp. 43–82; Schama, *Landscape and Memory*; Watkins (ed.), *European Woods and Forests*.
4. S. Daniels and C. Watkins, 'Picturesque Landscaping and Estate Management: Uvedale Price at Foxley, 1770–1829', *Rural History*, 2:2 (1991), pp. 141–70; J. M. Neeson, *Commoners: Common Right, Enclosure and Social Change in England, 1700–1820* (Cambridge: Cambridge University Press, 1993).
5. E. J. T. Collins, 'Woodlands and Woodland Industries in Great Britain During and After the Charcoal Iron Era', in J. P. Metaille (ed.) *Protoindustries et histoire des forets* (Toulouse: Groupement de recherche ISARD, Université de Toulouse-Le Mirai, 1992), pp. 109–20.
6. J. Main, *The Forest Planter and Pruner's Assistant* (London: Ridgway, 1839); J. West, *Remarks on the Management or rather the Mismanagement of Woods, Plantations and Hedgerow Timber* (Newark: J. Perfect, 1842); J. Standish and C. Noble, *Practical Hints on Planting Ornamental Trees* (London: Bradbury & Evans, 1852).
7. W. Ablett, *English Trees and Tree Planting* (London: Smith, Elder & Co., 1880), p. 402.
8. J. Nisbet (ed.) *The Forester: A Practical Treatise on Planting and Tending of Forest Trees and General Management of Woodland Estates by James Brown*, 6th edn, 2 vols (Edinburgh: W. Blackwood & Sons, 1894), vol. 1, p. 44.
9. J. Nisbet, *The Forester*, 2 vols (Edinburgh: William Black & Sons, 1905), vol. 1, p. 49.
10. H. FitzRandolph and M. Hay, *The Rural Industries of England and Wales* (Oxford: Oxford University Press, 1926); H. Edlin, *Woodland Crafts in Britain* (London: Batsford 1949); Eastnor Estate coppice sale records, Eastnor Estate Muniments (1933), Eastnor Castle Library, Herefordshire.
11. Daniels, 'The Political Iconography of Woodland', pp. 43–82; S. Seymour, 'Landed Estates, the "Spirit of Planting" and Woodland Management in Later Georgian Britain: A Case Study from the Dukeries, Nottinghamshire', in Watkins (ed.), *European Woods and Forests*, pp. 115–34.
12. R. Monteath, *The Forester's Guide and Profitable Planter*, 2nd edn (Edinburgh: Stirling & Kenney, 1824); A. C. Forbes, *English Estate Forestry* (London: Edward Arnold, 1904), pp. 16–17.
13. J. Brown, *The Forester*, 4th edn (Edinburgh: Blackwood, 1871), advertisement; Forbes, *English Estate Forestry*, p. 17.
14. Forbes, *English Estate Forestry*, p. 318.
15. J. Simpson, *The New Forestry* (Sheffield: Pawson & Brailsford, 1903).
16. W. Schlich, *A Manual of Foresty*, 5 vols (London: Bradbury, Agnew & Co., 1889), vol. 1, pp. v–vii.
17. A. C. Forbes, *The Development of British Forestry* (London: Edward Arnold, 1910), p. 252; N. D. G. James, *A History of English Forestry* (Oxford: Oxford University Press, 1981); N. D. G. James, 'A History of Forestry and Monographic Forestry Literature in Germany, France and the United Kingdom' in P. McDonald and J. Lassoie (eds) *The Literature of Forestry and Agroforestry* (Ithaca, NY; London: Cornell University Press, 1996), pp. 15–44; J. Tsouvalis and C. Watkins, 'Imagining and Creating Forests in Britain 1890–1939' in M. Agnoletti and S. Anderson (eds), *Forest History: International Studies on Socio-Economic and Forest Ecosystem Change* (Wallingford: Cabi International, 2000), pp. 371–86.
18. P. J. Jarvis, 'Plant Introductions to England and their Role in Horticultural and Silvicultural Innovation, 1500–1900' in H. S. A. Fox and R. A. Butlin (eds), *Change in the*

Countryside: Essays on Rural England 1500–1900, special publication, 10 (London: Institute of British Geographers, 1979), pp. 145–64, esp. p. 153.

19. M. Catesby, *Hortus Britanno-Americanus* (London: J. Ryall, 1763).

20. J. C. Loudon, *Arboretum et Fruticetum Britannicum*, 8 vols (London: for the author, 1838) vol. 1, pp. 12, 41, 45; J. Ray, *Historia Plantarum*, 2 vols (London: typis Mariae Clark, prostant apud Henricum Faithorne, 1686), vol. 2, pp. 1798–9; J. Britten, 'Banister, John (1650–1692)', rev. Marcus B. Simpson jun., *ODNB*.

21. D. Chambers, *The Planters of the English Landscape Garden: Botany, Trees and the Georgics* (New Haven, CT: Yale University Press, 1993), p. 3.

22. Chambers, *The Planters*, pp. 36, 45.

23. Chambers, *The Planters*, p. 111, quoting S. Switzer, *The Practical Husbandman*, 2 vols (London: S. Switzer, 1733) vol. 1, part 1, p. liv.

24. Loudon, *Arboretum Britannicum* (1844), vol. 1, p. 61.

25. Ibid., p. 54.

26. A. Murdoch, 'Campbell, Archibald, third duke of Argyll (1682–1761)', *ODNB*.

27. See M. Symes, A. Hodges and J. Harvey, 'The Plantings at Whitton', *Garden History*, 24:14 (1986), pp. 138–72; quotation from Pehr Kalm's *Kalm's Account of his Visit to England on his way to America in 1748*, trans. J. Lucas (London, 1892), pp. 31–2.

28. Chambers, *The Planters*, p. 92 quoting BL Add. MS 28727, fol. 5, 16 February 1747/8.

29. J. J. Cartwright (ed.), *The Travels through England of Dr. Richard Pococke Successively Bishop of Meath and of Ossory, during 1750, 1751, and Later Years*, Camden Society, New Series 44 (London: Camden Society, 1889) vol. 2, pp. 260–1.

30. Chambers, *The Planters*, p. 112, quoting a letter of 1 September 1741 from P. Collinson (1694–1768) to the American botanist and explorer J. Bartram (1699–1777) in W. Darlington, *Memorials of John Bartram and Humphry Marshall with Notices of their Botanical Contemporaries* (Philadelphia, PA: Lindsay & Blakiston, 1849), p. 145. See also D. Chambers, 'Collinson, Peter (1694–1768)', *ODNB*.

31. M. Symes and J. H. Harvey 'Lord Petre's Legacy: The Nurseries at Thorndon' *Garden History*, 24 (1996), pp. 272–82.

32. J. C. Loudon, 'Hints Respecting the Manner of Laying Out the Grounds of the Public Squares in London to the Utmost Picturesque Advantage', *Literary Journal*, 21 (1803), pp. 739–42; J. C. Loudon, 'A Short Account of the Life and Writing of John Claudius Loudon' in J. C. Loudon, *Self Instruction for Young Gardeners*, 2nd edn (London: Longman, Brown, Green, & Longmans, 1847), pp. xii–xiii.

33. Loudon, 'A Short Account of the Life and Writing', pp. xvii–xxiii; J. C. Loudon, *An Immediate and Effectual Mode of Raising the Rental of the Landed Property of England ... by a Scotch farmer, now farming in Middlesex* (London: Longman, Hurst, Rees & Orme, 1808); J. C. Loudon, *The Utility of Agricultural Knowledge to the Sons of the Landed Proprietors of England ... Illustrated by what has taken place in Scotland ... by a Scotch Farmer and Land Agent* (London: J. Harding, 1809); J. C. Loudon, *Observations on Laying out Farms in the Scotch Style Adapted to England* (London: J. Harding, 1812); M. L. Simo, *Loudon and the Landscape, From Country Seat to Metropolis, 1783–1843* (New Haven, CT: Yale University Press), pp. 17–26.

34. Loudon, *Observations on the Formation*.

35. Loudon, *Observations on ... Useful and Ornamental Plantations*; Loudon, 'A Short Account on the Life and Writing', pp. xiii–xvi.

36. Loudon, *Hints on the Formation of Gardens and Pleasure-Grounds*.

37. Loudon, *Arboretum Britannicum* (1844), vol. 1, pp. 1–2.
38. Ibid., vol. 1, pp. 221–2.
39. Ibid., vol. 1, pp. 226–7; vol. 3, pp. 1717–949; J. C. Loudon, *Encyclopaedia of Trees and Shrubs* (1842; London; Frederick Warne & Co., 1875), pp. 846–904.
40. Loudon, *Arboretum Britannicum* (1844), vol. 1, pp. 229–30.
41. Charles. R. Young, *The Royal Forests of Medieval England* (Philadelphia, PA: University of Pennsylvania Press, 1979); O. Rackham, *Trees and Woodland in the British Landscape* (London: Phoenix Press, 2001); James, *A History of English Forestry*.
42. S. Mastoris and S. Groves (eds), *Sherwood Forest in 1609: A Crown Survey by Richard Bankes*, Thoroton Society Record Series, 40 (1997).
43. Anon., *English Forests and Forest Trees, Historical, Legendary and Descriptive* (London: Ingram, Cooke, 1853) and H. Townley, *English Woodlands and their Story* (London: Methuen, 1910).
44. E. Taylor (ed.), *Chronicles of an Old English Oak; or, Sketches of English Life and History, as Reported by those who Listened to them* (London: Groombridge & Sons, 1860); J. G. Strutt, *Sylva Britannica or Portraits of Forest Trees Distinguished for their Antiquity, Magnitude or Beauty* (London: J. G. Strutt, 1826), introduction.
45. C. Hibbert, *Queen Victoria in Her Letters and Journals: A Selection* (London: John Murray, 1985), p. 273.
46. H. Rooke, *A Sketch of the Ancient and Present State of Sherwood Forest* (Nottingham: S. Tupman, 1799), pp. 18–19; Daniels, 'The Political Iconography of Woodland'.
47. W. Gilpin. *Remarks on Forest Scenery, and other Woodland Views* (London: R. Blamire, 1791), pp. 7, 8, 14.
48. U. Price, *Essays on the Picturesque* (Hereford: D. Walker, 1794), pp. 26–7; see also Daniels and Watkins, 'Picturesque Landscaping and Estate Management', pp. 156–8.
49. Gilpin, *Remarks on Forest Scenery*, p. 314, F. C. Laird, *A Topographical and Historical Description of the County of Nottingham* (London: Sherwood, Neely & Jones, George Cowie & Co., 1810), p. 67.
50. *Ivanhoe* was first published on 18 December 1819 in Edinburgh and 31 December 1819 in London – 'so close to the end of the year, Ivanhoe bore the date 1820 on its title-page', Walter Scott Digital Archive, University of Edinburgh: http://www.walterscott. lib.ed.ac.uk [accessed 5/01/2011].
51. M. Girouard, *The Return to Camelot: Chivalry and the English Gentleman* (New Haven, CT: Yale University Press, 1981); J. Fowler, *Landscapes and Lives: The Scottish Forest through the Ages* (Edinburgh: Canongate, 2002), pp. 101–7.
52. W. Irving, *Abbotsford and Newstead Abbey* (London: John Murray, 1835), pp. 233–4.
53. J. Carter, *A Visit to Sherwood Forest* (London: Longman & Co., 1850), p. 73; J. Searle, 'Leaves from Sherwood' in R. White, *Worksop, 'The Dukery' and Sherwood Forest* (London: Simpkin, Marshall, 1875), p. 244; E. Eddison, *History of Worksop; with Historical, Descriptive and Discursive Sketches of Sherwood Forest* (London: Longman, 1854), pp. 194–5.
54. F. Sissons, *Beauties of Sherwood Forest* (Worksop: Sissons & Sons, 1888), p. 58.
55. A. Jackson, 'Imagining Japan: The Victorian Perception of Japanese Culture', *Journal of Design History*, 5 (1996), pp. 245–56. See also S. Tachibana and C. Watkins, 'Botanical Transculturation: Japanese and British Knowledge and Understanding of Aucuba Japonica and Larix Leptolepis 1700–1920', *Environment and History*, 16 (2011), pp. 43–71.
56. S. Tachibana, S. Daniels and C. Watkins, 'Japanese Gardens in Edwardian Britain: Landscape and Transculturation', *Journal of Historical Geography*, 30 (2004), pp. 364–94.

57. E. Kaempfer (1651–1716), German traveller and naturalist; Carl Peter Thunberg (1743–1828), Swedish botanist; P. F. von Siebold (1796–1866), German doctor and botanist.

58. E. Kaempfer, *A History of Japan Together with a Description of the Kingdom of Siam*, trans. J. G. Scheuchzer (London: privately printed, 1727); E. Kaempfer, *A History of Japan*, 3 vols (Glasgow, 1906); C. P. Thunberg, *Travels in Europe, Africa, and Asia Made Between the Years 1770 and 1779* (London: F. & C. Rivington, 1795); C. P. Thunberg, *Flora Japonica* (Lipsiae: I. G. Mulleriano, 1784); C. P. Thunberg, *Plantarum Japonicarum Novae Species* (Upsaliae: Excudebant Palmblad, 1824); P. F. de Siebold and J. G. Zuccarini, *Flora Japonica sive Plantae* (translated into Japanese by M. Sekura, with a commentary by H. Oba), *Nihon no shokubutsu* (Tokyo: Kodansha, 1996). See also E. Charles Nelson, 'So Many Really Fine Plants: An Epitome of Japanese Plants in Western European Gardens', *Curtis's Botanical Magazine*, 16 (1999), pp. 52–68. J. G. Zuccarini (1797–1848) was professor of botany at the University of Munich between 1826 and 1848.

59. N. Kato, *MakinoHyohonkan shozou no Siebold Collection* (Siebold Collection at Makino Herbarium) (Kyoto: 2003); N. Kato, H. Kato, A. Kihara and M. Wakabayashi (eds), *MakinoHyohonkan shozou no Siebold Collection*, CD ROM Database (Tokyo: 2005).

60. C. Totman, *The Green Archipelago: Forestry in Preindustrial Japan* (Berkeley, CA: University of California, 1989), p. 261.

61. R. Ono and Y. Shimada, *Kai*, 2 vols (Tokyo: Yasaka Shobō, Shōwa, [1763] 1763).

62. B. Lindquist, 'Provenances and Type Variation in Natural Stands of Japanese Larch', *Acta Horti Gotoburgensis*, 20 (1955), pp. 1–34, p. 5.

63. H. J. Elwes and A. Henry, *The Trees of Great Britain and Ireland*, 7 vols (Edinburgh: privately printed, 1907) vol. 2, p. 384.

64. E. Baigent, 'John Gould Veitch (1839–1870), *ODNB*.

65. *Gardeners' Chronicle*, 15 December 1860, p. 1103.

66. R. Alcock, 'Narrative of a Journey in the Interior of Japan, Ascent of Fusiyama, and Visit to the Hot Sulphur Baths of Atami in 1860', read on 13 May 1861, *Journal of the Royal Geographical Society*, 31 (1861), pp. 321–55; R. Alcock, *The Capital of the Tycoon*, 2 vols (London: Longman, 1863), vol. 2, p. 402.

67. Elwes and Henry, *The Trees of Great Britain*, vol. 2, p. 385.

68. Larix leptolepis, Endlicher (1847); Larix japonica, Carriere (1855); Larix Kaempferi, C. S. Sargent, *The Silva of North America*, 14 vols (New York: Houghton Mifflin, 1891–1902); Pinus Larix, Thunberg (1784); Pinus Kaempferi, Lambert (1824); Abies Kaempferi, Lindley (1833); Abies leptolepis, Siebold et Zuccarini (1842); Pinus leptolepis, Endlicher (1847).

69. Elwes and Henry, *The Trees of Great Britain*, vol. 2, p. 384.

70. Ibid., pp. 385, 386.

71. Messrs. Dickson of Chester were said to have sold no less than 750,000 in the year 1905; Elwes and Henry, *The Trees of Great Britain*, vol. 2, pp. 386–7.

72. J. MacDonald, R. F. Wood, M. V. Edwards, and J. R. Aldhous, 'Exotic Forest Trees in Great Britain: Paper Prepared for the Seventh British Commonwealth Forestry Conference, Australia and New Zealand', *Forestry Commission Bulletin*, 30 (London: HMSO, 1957), p. 69.

73. W. Barron, *The British Winter Garden* (London: Bradbury & Evans, 1852).

74. Ibid., pp. 9–24.

75. W. Gilpin, *Remarks upon Forest Scenery*, 2 vols (Edinburgh: Fraser, 1834), vol. 1, p. 101, quoted in Loudon, *Arboretum Britannicum* (1844), vol. 4, pp. 2088–9.

76. J. E. Bowman, 'On the Longevity of the Yew, as ascertained from actual Section of its Trunk, and on the Origin of its Frequent Occurrence in Churchyards', *Magazine of Natural History*, 2nd series, 1 (1836), pp. 28–35, 85–7; W. T. Bree, 'Some Account of an Aged Yew Tree in Buckland Churchyard near Dover', *Magazine of Natural History*, 2nd series, 6 (1841), p. 47–51; Loudon, *Arboretum Britannicum* (1844), vol. 4, pp. 2069–72; C. A. Miles, *Christmas in Ritual and Tradition, Christian and Pagan* (London: T. Fisher Unwin, 1912), pp. 261–80.

77. Quoted in Loudon, *Arboretum Britannicum* (1844), vol. 4, pp. 2083–4; see also R. Blair, *The Grave*, ed. G Wright (London, 1786), p. 8.

78. Loudon, *Arboretum Britannicum* (1844), vol. 4, pp. 2073–82.

79. A. B. Lambert, *A Description of the Genus Pinus*, 1st edn (London: White, 1803); J. Forbes, *Hortus Woburnensis* (London: Ridgway, 1833); Loudon, *Arboretum Britannicum* (1844), vol. 1, pp. 70–1; M. Symes, 'A. B. Lambert and the conifers at Painshill', *Garden History*, 16 (1998), pp. 24–40 (pp. 24, 27).

80. *Derby Mercury*, 25 May 1852; *Derby Mercury*, 16 June 1852; Derby Temperance Society, Annual Reports (1852–76), Local Studies Library, Derby (BA 178).

2 Trees and Taxonomy

1. G. C. Gorham, *Memoirs of John Martyn FRS and Thomas Martyn, BD, FRS, FLS, Professors of Botany in the University of Cambridge* (London: Hatchard & Son, 1830), p. 137.

2. J. von Sachs, *History of Botany (1530–1860)* (Oxford: Clarendon Press, 1890); J. Reynolds Green, *A History of Botany in the United Kingdom* (London: Dent, 1914); Allen, *The Naturalist in Britain*; L. Brockway, *Science and Colonial Expansion: The Role of the British Botanic Gardens* (New Haven, CT: Yale University Press, 1979); Russell, *Science and Social Change*; Inkster and Morell (eds), *Metropolis and Provinces*; Young, *Darwin's Metaphor*; Merrill, *The Romance of Victorian Natural History*; Bowler, *History of the Environmental Sciences*; Stevens, *The Development of Biological Systematics*; Desmond, *Kew*; Jardine et al. (eds), *Cultures of Natural History*; Shteir, *Cultivating Women, Cultivating Science*; Lightman (ed.), *Victorian Science in Context*; D. P. McCracken, *Gardens of Empire: Botanical Institutions of the Victorian British Empire* (Leicester: Leicester University Press, 1997); Drayton, *Nature's Government*; Merchant, *Reinventing Eden*; B. Elliott, *The Royal Horticultural Society: A History, 1804–2004* (Chichester: Phillimore, 2004); Yanni, *Nature's Museums*; Daunton (ed.), *The Organisation of Knowledge*; P. Ayers, *The Aliveness of Plants: The Darwins at the Dawn of Plant Science* (London: Pickering and Chatto, 2008); Endersby, *Imperial Nature*.

3. H. Ritvo, 'Zoological Nomenclature and the Empire of Victorian Science', in Lightman (ed.), *Victorian Science in Context*, pp. 334–53, pp. 337–8; J. Endersby, 'Classifying Sciences: Systematics and Status in mid-Victorian Natural History', in Daunton (ed.), *The Organisation of Knowledge*, pp. 61–86.

4. Endersby, 'Classifying Sciences', pp. 61–86.

5. For convenience the term botanist is employed, although it was regarded as part of natural history. The concepts of 'natural' and 'natural system' were, of course, highly contested and unless otherwise stated, refer to Jussieuian botany in distinction to the 'artificial' Linnaean system. Of course, all botanists in the period would probably have claimed that their systems were founded upon real experience and study of the external world.

6. Sachs, *History of Botany*, pp. 7–10, 79–154; Green, *A History of Botany*, pp. 207–353; C. E. Raven, *John Ray, Naturalist: His Life and Works* (Cambridge: Cambridge University Press, 1950), pp. 181–307; E. Mayr, *The Growth of Biological Thought* (Cambridge, MA: Harvard University Press, 1982), pp. 177–80, 196–202; Stevens, *Development of Biological Systematics*; P. Smith (ed.), *Memoir and Correspondence of the late Sir James Edward Smith*, 2 vols (London: Longman, Rees, Orme, Brown, Green, & Longman, 1832); M. Walker, *Sir James Edward Smith* (London: Linnean Society of London, 1988); J. E. Smith, *An Introduction to Physiological and Systematical Botany*, 5th edn (London: Longman, Hurst, Rees, Orme, Brown, & Green, 1825), p. 288.

7. C. Linnaeus, A *System of Vegetables ...Translated from the Thirteenth Edition of the' Systema Vegetabilium' of ... Lineus* [sic] *... by a Botanical Society at Lichfield*, 2 vols (Lichfield: Leigh & Sotheby, 1783); C. Linnaeus, *The Families of Plants ... Translated from the Last Edition of the 'Genera Plantarum' ... by a Botanical Society at Lichfield*, 2 vols (Lichfield: John Jackson, 1787); Loudon, *Encyclopaedia of Gardening* (1830), pp. 128–30; J. C. Loudon, *Encyclopaedia of Plants*, new edn (London: Longman et. al., 1855), pp. xxi–xxii, 1–2, 1051–2; Bowler, *History of the Environmental Sciences*, pp. 155–66; Endersby, 'Classifying Sciences', pp. 61–86; J. Lindley, *The Vegetable Kingdom* (London: Bradbury & Evans, 1846), pp. xxiii, 33–4; Sachs, *History of Botany*, pp. 79–107; Mayr, *The Growth of Biological Thought*, pp. 172–3; A. Wolf, *A History of Science, Technology and Philosophy*, 2 vols (New York: Harper Torch Books, 1963) vol. 2, pp. 426–77; N. Jardine, 'Inner History; or, How to End Enlightenment', in J. Clark, J. Golinski and S. Schaffer (eds), *The Sciences in Enlightened Europe* (Chicago, IL: Chicago University Press, 1999), pp. 477–94; Allen, *Naturalist in Britain*, pp. 36–51; Thomas, *Man and the Natural World*, pp. 65–9.

8. Smith, *An Introduction to Physiological and Systematical Botany* (1825), pp. 288–98, 290, 296–8, 316–30.

9. P. W. Watson, *Dendrologia Britannica*, 2 vols (London: for the author, 1824–5), vol. 1, p. xvii.

10. A. P. de Candolle, *Theorie elementaire de la botanique*, 2 vols, 1st edn (Paris: Deterville, 1813); A. P. Decandolle and K. Sprengel, *Elements of the Philosophy of Plants*, trans. (Edinburgh: William Blackwood, 1821), pp. 1–172; Lindley, *The Vegetable Kingdom*, pp. xxxv–xxxvi; Sachs, *History of Botany*, pp. 126–38; Stevens, *The Development of Biological Systematics*, pp. 79–91.

11. Candolle, *Theorie elementaire*, p. 206, quoted in Lindley, *The Vegetable Kingdom*, p. xxxv.

12. Sachs, *History of Botany*, pp. 127–30, 136–7; Stevens, *The Development of Biological Systematics*.

13. R. Brown, *Prodromus Florae Naovae Hollandiae et Insulae Van Diemen* (London: Johnson & Co., 1810); Sachs, *History of Botany*, pp. 139–45; Green, *A History of Botany*, pp. 309–55; D. J. Mabberley, *Jupiter Botanics: Robert Brown of the British Museum* (London: Lubrecht & Cramer, 1985); Stevens, *The Development of Biological Systematics*, pp. 98–102; P. F. Stevens, 'On Amateurs and Professionals in British botany in 1858 – J. D. Hooker on Bentham, Brown and Lindley', *Harvard Papers in Botany*, 2 (1997), pp. 125–32; W. T. Stearn (ed.), *John Lindley, 1799–1865: Gardener-Botanist and Pioneer Orchidologist* (London: Woodbridge Antique Collectors Club, 1999); Endersby, *Imperial Nature*, pp. 213-6.

14. W. J. Hooker and T. Tylor, *Muscologia Britannica* (London: Longman, Hurst, Rees, Orme & Brown, 1818); W. J. Hooker, *Flora Scotica* (Edinburgh: Archibald Constable & Co., 1821).

15. R. Sweet, *Hortus Suburbanus Londoniensis; or a Catalogue of Plants Cultivated in the Neighbourhood of London* (London, 1818), pp. v–vi.
16. S. F. Gray, *Natural Arrangement of British Plants*, 2 vols (London: Baldwin, Cradock & Joy, 1821); M. Allan, *Darwin and his Flowers* (London: Allen Lane, 1977), pp. 41–2; Endersby, 'Classifying Sciences', pp. 61–86.
17. Candolle, *Theorie elementaire* (1813); Hooker, *Flora Scotica*; Loudon, *Encyclopaedia of Plants* (1855), p. 1054; A. Desmond and J. Moore, *Charles Darwin* (London: Michael Joseph, 1991), p. 43.
18. J. Lindley, *Observations on the Structure of Fruits and Seeds*, trans. from L.-C. Richard, *Analyse du Fruit* Richard (London: John Harding, 1819), pp. vi–vii.
19. J. C. Loudon, *Encyclopaedia of Plants*, 2nd edn (London: Longman, Brown, Green, & Longmans, 1841, pp. iii–iv.
20. Ibid., pp. iii–iv, xiii, xxi.
21. J. Lindley, *Synopsis of British Flora According to the Natural Orders* (London: Longman, Rees, Orme, Brown & Green, 1829), preface, p. xi.
22. Loudon, *Encyclopaedia of Plants* (1841), p. 2.
23. Ibid., pp. 1053–4.
24. Loudon, *Arboretum Britannicum* (1838).
25. Lindley, *The Vegetable Kingdom*, p. xxiii.
26. Watson, *Dendrologia*, vol. 1, pp. xv–xvi.
27. Lindley, *The Vegetable Kingdom*, pp. xxiii–iv.
28. R. Leakey and R. Lewontin, *The Sixth Extinction: Biodiversity and its Survival* (London: Phoenix, 1998), pp. 103–4, 120–1.
29. Jane Loudon (ed.), *Encyclopaedia of Plants* (London: Longman et al., 1855), pp. 8–29, 76–108, 108–236, 296–9, 408–56, 456–90, 598–649, 660–747.
30. Ibid., pp. 768–874.
31. H. Edlin, *Trees, Woods and Man*, 3rd edn (London: Collins, 1970), pp. 1–5; G. Aas and A. Riedmiller, *Trees of Britain and Europe*, trans. M. Walters (London: HarperCollins, 1994); J. Hillier and A. Coombes, *The Hillier Manual of Trees and Shrubs* (Newton Abbott: David & Charles, 2002); P. Sterry, *Collins Complete British Trees* (London: Collins, 2007), pp. 7–14.
32. Price, *Essay on the Picturesque* (1794), p. 286.
33. Edlin, *Trees, Woods and Man*, pp. 1–6; Sterry, *Collins Complete British Trees*, pp. 7–14.
34. Edlin, *Trees, Woods and Man*, p. 9.
35. Sterry, *Collins Complete British Trees*, pp. 7–14.
36. Edlin, *Trees, Woods and Man*, pp. 10–11; Sterry, *Collins Complete British Trees*, pp. 7–14.
37. Watson, *Dendrologia* (1824), vol. 2, p. 30, tables.
38. J. E. Smith, *A Grammar of Botany* (New York; London: Longman, Hurst, Rees, Orme & Brown, 1822), pp. 214–66; J. Lindley, *An Introduction to the Natural System of Botany*, 1st edn (London: Longman, Rees, Orme, Brown & Green, 1830), pp. iii–iv.
39. R. Preston, *North American Trees*, 3rd edn (Cambridge, MA: MIT Press, 1976), pp. xv–xxi.
40. Edlin, *Trees, Woods and Man*, pp. 11–12.
41. Ibid., p. 12.
42. Ibid., pp. 6–8.
43. Depleted populations caused by enclosure and Highland clearances in the name of estate improvement presented an opportunity for landlords to plant trees, especially where the land was too poor in quality for agriculture.

44. Stevens, *The Development of Biological Systematics*, pp. 384–5.
45. Smith, An *Introduction to Physiological and Systematical Botany* (1825), pp. 10, 14, 16–24, 25–30; N. Grew, *An Idea of a Phytological History Propounded* (London: Richard Chiswell, 1673); J. Evelyn, *Silva: or a Discourse of Forest Trees* (London: J. Martyn & J. Allestry, 1664); S. Hales, *Statical Essays,* 2nd edn, 2 vols (London: W. Innys, 1731–3); E. Darwin, *Phytologia: or the Philosophy of Agriculture and Gardening* (London: J. Johnson, 1800); T. A. Knight, *A Selection from the Physiological and Horticultural Papers ... by the late Thomas Andrew Knight* (London: Longman, Orme, Brown, Green & Longmans, 1841), p. 11; H. L. Duhamel du Monceau, *Traite des arbres et arbustes qui se cultivent en France,* 2 vols (Paris: H. L . Guerin & L. F. Delatour, 1755); H. L. Duhamel du Monceau, *La physique des arbres, ou il est traite de l'anatomie des plantes et de l'economie vegetale,* 2 vols (Paris: H. L Guerin & L. F. Delatour, 1758); C. F. B. de Mirbel, *Traité d'anatomie et de physiologie végétales,* 2 vols (Paris, 1802); C. F. B. de Mirbel, *Éléments de physiologie végétale et de botanique,* 3 vols (Paris, 1815).
46. Watson, *Dendrologia* (1825), vol. 1, p. vii.
47. Watson, *Dendrologia* (1824), vol. 1, p. ii.
48. Ibid., vol. 1, preface.

3 British Arboriculture c. 1800–35

1. S. Naylor, 'Provincial Authorities and Botanical Provinces: Elizabeth Warren's Hortus Siccus of the Indigenous Plants of Cornwall', in Elliott et al. (eds), 'Cultural and Historical Geographies of the Arboretum', pp. 84–95.
2. Loudon, *Encyclopaedia of Gardening* (1830), pp. 86–7; Loudon, *Arboretum Britannicum* (1844), vol. 1, pp. 129–30; Sachs, *History of Botany*; Green, *A History of Botany*; Allen, *The Naturalist in Britain*; Brockway, *Science and Colonial Expansion*; Russell, *Science and Social Change*; Young, *Darwin's Metaphor*; A. T. Gage and W. T. Stearn, *A Bicentenary History of the Linnean Society of London* (London: Academic Press, 1988); Bowler, *History of the Environmental Sciences*; Stevens, *The Development of Biological Systematics*; Jardine et al. (eds), *Cultures of Natural History*; Shteir, *Cultivating Women, Cultivating Science*; Lightman (ed.), *Victorian Science in Context*; McCracken, *Gardens of Empire*; Drayton, *Nature's Government*; Merchant, *Reinventing Eden*; Elliott, *The Royal Horticultural Society*; Endersby, *Imperial Nature*; P. A. Elliott, *Enlightenment, Modernity and Science: Geographies of Scientific Culture and Improvement in Georgian England* (London: I. B. Tauris, 2010), pp. 125–66.
3. Loudon, *Encyclopaedia of Gardening* (1830), pp. 1057–9.
4. Watson, *Dendrologia* (1825), vol. 1, p. xvi.
5. Loudon, *Encyclopaedia of Gardening* (1830), p. 1059.
6. Ibid.
7. J. E. Smith and J. Sowerby, *English Botany,* 36 vols (London: J. Davis, 1790).
8. H. C. Andrews, *The Botanist's Repository for New and Rare Plants,* 10 vols (London: T. Bensley, 1797).
9. Watson, *Dendrologia* (1825), vol. 1, p. x.
10. J. Williamson, 'A Narrative of Experiments Made on Trees', Royal Botanic Garden library, Edinburgh, MSS. F39; M. McCoig (attrib.), 'Observations on the Vernal Area', Hope Papers, GD 253/144/12/8; Green, *A History of Botany,* pp. 111–4, 185–8, 276–83, 393–5, 572–6; H. R. Fletcher and W. H. Brown, *The Royal Botanic Garden Edinburgh, 1670–1970* (Edinburgh: HMSO, 1970), pp. 57–67, 80–112, 179–89, 281;

M. V. Mathew, *The History of the Royal Botanic Garden Library Edinburgh* (Edinburgh: HMSO, 1987).

11. Green, *A History of Botany*, pp. 111–4, 185–8, 276–83, 393–5, 572–6; Fletcher and Brown, *Royal Botanic Garden Edinburgh*, pp. 80–112; J. H. Balfour, *Guide to the Royal Botanic Garden, Edinburgh* (Edinburgh: Edmonston & Douglas, 1873); Elliott, *Enlightenment Botany*; 'The Garden at Elvaston Castle'; 'William Barron', obituary, *Gardener's Chronicle* (1891), pp. 522–4; C. Byrom, 'The Pleasure Grounds of Edinburgh New Town', *Garden History*, 23 (1995), pp. 69–90, on pp. 80–2.

12. Loudon, *Encyclopaedia of Gardening* (1830), pp. 86–7; Loudon, *Arboretum Britannicum* (1844), vol. 1, pp. 129–30; N. C. Johnson, 'Cultivating Science and Planting Beauty: The Spaces of Display in Cambridge's Botanical Garden', *Interdisciplinary Science Reviews*, 31 (2006), pp. 42–57; Elliott, *Enlightenment, Modernity and Science*, pp. 125–66.

13. Hull Botanic Garden, *The Addresses of the President and Treasurer at the first General Meeting of the Subscribers to the Hull Botanic Garden; and the Report of the Provisional Committee ... with the Laws of the Institution and a List of the Subscribers* (Hull: Perkins, 1812), p. 7.

14. T. Sherrard, 'A Survey of the Botanic Garden at Glasnevin in the County of Dublin', engraved by J. Taylor, *Transactions of the Dublin Society ... for the Year 1799* (Dublin, 1800); Dublin Botanic Garden, 'Catalogue of Plants in the Dublin Society's Botanic Garden at Glasnevin', *Transactions of the Royal Dublin Society*, 4 (1800) W. Wade, *Plantae Rariores in Hibernia Inventae* (Dublin: Graisbury & Campbell, 1804); E. McCracken and E. C. Nelson, *The Brightest Jewel: History of the National Botanic Gardens, Glasnevin, Dublin* (Kilkenny: Boethius Press, 1987), pp. 21–82.

15. Dublin Botanic Garden, *Catalogue of Plants in the Dublin Society's Botanic Garden at Glasnevin*; Wade, *Plantae Rariores*; Loudon, *Encyclopaedia of Gardening* (1830), pp. 1094–5; Loudon, *Arboretum Britannicum* (1844), vol. 1, pp. 129–30; McCraken and Nelson, *The Brightest Jewel*, pp. 21–82; F. O'Kane, *Landscape Design in Eighteenth-Century Ireland* (Cork: Cork University Press, 2004); F. O'Kane, 'Educating a Sapling Nation: The Irish Nationalist Arboretum', in Elliott et al. (eds), 'Cultural and Historical Geographies of the Arboretum', pp. 185–95; N. Johnson, 'Names, Labels and Planting Regimes: Regulating Trees at Glasnevin Botanic Gardens, Dublin, 1795–1850', in Elliott et al. (eds), 'Cultural and Historical Geographies of the Arboretum', pp. 53–70.

16. G. Jackson, *Hull in the Eighteenth Century: A Study in Economic and Social History* (Hull: Hull Academic Press, 1972), pp. 66–9; H. Calvert, *A History of Kingston-Upon-Hull* (Chichester: Phillimore, 1978); K. J. Allison (ed.), *Victoria History of the County of York: East Riding vol. I: The City of Kingston-Upon-Hull* (Oxford: Oxford University Press, 1969), pp. 174–214; G. Kitteringham, 'Science in Provincial Society: The Case of Liverpool in the Early Nineteenth Century', *Annals of Science*, 39 (1982), pp. 329–448; I. Inkster, 'Scientific Culture and Scientific Education in Liverpool prior to 1812 – A Case Study in the Social History of Education', in I. Inkster, *Scientific Culture and the Urbanisation in Industrialising Britain* (Aldershot: Ashgate, 1997), pp. 28–47; A. Wilson, 'The Cultural Identity of Liverpool, 1790–1850: The Early Learned Societies', *Transactions of the Historic Society of Lancashire and Cheshire*, 147 (1997), pp. 58–73; J. Stobart, 'Culture Versus Commerce: Societies and Spaces for Elites in Eighteenth-Century Liverpool', *Journal of Historical Geography*, 28 (2002), pp. 471–85; Elliott, *Enlightenment, Modernity and Science*, pp. 152–63.

17. W. Roscoe, *An Address Delivered before the Proprietors of the Botanic Garden in Liverpool ... to which are added the Laws of the Institution of a List of the Proprietors* (Liverpool: J

M'Reery, 1802); Loudon, *Encyclopaedia of Gardening* (1830), pp. 1080–1; Liverpool Botanic Garden, *A Catalogue of Plants in the Botanic Garden at Liverpool* (Liverpool: James Smith, 1808), pp. iii–vii.

18. Roscoe, *An Address*, pp. 16–37; H. Roscoe, *Life of William Roscoe*, 2 vols (London: Cadell, 1833), vol. 1, pp. 253–65; Liverpool Botanic Garden, *Catalogue of Plants*, pp. 22–46.

19. Watson, *Dendrologia* (1825), vol. 1, pp. xiii–xiv.

20. Hull Botanic Garden, *The Addresses of the President and Treasurer at the first General Meeting of the Subscribers to the Hull Botanic Garden* (Hull: J. Perkins, 1812); Watson, *Dendrologia*, vol. 1, pp. xii–xvi; W. T. Stearn, 'Adrian Hardy Haworth, 1768–1833', in A. H. Haworth, *Complete Works on Succulent Plants*, 5 vols (London: Gregg Press, 1965), vol. 1, pp. 9–57; Allen, *Naturalist in Britain*, pp. 94–5; *Hull Packet*, 30 June 1812; Loudon, *Encyclopaedia of Gardening* (1830), p. 1079; The botanic garden was moved to a much larger forty-nine-acre site around 1880 near Spring Bank but was sold to the corporation for Hymen's College around 1890; E. Wrigglesworth, *Brown's Illustrated Guide to Hull* (Hull: A. Brown, 1891), p. 134; Allison, *Victoria History*, p. 382.

21. *Hull Packet*, 20 December 1814.

22. D. William, *The Glasgow Guide* (Edinburgh: Canongate Books, 1999), pp. 227–30; Fowler, *Landscapes and Lives*, pp. 111–15.

23. Loudon, *Encyclopaedia of Gardening* (1830), pp. 1089–90.

24. *Bristol Mercury*, 17 October 1835; *Bristol Mercury*, 14 May 1836.

25. T. Appleby, *The Cottage Gardener and Country Gentleman's Companion* (1856) quoted in J. Carder, *The Sheffield Botanical Gardens: A Short History* (Sheffield: Sheffield Council, 1986), pp. 26–7; R. Alison Hunter, *Sheffield Botanical Gardens; People, Plants and Pavilions* (Sheffield: Friends of the Botanical Gardens, 2007), pp. 4–11; R. Harman and J. Minnis, *Sheffield: Pevsner City Guides* (New Haven, CT: Yale University Press, 2004), pp. 262–3.

26. Printed reports of the Garden Committee, numbers 1–5 (1823–7); and Manuscript minutes of the Gardening Committee (1818–30) (M12/01), Royal Horticultural Society, Lindley Library, London; Loudon, *Encyclopaedia of Gardening* (1830), pp. 1059–60; Fowler, *Landscapes and Lives*, pp. 101–7; Elliott, *Royal Horticultural Society*.

27. Loudon, *Encyclopaedia of Gardening* (1830), pp. 1059–60; Loudon, *Arboretum Britannicum* (1844), vol. 1, pp. 129–30; *Gardener's Magazine*, 5 (1830), p. 326, Fig. 79; *Gardener's Magazine*, 6 (1830), p. 250, Fig. 44.

28. A. B. Lambert, *A Description of the Genus Cinchona*, 1st edn (London: Benjamin & John White 1797); see also, A. B. Lambert, *An Illustration of the Genus Cinchona* (London: J. Searle, 1821).

29. H. S. Miller, 'The Herbarium of Aylmer Bourke Lambert', *Taxon*, 19 (1970), pp. 489–553; W. Darlington, *Reliquiae Baldwinianae: Selections from the Correspondence of the late William Baldwin, M.D* (1843; New York: Hafner, 1969); E. L. Little, Jr, 'Lambert's "Description of the *Genus Pinus*", 1832 edition', *Madrono*, 10 (1949), 33–47; H. W. Renkema and J. Ardagh, 'Aylmer Bourke Lambert and his "Description of the Genus Pinus"', *Linnean Society Journal* – Botany, 48 (1930), pp. 439–66. Don has been credited with making the first descriptions of several major conifers including the Coast redwood *Taxodium sempervirens*, now named *Sequoia sempervirens*, and the Grand fir, *Pinus grandis*.

30. Watson, *Dendrologia* (1824), preface.

31. Ibid.

32. Ibid.

33. Watson, *Dendrologia* (1825), vol. 1, p. vii.

34. Ibid., vol. 1, p. v.

35. Watson, *Dendrologia* (1824), vol. 1, p. i.

36. A. B. Lambert, *A Description of the Genus Pinus, Illustrated with Figures,* 2nd edn, 2 vols (London: John Gale, 1828), pp. iii–iv. [Unless otherwise stated, reference to the *Genus Pinus* is taken from the 1832 new edition, printed by Weddell].

37. Watson, *Dendrologia* (1824), vol. 1, p. i.

38. G. S. Boulger, rev. Giles Hudson, 'P. S. Watson', *ODNB*; Loudon, *Arboretum Britannicum* (1844), vol. 1, p. 188.

39. A. Bermingham, *Learning to Draw: Studies in the Cultural History of a Polite and Useful Art* (New Haven, CT: Yale University Press, 2000); Endersby, *Imperial Nature*, pp. 124–31; J. Bonehill and S. Daniels (eds), *Paul Sandby: Picturing Britain* (Nottingham and London: Royal Academy of Arts, 2009), pp. 230–3.

40. Watson, *Dendrologia* (1825), vol. 1, p. viii.

41. Ibid., vol. 1, pp. iv–x, 1–72.

42. Ibid., vol. 1, p. vii.

43. Watson, *Dendrologia* (1824), vol. 1, preface.

44. Loudon, *Arboretum Britannicum* (1844), vol. 1, p. 188.

45. Lambert, *Genus Pinus* (1828), vol. 1, p. vi.

46. A. B. Lambert, *A Description of the Genus Pinus*, 1st edn (London: White, 1803), pp. 64–82.

47. Watson, *Dendrologia* (1825), vol. 1, pp. xv–xvi.

48. Lambert, *Genus Pinus* (1824), vol. 1, p. v.

49. Ibid.

50. Ibid., vol. 2, pp. 33–4.

51. Ibid., vol. 1, pp. 54–5.

52. Lambert, *Genus Pinus* (1828), vol. 1, p. v; *Linnean Transactions*, 15 (1824), p. 498.

53. Lambert, *Genus Pinus* (1824), vol. 1, p. v.

54. Ibid., vol. 1, p. vi.

4 John Claudius Loudon's Arboretum

1. Loudon, 'A Short Account of the Life and Writings', pp. ix–li; J. C. Loudon, *The Derby Arboretum: ... A Catalogue of the Trees and Shrubs* (London: Longman, Ormer, Brown, Green & Longmans, 1840); Chambers, 'Biographical Sketches; John Claudius Loudon', pp. 284–86; L. Fricker, 'John Claudius Loudon, the Plane Truth?', in P. Wallis (ed.), *Furor Hortensis: Essays on the History of the English Landscape in Memory of H. F. Clark* (Edinburgh: Elysium Press, 1974), pp. 76–88; P. Boniface, *The Travels of John Claudius Loudon and his wife Jane* (London: Guild Publishing, 1987); E. B. MacDougall (ed.), *John Claudius Loudon and the Early Nineteenth Century in Great Britain* (Washington, DC: Dumbarton Oaks, 1980); Elliott, *Victorian Gardens*; Simo, *Loudon and the Landscape*; H. Conway, *People's Parks: The Design and Development of Victorian Parks in Britain* (Cambridge: Cambridge University Press, 1991); B. Hartley, 'Sites of Knowledge and Instruction: Arboretums and the *Arboretum et Fruticetum Britannicum*', in Elliott, et al. (eds), *Cultural and Historical Geographies of the Arboretum*, pp. 28–52. Loudon's personal papers, unfortunately, have not survived, having apparently been destroyed during the Second World War.

2. Adamson, *English Education*; Allen, *The Naturalist in Britain*; Russell, *Science and Social Change*; Inkster and Morell (eds), *Metropolis and Provinces*; Young, *Darwin's Metaphor*; Merrill, *The Romance of Victorian Natural History*; Bowler, *History of the Environmental Sciences*; Stevens, *The Development of Biological Systematics*; Jardine et al. (eds), *Cultures of Natural History*; Shteir, *Cultivating Women, Cultivating Science*; Lightman (ed.), *Victorian Science in Context*; Drayton, *Nature's Government*; Merchant, *Reinventing Eden*; Yanni, *Nature's Museums*; Daunton (ed.), *The Organisation of Knowledge*.

3. J. B. Morrell and A. Thackray, *Gentlemen of Science: Early Years of the British Association for the Advancement of Science* (Oxford: Clarendon Press, 1981).

4. Green, *A History of Botany*, p. 379. The main exception is Simo's *Loudon and the Landscape*.

5. Simo, *Loudon and the Landscape*.

6. D. E. Allen, *The Fern Craze: A History of Pteridomania* (London: Hutchinson, 1969), pp. 7–15; Allen, *Naturalist in Britain*, 96–8; Simo, *Loudon and the Landscape*, pp. 14, 132–3, 139–41, 275–6.

7. Sachs, *History of Botany*, pp. 147–54; Green, *A History of Botany*, pp. 336–53; Adamson, *English Education*, pp. 89–95; Stevens, *The Development of Biological Systematics*, pp. 105–7; Drayton, *Nature's Government*, pp. 144–69; L. Brockway, *Science and Colonial Expansion*; Desmond, *Kew*; Shteir, *Cultivating Women*, pp. 149–65; McCracken, *Gardens of Empire*; Elliott, *The Royal Horticultural Society*; J. Endersby 'Classifying Sciences', pp. 61–85; Ritvo, 'Zoological Nomenclature'.

8. Loudon, *Self Instruction for Young Gardeners*.

9. Loudon, *Encyclopaedia of Gardening* (1830), p. 1035; Simo, *Loudon and the Landscape*, pp. 150–1.

10. *Gardener's Magazine*, 2 (1827), pp. 359–60.

11. *Gardener's Magazine*, 9 (1833), pp. 471–2.

12. Ibid., pp. 467–9.

13. J. C. Loudon, *Hortus Britannicus*, 2nd edn (London, 1830), pp. 491–2; J. C. Loudon, *Illustrations of Landscape Gardening and Garden Architecture* (London, 1830), plate II, sheet 4.

14. Loudon, 'A Short Account of the Life and Writings', p. xxiv.

15. Loudon, *Encyclopaedia of Gardening* (1830), p. 807; J. C. Loudon, *Suburban Gardener and Villa Companion* (London; Edinburgh: Longman, Orme, Brown, Green, Longmans, & W. Black, 1838), pp. 276–7.

16. Loudon, *Suburban Gardener*, pp. 276–7.

17. Ibid., p. 278.

18. Ibid., pp. 562–8, 654–61.

19. Ibid., pp. 345–6.

20. Ibid., p. 340.

21. Ibid., pp. 347–9.

22. J. C. Loudon, *Encyclopaedia of Cottage, Farm and Villa Architecture* (London: Longman, Rees, Ormer, Brown, Green, & Longmans, 1833), p. 790.

23. Ibid., p. 790.

24. A. C. Quatremere de Quincy, *An Essay on the Nature, the End and the Means of Imitation in the Fine Arts*, trans. J. C. Kent (London, 1837); Simo, *Loudon in the Landscape*, pp. 172–3.

25. *Gardener's Magazine*, 2 (1827), pp. 237–8.

26. Ibid., pp. 33–5; J. C. Loudon, *Hortus Britannicus*, 1st edn (London, 1826).

27. *Gardener's Magazine*, 6 (1830), pp. 582, 718.
28. Loudon, *Arboretum Britannicum* (1838), vol. 1, p. vii.
29. Loudon, 'A Short Account of the Life and Writings', pp. x, xvii, xix, xxiv, xxx–xxxi, xxxv, xxxviii–xxxix.
30. Loudon, *Arboretum Britannicum* (1844), vol. 1, p. xv; vol. 4, appendix 1. The letters were later passed to Jane Jukes (1791–1873), a Birmingham geologist, and later obtained by the Yale Centre for British Art, New Haven, Connecticut; W. Roberts, 'The Centenary of Loudon's Arboretum', *Journal of the Royal Horticultural Society*, 61 (1936), pp. 277–84; Simo, *Loudon and the Landscape*, pp. 168–9, 318; D. E. Allen, 'Natural History in Britain in the Eighteenth Century', *Archives of Natural History*, 20 (1993), pp. 333–47; V. Jankovic, 'The Place of Nature and the Nature of Place: The Chorographic Challenge to the History of British Provincial Science', *History of Science*, 38 (2000), pp. 79–113.
31. Loudon, *Arboretum Britannicum* (1844), vol. 1, p. viii.
32. Ibid., vol. 1, p. ix.
33. Ibid., vol. 1, pp. 4, 1–191.
34. Ibid., vol. 1, pp. 4, 192–230.
35. Ibid., vol. 1, pp. 239–494; see also vols 2, 3 and 4.
36. Ibid., vol. 1, p. 6.
37. Ibid., vol. 1, pp. 219–30.
38. Ibid., vol. 1, pp. 192–230, p. 202.
39. Ibid., vol. 4, pp. 2293–329.
40. Ibid., vol. 1, p. 222.
41. Ibid., vol. 1, pp. 223–4.
42. Ibid., vol. 1, pp. 223–6.
43. Ibid., vol. 1, pp. 227–9; vol. 3, pp. 1383–90.
44. Ibid., vol. 1, pp. 230; vol. 3, p. 967.
45. Ibid., vol. 1, p. 223; Hartley, 'Sites of Knowledge and Instruction', pp. 42–50.
46. A collection of MSS and other material relating to the Sowerbys and their works is held by the Natural History Museum, London, which includes sixty-seven pencil and watercolour drawings and notes for the *Arboretum Britannicum* (L MSS Sowerby coll/C48 box no. 40 1838); R. J. Cleevely, 'The Sowerbys and their Publications in the light of the Manuscript Material in the British Museum (Natural History)', *Journal of the Society for the Bibliography of Natural History*, 7 (1976), pp. 343–68; J. B. MacDonald, 'The Sowerby Collection in the British Museum (Natural History): A Brief Description of its Holdings and a History of its Acquisition from 1821–1971', *Journal of the Society for the Bibliography of Natural History*, 6:6 (1975), pp. 380–401; G. Meynell, 'The Royal Botanic Society's Garden, Regent's Park', *London Journal*, 6 (1980), pp. 135–46.
47. Loudon, *Arboretum Britannicum* (1844), vol. 1, p. v.
48. Ibid., vol. 1, p. vi.
49. Ibid., vol. 1, pp. xii–xiii.
50. Ibid., vol. 1, preface.
51. Loudon, 'A Short Account of the Life and Writings', pp. xxxvii–xxxix.
52. Ibid., pp. xxxv, xlv–xlvii.

5 The Botany of the *Arboretum Britannicum*

1. Sachs, *History of Botany*, pp. 108–54; Allen, *The Naturalist in Britain: A Social History*; L. Barber, *The Heyday of Natural History* (London: Jonathan Cape, 1980); Russell, *Science and Social Change*; Merrill, *The Romance of Victorian Natural History*, pp. 29–34; Shteir, *Cultivating Women, Cultivating Science*, pp. 149–69; Drayton, *Nature's Government*, pp. 144–8; A. B. Shteir, 'Elegant Recreations? Configuring Science Writing for Women', in Lightman (ed.), *Victorian Science in Context*, pp. 236–43; Yanni, *Nature's Museums*, pp. 14–61; R. Watts, *Women in Science: A Social and Cultural History* (London: Routledge, 2007), pp. 103–6; P. F. Stevens, 'J. D. Hooker, George Bntham, Asa Gray and Ferdinand Mueller on Species Limits in Theory and Practice: A mid-Nineteenth Century Debate and its Repercussions', *Historical Records of Australian Science*, 11 (1997), pp. 345–70; C. Bonneuil, 'The Manufacture of Species: Kew Gardens, the Empire and the Standardisation of Taxonomic Practise in late Ninteenth-Century Botany', in M. N. Bourget, C. Licoppe and O. Sibum (eds), *Instruments, Travel and Science: Itineraries of Precision from the 17ᵗʰ to the 20ᵗʰ Century* (London: Routledge, 2000), pp. 189–215; G. McOuat, 'Cataloguing Power: Delineating Comptetent Naturalists and the Meaning of Species at the British Museum', *British Journal for the History of Science*, 34 (2001), pp. 1–28; M. Winsor, 'Non-essentialist Methods in pre-Darwinian Taxonomy', *Biology and Philosophy*, 18 (2003), pp. 387–400; D. Allen, 'George Bentham's *Handbook for the British Flora*: from Controversy to Cult', *Archives of Natural History*, 30 (2003), pp. 224–36; R. Bellon, 'Joseph Hooker Takes a Fixed Post: Transmutation and the Present Unsatisfactory State of Systematic Botany, 1844–60', *Journal of the History of Biology*, 39 (2006), pp. 1–39; Endersby, *Imperial Nature*.

2. Loudon, 'A Short Account of the Life and Writings', pp. ix–li; Loudon, *The Derby Arboretum*; Chambers, 'Biographical Sketches; John Claudius Loudon'; Fricker, 'John Claudius Loudon, the Plane Truth?', pp. 76–88; Boniface, *The Travels of John Claudius Loudon*; MacDougall (ed.), *John Claudius Loudon*; Simo, 'Review Essay: John Claudius Loudon'; Elliott, *Victorian Gardens* (1986); Simo, *Loudon and the Landscape*; Conway, *People's Parks*; Hartley, 'Sites of Knowledge and Instruction, pp. 28–52.

3. Lindley, *Introduction to the Natural System of Botany* (1830); J. Lindley, *Introduction to the Natural System of Botany*, 2nd edn (London: Longman, Rees, Orme, Brown & Green, 1836); Lindley, *The Vegetable Kingdom*.

4. Loudon, *Arboretum Britannicum* (1844), vol. 1, pp. viii–ix; C. Morris, 'The Diffusion of Useful Knowledge: John Claudius Loudon and his Influence in the Australian Colonies', *Garden History*, 32 (2004), pp. 101–23.

5. Loudon, *Arboretum Britannicum* (1844), vol. 1, pp. viii–ix; for Denson: S. G. Sealy, 'Cuckoos and their fosterers: uncovering details of Edward Blyth's field experiments', *Archives of Natural History*, 36 (2009), pp. 129–35; M. R. D. Seaward, 'William Borrer (1781–1862), father of British Lichenology', *Bryologist*, 105 (2002), pp. 70–7.

6. Loudon, *Arboretum Britannicum* (1844), vol. 1, pp. 211–12.

7. Ibid., vol. 1, p. 212.

8. Ibid., vol. 1, p. 211.

9. Ibid., vol. 1, pp. 7–8.

10. Ibid., vol. 1, p. 8.

11. Ibid., vol. 1, p. 7.

12. Ibid., vol. 1, pp. 8–9.

13. Ibid., vol. 1, p. 212.

14. Ibid., vol. 1, p. 217.
15. Ibid., vol. 1, pp. 212–13.
16. Ibid., vol. 1, pp. 9–10; J. C. Loudon, *Encyclopaedia of Agriculture*, second edition (London, 1844); W. J. Dempster, *Evolutionary Concepts in Nineteenth-Century Britain: Natural Selection and Patrick Matthew* (Edinburgh: Pentland Press, 1996).
17. Lindley, *An Introduction to the Natural System of Botany* (1836), p. 365.
18. Loudon, *Arboretum Britannicum* (1844), vol. 1, p. 213.
19. Ibid., vol. 1, p. 213.
20. Ibid., vol. 1, p. 214; Lindley, *Introduction to Botany* (1836), pp. 306, 366–7; Sachs, *History of Botany*, pp. 108–54; Green, *A History of Botany*, pp. 336–53.
21. Loudon, *Arboretum Britannicum* (1844), vol. 1, p. 215; A. de Candolle, *Physiologie végétale*, 3 vols (Paris: Béchet, 1832), vol. 2, p. 689.
22. Loudon, *Arboretum Britannicum* (1844), vol. 1, p. 215.
23. Ibid., vol. 1, pp. 215–16.
24. Ibid., vol. 1, p. 216.
25. Ibid.
26. Ibid.
27. Ibid., vol. 1, pp. 216–17.
28. Ibid., vol. 1, p. 217; vol. 2, pp. 1213–41.
29. Ibid., vol. 1, pp. 217–18.
30. Ibid., vol. 1, p. 218; vol. 2, pp. 1231–3.
31. Ibid., vol. 1, p. 218.
32. Ibid., vol. 1, pp. 218–19.
33. W. J. Bean, *Trees and Shrubs Hardy in the British Isles*, 7th edn, 3 vols (London: John Murray, 1951), vol. 1, preface, p. vii; Drayton, *Nature's Government*, pp. 221–68.
34. M. T. Masters, 'List of Conifers and Taxads in Cultivation in the Open Air in Great Britain and Ireland', *Journal of the Royal Horticultural Society*, 14 (1892), pp. 1–18; M. T. Masters, 'Notes on the Genera of Taxaceae and Coniferae', *Journal of the Linnean Society*, 30 (1893), pp. 1–42; A. Kent, *Veitch's Manual of Coniferae*, 2nd edn (London: J Veitch & Sons, 1900), pp. 102–5.
35. J. Hoopes, *Book of Evergreens* (New York: Orange Judd & Co., 1868), pp. 239–44; G. Gordon, *The Pinetum: Being a Synopsis of All the Coniferous Plants at Present Known*, 2nd edn (London: Henry G. Bohn, 1875), pp. 414–16; Bean, *Trees and Shrubs*, vol. 3, pp. 303–5.
36. G. Bentham, *Handbook of the British Flora*, 7th edn, rev. J. D. Hooker and A. B. Rendle (Ashford: L. Reeve & Co. Ltd, 1924), pp. xxxiii, xxxv.
37. Bentham, *Handbook*, pp. xxxviii–xl.
38. Ibid., p. xi.
39. Loudon, *Arboretum Britanniucm* (1844), vol. 1, p. 220.
40. Ibid., vol. 1, pp. 220–1; vol. 4, pp. 2159–67.
41. Ibid., vol. 1, p. 221.
42. Ibid., vol. 1, p. 1; J. Browne, *The Secular Ark: Studies in the History of Biogeography* (New Haven, CT: Yale University Press, 1985); C. Lever, *They Dined on Eland: The Story of the Acclimatisation Societies* (London: Quiller Press, 1992); R. H. Grove, *Green Imperialism: Colonial Expansion, Tropical Island Edens and the Origins of Environmentalism, 1600–1860* (Cambridge: Cambridge University Press, 1995).
43. Loudon, *Arboretum Britannicum* (1844), vol. 1, pp. 3–4.
44. Ibid., vol. 1, pp. 4, 15–186.

45. Ibid., vol. 1, p. 5; vol. 3, pp. 1926–49.
46. Ibid., vol. 1, p. 6.
47. Ibid., vol. 1, p. 5.
48. Ibid., vol. 1, p. 226.
49. Ibid., vol. 1, p. v.
50. Ibid., vol. 1, p. vii.
51. Ibid., vol. 1, p. xii.
52. Ibid., vol. 1, pp. 10–11.
53. Ibid., vol. 1, p. 11; Loudon, 'Recollections of a Tour', *Gardener's Magazine*.
54. Loudon, *Arboretum Britannicum* (1844), vol. 1, p. 2.
55. Ibid., vol. 1, p. 2; vol. 2, pp. 748–812.
56. Ibid., vol. 1, p. 9.
57. Ibid., vol. 1, pp. 11–12.
58. Ibid., vol. 1, pp. 12–13.
59. Ibid., vol. 1, pp. 13–14; Gilpin, *Remarks on Forest Scenery*; E. Kennion, *An Essay on Trees in Landscape* (London: C. J. Kennion, 1816); Bonehill and Daniels (eds), *Paul Sandby: Picturing Britain*, pp. 230–3; Ruskin, *Modern Painters*, vol. 1 pp. 379–402; J. Ruskin, *Selections*, pp. 81–90.
60. Loudon, *Arboretum Britannicum* (1844), vol. 1, p. vii.

6 The Derby Arboretum

1.   Not being owned and managed by the council, Derby Arboretum was not fully a municipal park.
2.   Loudon, *The Derby Arboretum*; Chambers, 'Biographical Sketches; John Claudius Loudon'; Fricker, 'John Claudius Loudon, the Plane Truth?', pp. 76–88; Boniface, *Travels of John Claudius Loudon*; MacDougall (ed.), *John Claudius Loudon*; Elliott, *Victorian Gardens* (1986); Simo, *Loudon and the Landscape*; Conway, *People's Parks*; C. Thacker, *The Genius of Gardening: The History of Gardens in Britain* (London: Weidenfeld & Nicolson, 1994); Taylor, 'Urban Public Parks, 1840–1900: Design and Meaning', *Garden History*, 23 (1995), pp. 201–21; Hartley, 'Sites of Knowledge and Instruction', pp. 28–52; see also the website maintained by Chris Harris about the Derby Arboretum at: http://www.derbyarboretum.co.uk/.
3.   P. A. Elliott, *The Derby Philosophers: Science and Urban Culture in British Society c. 1700–1850* (Manchester: Manchester University Press, 2009).
4.   E. Higginson, *Doctrines and Duties of Unitarians* (Lincoln: W. Brooke, 1820), pp. 14–15.
5.   N. Jones, *Life and Death: Discourse on Occasion of the Lamented Death of Joseph Strutt* (London: Green, 1844), pp. 14–16.
6.   J. Farey, *General View of the Agriculture and Minerals of Derbyshire*, 3 vols (London: Sherwood, Neely & Jones, 1811–17), vol. 1, pp. v–xiv, xvii–xxv; vol. 2, pp. 219–340, quotation, p. 239; vol. 3, pp. 685–7; J. C. Loudon, *Encyclopaedia of Agriculture*, 5th edn (London: Longman, Brown, Green, & Longmans, 1844), pp. 1152–4.
7.   J. C. Loudon, *A Short Treatise on Several Improvements Recently made in Hot-Houses* (Edinburgh: for the author, 1805).
8.   Loudon, *Cottage, Farm and Villa Architecture*, p. 1276; A. Ure, 'Official Report ... upon Bernhardt's Stove-Furnaces', *Architectural Magazine*, 5 (1836), pp. 31–42.

9.   W. Strutt, 'Letter to Anne Strutt', 9 May 1823, Strutt Family Correspondence, Local Studies Library, Derby, DLSL, D125/-.

10. G. F. Chadwick, *The Park and Town: Public Landscape in the 19th and 20th Centuries* (London: Architectural Press, 1966), p. 44; S. Lasdun, *English Park* (London: Andre Deutsch, 1991), pp. 119–35.

11. J. V. Beckett et. al. (eds), *Centenary History of Nottingham* (Manchester: Manchester University Press, 1997), pp. 204–11; P. Elliott, C. Watkins and S. Daniels, 'Nottingham Arboretum: Natural History, Leisure and Public Culture in a Victorian Regional Centre', *Urban History*, 35 (2008), pp. 48–71.

12. Loudon, *The Derby Arboretum*, p. 83; Simo, *Loudon and the Landscape*, p. 236; Conway, *People's Parks*, pp. 34–40; Lasdun, *English Park*, pp. 136–52.

13. A. Delves, 'Popular Recreation and Social Conflict in Derby, 1800–1850', in E. Yeo and S. Yeo (eds), *Explorations in the History of Labour and Leisure* (Brighton: Harvester Press, 1981), pp. 89–127, 104; E. Higginson, letter to E. Higginson junior, 26 April 1825, James Martineau papers, Harris Manchester College, Oxford, Oxfordshire; S. Glover, *History of the County of Derby*, 2 vols (Derby: H. Mozley, 1833) vol. 2, part I, p. 406.

14. Simo, *Loudon and the Landscape*, pp. 5–6, 231–3, 247–8.

15. J. C Loudon, 'Recollections of a Tour', *Gardener's Magazine*, 15 (1839), p. 449.

16. *Gardener's Magazine*, 14 (1838), pp. 29, 79; R. S. Fitton and A. P. Wadsworth, *The Strutts and the Arkwrights, 1758–1830* (Manchester: Manchester University Press, 1958).

17. Loudon, *The Derby Arboretum*; *Derby Mercury* (22 January, 9, 16, 23 September 1840).

18. J. C. Loudon, 'Remarks on Laying out Public Gardens and Promenades', *Gardener's Magazine*, 11 (1835), pp. 644–59; Chadwick, *Park and the Town*, p. 57; Simo, *Loudon and the Landscape*.

19. Loudon, 'Remarks on Laying out Public Gardens and Promenades'.

20. Loudon, *Arboretum Britannicum* (1844), vol. 1, pp. 199–200.

21. Ibid., vol. 1, pp. 200–2.

22. Ibid.

23. Chambers, 'Biographical Sketches: John Claudius Loudon', p. 285; Loudon, *The Derby Arboretum*, p. 74; Conway, *People's Parks*, p. 44.

24. Loudon, *Arboretum Britannicum* (1838), vol. 1, pp. xv–xxiii; vol. 4, p. 2609.

25. Loudon, *Trees and Shrubs* (1842); J. Lindley, Review of J. C. Loudon's *Encyclopaedia of Trees and Shrubs*, from the *Gardener's Chronicle*, quoted in the Catalogue of the Encyclopaedias and Dictionaries of Longman, Brown, Green & Longmans (London, 1845), p. xvi.

26. Loudon, *The Derby Arboretum*, p. 7; Simo, *Loudon and the Landscape*, pp. 166–7.

27. J. C. Loudon, 'Remarks on Laying out Public Gardens and Promenades', *Gardener's Magazine*, 2 (1835), p. 650; Conway, *People's Parks*, p. 81.

28. Loudon, *The Derby Arboretum*, pp. 7–8.

29. E. Darwin, *The Botanic Garden: A Poem in Two Parts; The Loves of the Plants and the Economy of Vegetation*, 2 vols (London: J. Johnson, 1791); Elliott, *Enlightenment, Modernity and Science*, pp. 48–76.

30. *Derby Mercury*, 22 April 1835.

31. *Catalogue of the Library belonging to the Derby Philosophical Society*; *Derby Mechanics' Institute: Catalogue of Books* (Derby, 1851).

32. Loudon, *The Derby Arboretum*, pp. 71–3.

33. C. M. Hovey, *Magazine of Horticulture*, 11 (1845), pp. 122–8.

34. Report from the *Gardener's Journal* reproduced in the *Derby Mercury* (28 June, 1848).

35. Simo, *Loudon and the Landscape*, p. 192; Loudon, 'Recollections of a Tour'; Loudon, *The Derby Arboretum*, p. 77.

36. Ibid., pp. 90–5; *Westminster Review*, 35 (January 1841), pp. 422–31.

37. W. Pontey, *Forest Pruner, or Timber Owner's Assistant*, 4th edn (Leeds: T. Inchbold, 1826).

38. Cash book of the Derby Philosophical Society (1813–47), Local Studies Library, Derby (BA 9229–30); Elliott, *Derby Philosophers*.

39. 'Review of J. C. Loudon, *Arboretum Britannicum*', *Quarterly Review*, 144 (1838), pp. 181–96; 'Review of J. C. Loudon, *Arboretum Britannicum*', *Edinburgh Review*, 69 (1839), p. 384; A. Gorrie, 'Review of J. C. Loudon, *Arboretum Britannicum*', *Quarterly Journal of Agriculture*, 9 (1838–9), pp. 55–64; *Edinburgh New Philosophical Journal*, 28 (1839–40); 'Review of J. C. Loudon, *Arboretum Britannicum*', *The Times*, 3 August 1838; J. C. Loudon, 'Opinions of the Public Press' appended to Loudon (ed.), *The Landscape Gardening and Landscape Architecture of the Late Humphry Repton Esquire* (London, 1840), pp. 622–3.

40. *Allgemeine Gartenzeitung*, 6 October 1838, quoted in Loudon, 'Opinions of the Public Press', pp. 622–3; *Biblioteche Universelle de Geneva* 20 (1839), p. 195, quoted in 'Opinions of the Public Press', pp. 622–3.

41. J. Lindley, 'Review of the *Arboretum Britannicum*, *Annals of Natural History*', 3 (May 1839), p. 188–9, on p. 188.

42. Jones, *Discourse on Occasion*, p. 16.

43. Appendix, Part II, to the *Second Report of the Commissioners of Inquiry into the State of Large Towns and Populous Districts* (London: HMSO, 1845), p. 274.

44. J. R. Martin, 'Report on the State of Nottingham, Coventry, Leicester, Derby, Norwich, and Portsmouth', in the *Second Report of the Commissioners for Inquiry into the State of Large Towns and Populous Districts*, 2 vols (London: HMSO, 1845), vol. 2.

45. *Derby Mercury*, 20 August 1851.

46. Simo, *Loudon and the Landscape*, p. 200.

47. *Westminster Review*, 35 (1841), pp. 418–57.

48. E. Chadwick, 'Effects of Public Walks and Gardens on the Health and Morals of the Lower Classes of the Population', in the *Report on the Sanitary Condition of the Labouring Population of Great Britain* (London: HMSO, 1842), pp. 275–6.

49. W. Jerdan, 'Characteristic Letters', *The Leisure Hour* (1 February, 1869), p. 140, quoted in Simo, *Loudon and the Landscape*, p. 201.

50. Hovey, *Magazine of Horticulture*, pp. 122–8, quoted in Simo, *Loudon and the Landscape*, pp. 202–3.

51. A. J. Downing, *Rural Essays* (New York: Leavitt & Allen, 1856), pp. 497–557; D. Schuyler, *Apostle of Taste: Andrew Jackson Downing, 1815–1852* (Baltimore, MD: Johns Hopkins University Press, 1999), pp. 187–203; J. K. Major, *To Live in the New World: Andrew Jackson Downing and American landscape Gardening* (Cambridge, MA: MIT Press, 1997), pp. 137–40; T. Schlereth, 'Early North-American Arboreta', in Elliott et al. (eds), *Cultural and Historical Geographies of the Arboretum*, pp. 196–216.

52. Downing, *Rural Essays*, pp. 516–9.

7 Estate Arboretums

1. Loudon (ed.), *The Landscape Gardening and Landscape Architecture*, pp. 549–50; on collecting in natural history see: J. Elsner and R. Cardinal, *Cultures of Collecting* (Cambridge MA: Harvard University Press); S. C. Parrish, *American Curiosity: Cultures of*

Natural History in the Colonial British Atlantic World (Chapel Hill, NC: University of North Carolina Press, 2006).

2. W. D. Rubinstein, *Men of Property: The Very Wealthy in Britain since the Industrial Revolution* (London: Social Affairs Unit, 1981); D. Cannadine, *The Decline and Fall of the British Aristocracy* (London: Macmillan, 1992).

3. Loudon, *Arboretum Britannicum* (1838), vol. 1, pp. 127–8; B. Hofland and T. C. Hofland, *A Descriptive Account of the Mansion and Gardens of Whiteknights, a Seat of His Grace, the Duke of Marlborough* (London: privately printed, 1819); E. Smith, *A History of Whiteknights* (Reading: University of Reading, 1957); B. V. J. Thompson, *Whiteknights: A History of the University Site* (Reading: University of Reading, 1986); A. Hankin, 'From Aristocrats to Academics: 200 Years in the History of Whiteknights Park', unpublished manuscript history, QR/NW/PG 18, Reading Library; M. Soames, *The Profligate Duke* (London: Collins, 1987).

4. Hofland and Hofland, *A Descriptive Account*, pp. 85–6.

5. Ibid.

6. Ibid., p. 109.

7. Smith, *History of Whiteknights*, pp. 22–3.

8. Ibid., pp. 22–7.

9. Loudon, *Arboretum Britannicum* (1844), vol. 1, pp. 127–8.

10. J. Pearson, *Stags and Serpents: The Story of the House of Cavendish* (London: Macmillan, 1983), p. 132.

11. Pearson, *Stags and Serpents*, pp. 133–4; J. Lees-Milne, *The Bachelor Duke: 6th Duke of Devonshire 1790–1858* (London: John Murray, 1998), pp. 42–3, 164–9; W. Spencer, Sixth Duke of Devonshire, *Handbook of Chatsworth and Hardwicke* (London: privately printed, 1844).

12. S. Piebenga and S. Toomer, 'Westonbirt Arboretum: from Private, Nineteenth-Century Estate Collection to National Arboretum' in Elliott et al. (eds), 'Cultural and Historical Geographies of the Arboretum', pp. 113–28.

13. A. B. Jackson, *Catalogue of the Trees and Shrubs at Westonbirt* (London: Oxford University Press, 1927); M. Woodward, *The Trees at Westonbirt* (London: Westminster Press, 1933); S. Y. Barkley, *Trees of Westonbirt School* (Reading: Bradley & Son, 1952), pp. 5–6; C. Sebag-Montefiore, 'Holford, Robert Stayner (1808–1892)', *ODNB*; D. Bown, *Westonbirt: The Forestry Commission's Finest Arboretum* (Baltonsborough: Julian Holland, 1990); M. Symes, 'Westonbirt Gardens: A Victorian Elysium', *Garden History*, 18 (1990), pp. 155–73; S. Piebenga, 'William Sawry Gilpin (1762–1843): Picturesque Improver', *Garden History*, 22 (1994), pp. 175–96; Nicholas Pearson Associates, *Westonbirt Arboretum and Gardens: Historic Landscape Survey and Restoration Plan* (2003); Piebenga and Toomer, 'Westonbirt Arboretum', pp. 113–28. We are grateful to Sophie Piebenga, Simon Toomer and Sally Day for offering considerable help and information concerning Westonbirt.

14. 'The Garden at Elvaston Castle', *Gardener's and Farmer's Journal*, reproduced, *Derby Mercury*, 19 May 1852; 'William Barron', obituary, *Gardener's Chronicle* (1891), pp. 522–4; Byrom, 'The Pleasure Grounds of Edinburgh New Town', pp. 69–90, on pp. 80–2; Fowler, *Landscapes and Lives*.

15. Girouard, *The Return to Camelot*, pp. 88–9; K. D. Reynolds, 'Charles Stanhope, fourth Earl of Harrington (1780–1851)', J. Knight (K. D. Reynolds, rev.), 'Maria Foote (Stanhope) (1797–1867)', *ODNB*.

16. J. C. Loudon, 'Elvaston Castle: The Seat of the Earl of Harrington', *Gardener's Magazine*, 15 (1839), pp. 458–60; P. Boniface, *In Search of English Gardens: The Travels of John Claudius Loudon and his Wife Jane* (London: Century, 1987), pp. 182–8.

17. S. M. Farrell, 'Charles Stanhope, third Earl of Harrington (1753–1829), *ODNB*; Barron, *The British Winter Garden*; T. Baines, 'Eastnor Castle Ledbury, the Seat of Earl Somers', *Gardener's Chronicle* (December 1876), pp. 807, 838; R. Glendinning, 'Elvaston Castle, the seat of the Earl of Harrington', *Gardener's Chronicle*, 49 (1849), pp. 773, 789, 805, 820; (1850), pp. 4, 21, 36, 50, 53, 69, 84, 100; Loudon, 'Elvaston Castle'; Girouard, *Return to Camelot*, pp. 88–9; Elliott, *Victorian Gardens*, pp. 83–7.

18. Glendinning, 'Elvaston Castle', p. 4.

19. 'The Gardens at Elvaston Castle', *Gardener's and Farmer's Journal* (1852), reproduced in the *Derby Mercury*, 19 May 1852.

20. Glendinning, 'Elvaston Castle', pp. 36–7.

21. W. Barron and Son, *Principal Works Carried out by William Barron and Son Ltd* (Derby: c.1930), p. 6; Select Committee on Forestry, *Report from the Select Committee on Forestry* (London: HMSO, 1887), pp. 1, 9.

22. E. Kemp, *Description of the Gardens at Biddulph Grange* (London: Bradbury & Evans, 1862); Elliott, *Victorian Gardens*, pp. 102–6; P. Hayden, *Biddulph Grange, Staffordshire: A Victorian Garden Rediscovered* (London: George Philip with National Trust, 1989).

23. Select Committee on Forestry, *Report*, pp. 1–11.

24. Symes, 'Westonbirt Gardens', pp. 170–1; Fowler, *Landscapes and Lives*, pp. 109–42; D. Whitehead, 'Veterans in the Arboretum: Planting Exotics at Holme Lacy, Herefordshire, in the late Nineteenth Century', in Elliott et al. (eds), 'Cultural and Historical Geographies of the Arboretum', pp. 96–112.

25. See R. S. Holford's manuscript memorandum book (*c.* 1880–91), pp. 18, 25, 27–8, 35, 37, at the National Arboretum, Westbirton, Tetbury, Gloucestershire. Probably Thomas Henry Sutton Southerton Estcourt (1801–76) of the Estcourt estate, Tetbury, Gloucestershire.

26. Holford, manuscript memorandum book (*c.* 1880–91), pp. 18, 25, 27, 35, 37.

27. See the Diary of Jonah Neale, pp. 56, 62, 81–6, 87–8, at the National Arboretum, Westbirton, Tetbury.

28. Diary of Jonah Neale, pp. 1–4, 7–9, 99–104.

29. Baines, 'Eastnor Castle Ledbury, the Seat of Earl Somers', p. 108.

30. Barron and Son, *Principal Works*; Select Committee on Forestry, *Report*, pp. 3, 6–7; P. Bowe, 'Puckler-Muskau's Estate and its Influence on American Landscape Architecture', *Garden History*, 23 (1995), pp. 192–200.

31. Gordon, *The Pinetum*, pp. 430, 424–5, 429.

32. Loudon, *Arboretum Britannicum* (1844), vol. 1, pp. 130–1.

33. See R. S. Holford's manuscript notebooks held at the National Arboretum, Westbirton, Tetbury, Gloucestershire; Diary of Jonah Neale, pp. 57, 95–6, 99.

34. 'British Gardeners – XXVIII: William Coleman', *Gardener's Chronicle*, 25 October 1875, p. 517.

35. G. Mullins, 'Eastnor Castle: Gardener and Gardens', *Gardener's Chronicle*, 29 August 1903, p. 155.

36. Garden expenditure books kept by W. Coleman (1860–71), Eastnor Castle Library, Herefordshire.

37. J. C. Loudon, 'General Results of a Gardening Tour', *Gardener's Magazine*, 7 (1831), pp. 395–7; J. Paxton, 'Some Account of the Arboretum lately Commenced by His Grace

the Duke of Devonshire at Chatsworth in Derbyshire', *Gardener's Magazine*, 11 (1835), pp. 385–95; 'Paxtonia rosea', *Gardener's Magazine*, 15 (1839), p. 300; Loudon, 'Recollections of a Tour', pp. 450–3; W. Adam, *Gem of the Peak* (London: Longman, 1840), pp. 126–32; G. F. Chadwick, *The Works of Joseph Paxton* (London: Architectural Press, 1963); K. Colquhoun, *A Thing in Disguise: The Visionary Life of Joseph Paxton* (London: Fourth Estate, 2003); J. Barnatt and T. Williamson, *Chatsworth: A Landscape History* (Macclesfield: Windgather Press, 2005), pp. 135–7.

38. Paxton, 'Some Account of the Aboretum'; Loudon, 'Recollections of a Tour', pp. 450–3; Adam, *Gem of the Peak*, pp. 126–32; Chadwick, *The Works of Joseph Paxton*; Colquhoun, *A Thing in Disguise*; Barnatt and Williamson, *Chatsworth*, pp. 135–7.

39. Adam, *Gem of the Peak*, pp. 126–32; Chadwick, *The Works of Joseph Paxton*; Colquhoun, *A Thing in Disguise*; Barnatt and Williamson, *Chatsworth*, pp. 135–7; Hayden, *Biddulph Grange*.

40. Baines, 'Eastnor Castle, Ledbury, the Seat of Earl Somers', p. 76.

41. Glendinning, 'Elvaston Castle', pp. 773, 789.

42. A. Cramb, 'Travelling Notes on Gardens in the Midland Counties', *Gardener's Chronicle*, April 1870, pp. 39–41, 165–8.

43. Baines, 'Eastnor Castle, Ledbury: The seat of Earl Somers', pp. 107–9.

44. Cramb, 'Travelling Notes'.

45. Baines, 'Eastnor Castle, Ledbury: The Seat of Earl Somers', pp. 76–8; A. Barker, 'Eastnor Castle', *Garden* (21 April 1888), pp. 357–8; Mullins, 'Eastnor Castle: Gardener and Gardens', p. 155; Septuagenarian, 'The late William Coleman', 'Obituary: William Coleman', *Gardener's Chronicle*, 29 February 1908, pp. 141, 143; P. Fothergill, *Eastnor Castle Estate: A List of Trees in the Castle Grounds* (Eastnor Castle, Ledbury, 1982).

46. Symes, 'Westonbirt Gardens', p. 164.

47. See J. Neale, 'Westonbirt near Tetbury, Gloucestershire ... proofs selected from twenty seven years practical experience on plants and flowers' and R. S. Holford's notebook, *c.* 1860–91, both held at the National Arboretum, Westonbirt, Tetbury.

48. Symes, 'Westonbirt Gardens', pp. 168–9.

49. Barkley, *Trees of Westonbirt School*, p. 7.

50. C. McIntosh, *Book of the Garden*, 2 vols (Edinburgh, 1853), vol. 1, pp. 584–7.

51. Ibid., vol. 1, pp. 587–90.

52. Ibid.

53. Loudon, *Encyclopaedia of Gardening* (1830), pp. 998–9.

54. Barron and Son, *Principal Works*, pp. 27–46; Elliott, *Victorian Gardens*, pp. 16–20.

55. *Derbyshire Courier*, 15 April 1830; Chadwick, *The Works of Joseph Paxton*, p. 25.

56. H. Steuart, *The Planter's Guide* (Edinburgh: J. Murray, 1828); Loudon, *Encyclopaedia of Gardening* (1830), pp. 291, 402–6; Loudon, *Encyclopaedia of Agriculture* (1844), pp. 642–3.

57. Barron, *The British Winter Garden*, pp. 25–48.

58. Barron and Son, *Principal Works*, pp. 27–46; Elliott, *Victorian Gardens*, p. 250.

59. Barron and Son, *Select Catalogue of Coniferae and other Ornamental Plants ... offered for sale by William Barron and Son* (London and Derby: Bemrose & Sons, 1875), p. 24.

60. Holford, manuscript memorandum book, pp. 17–18.

61. Barron, *The British Winter Garden*, pp. 21–2.

62. Chadwick, *The Works of Sir Joseph Paxton*, pp. 19–25, 72–103; Lees-Milne, *The Bachelor Duke*, pp. 113–5, 130–1; Colquhoun, *A Thing in Disguise*, pp. 80–9, 99.

63. Chadwick, *The Works of Sir Joseph Paxton*, pp. 19–25, 72–103; Lees-Milne, *The Bachelor Duke*, pp. 113–5, 130–1; Colquhoun, *A Thing in Disguise*, pp. 80–9, 99.
64. Cramb, 'Travelling Notes'.
65. H. M. Stephens, 'Cocks, Arthur Herbert (1819–1881)', rev. Katherine Prior, *ODNB*; I. Tyrrell, 'Somerset, Lady Isabella Caroline [Lady Henry Somerset] (1851–1921)', *The Bengal Obituary* (London and Calcutta, 1851), pp. 31, 316.
66. J. Forbes, *Pinetum Woburnense* (London: J. Moyes, 1839); Jackson, *Catalogue of the Trees and Shrubs*; H. Clinton-Baker, *Illustrations of Conifers*, 2 vols (London: Simpson, 1909).
67. Colquhoun, *A Thing in Disguise*, p. 57.

8 Public Urban Arboretums

1. Public Walks, Report from the Select Committee on Public Walks (1833), *Parliamentary Papers*, 15 (Dublin: Irish University Press), pp. 2–11; Simo, *Loudon and the Landscape*, p. 236; Conway, *People's Parks*, pp. 34–40; R. Rodger, *Housing in Urban Britain, 1780–1915* (Basingstoke: Macmillan, 1989); *Local Reports on the Sanitary Condition of the Labouring Population of England* (London, 1842); *Second Report of the Commissioners of Inquiry into the State of Large Towns and Populous Districts*; R. W. Malcolmson, *Popular Recreations in English Society, 1700–1850* (Cambridge: Cambridge University Press, 1973); P. C. Bailey, A Mingled Mass of Perfectly Legitimate Pleasures: The Victorian Middle Class and the Problem of Leisure', *Victorian Studies*, 21 (1977), pp. 7–28; J. M. Golby and A. W. Purdue, *The Civilisation of the Crowd: Popular Culture in England, 1750–1900* (London: Batsford Academic and Educational, 1984), pp. 144–63; B. H. Harrison, 'Religion and Recreation in Nineteenth Century England', *Past and Present*, 38 (1967), pp. 98–125; J. Lowerson, *Sport and the English Middle Classes, 1870–1914* (Manchester: Manchester University Press, 1993), pp. 26–7; Russell, *Science and Social Change*, pp. 54–73.
2. C. Kingsley, *Letters and Memories*, vol. 1, pp. 138–9, quoted in W. Houghton, *The Victorian Frame of Mind, 1830–1870* (New Haven, CT: Yale University Press, 1964), p. 80.
3. A. Briggs, *Victorian Cities* (London: Pelican, 1963); H. J. Dyos and M. Wolff (eds), *The Victorian City: Images and Realities*, 2 vols (London: Routledge & Regan Paul, 1973), vol. 2, pp. 603–738; D. Fraser, *Urban Politics in Victorian England: The Structure of Politics in Victorian Cities* (Leicester: Leicester University Press, 1983); D. Fraser (ed.), *Municipal Reform and the Industrial City* (Leicester: Leicester University Press, 1982); P. Clark (ed.), *The Cambridge Urban History of Britain*, 2 vols (Cambridge: Cambridge University Press, 2000); M. Daunton (ed.), *The Cambridge Urban History of Britian*, 3 vols (Cambridge: Cambridge University Press, 2000), vol. 3, pp. 95–132, 207–28, 315–50, 351–94; T. Hunt, *Building Jerusalem: The Rise and Fall of the Victorian City* (London: Weidenfeld & Nicolson, 2004); Russell, *Science and Social Change*, pp. 154–77; Inkster and Morrell (eds), *Metropolis and Province*; E. Hooper-Greenhill, *Museums and the Shaping of Knowledge* (London; New York: Routledge, 1992); Yanni, *Nature's Museums*; Drayton, *Nature's Government*; J. Rose, *The Intellectual Life of the British Working Classes* (New Haven, CT: Yale University Press, 2002), pp. 58–91; Alberti, *Nature and Culture*.
4. *Derby Mercury*, 16 July 1851.
5. *Derby Mercury*, 1 March 1848.
6. *Nottingham Journal*, 18 August 1843; 19 July 1844; 2 August 1844; 1 August 1845; 31 August 1846; 16 June 1848; 5 July 1850; *Leicester Journal*, 18 August 1843, 14 August

1846, 24 July 1846, 30 July 1847; *Leicester Journal*, 14 August 1846; *Nottingham Journal*, 14 May 1852; *Nottinghamshire Review*, 14 May 1852; *Illustrated London News*, 15 May 1852; W. K. H. White, *White's Nottingham Directory and Borough Register* (Nottingham: White & Co., 1854), pp. 33–5; R. Illiffe and W. Baguley, *Victorian Nottingham: A Story in Pictures, Vol. 10* (Nottingham: Nottingham Historical Film Unit, 1973), pp. 31–49.

7. Chadwick, *The Works of Sir Joseph Paxton*, p. 47; UK database of Historic Parks and Gardens.

8. Elliott et al. 'Nottingham Arboretum'.

9. *Worcester Journal*, 2 May 1857; F. Covins, *The Arboretum Story* (Worcester: Arboretums Residents Association, 1989), p. 22.

10. *Worcester Journal*, 2 May 1857; Covins, *The Arboretum Story*, p. 22; W. Barron, Plan of the Worcester Pleasure Grounds (1864), Worcester Record Office, 372/1, Worcester, Worcestershire; see also in the Worcester Records Office: 5003, 8018/6ii, 8018/30 and 899:2.

11. J. S. Henslow, *Address Delivered in Ipswich Museum on the 9th March 1848* (Ipswich: S. Piper, 1848); L. Jenyns, *Memoir of the Revd. John Stevens Henslow* (London: John van Voorst, 1862); R. Gowing, *Descriptive Handbook of Ipswich* (Ipswich, 1864), pp. 72–4; J. E. Taylor, *Descriptive Handbook of Ipswich and Neighbourhood* (Ipswich: William Hunt, 1873), pp. 34–7; S. M. Walters, *The Shaping of Cambridge Botany* (Cambridge: Cambridge University Press, 1981); R. A. D. Markham, *A Rhino in the High Street: Ipswich Museum – the Early Years* (Ipswich: Ipswich Borough Council, 1990); S. M. Walters and E. A. Stow, *Darwin's Mentor: John Stevens Henslow, 1796–1861* (Cambridge: Cambridge University Press, 2001).

12. Ipswich Borough Council, Council Minute Book: 1 January 1844; 24 April 1844; 23 February 1848; 27 April 1848; 28 May 1848; 9 November 1853 (DA 8 151/3); also see the Minutes of the Estate Committee: 4 March 1881; 1 April 1881; 7 October 1881 (DB11/2), held at the Suffolk County Record Office, Ipswich; W. K. H. White, *History, Directory and Gazetteer of Suffolk* (Sheffield: White & Co., 1855), pp. 88, 130; Gowing, *Descriptive Handbook*, pp. 82–3; Ipswich, *The Visitors and General Guide to Ipswich* (Ipswich, 1869), p. 23; Gowing, *Descriptive Handbook*, p. 39; J. E. Taylor, *Pawsey and Hayes' Illustrated Guide to Ipswich and the Neighbourhood* (Ipswich, 1890), pp. 62–4.

13. See the Corporation Committee minutes of 7 June 1868, held at the Lincolnshire County Archives, L1/1/8/4; Report of the Derby Arboretum committee, minutes of the Borough of Derby, 9 November 1847.

14. Obituary of Samuel Curtis, additional page appended to *Curtis's Botanical Magazine*, 3rd series, vol. 16 (1860); 'Samuel Curtis', *Journal of the Royal Horticultural Society*, 58 (1933); M. Hadfield, R. Harling and L. Highton, *British Gardeners: A Biographical Dictionary* (London: Zwemmer, 1980), pp. 80–7; S. Curtis, *A Monograph on the Genus Camellia* (London: J. and A. Arch, 1819); S. Curtis, *Beauties of Flora: Being a Selection of Flowers Painted from Nature by Eminent Artists* (Gamston, Nottingham, 1820).

15. *Nottingham Journal*, 14 May 1852; *Nottinghamshire Review*, 14 May 1852; *Illustrated London News*, 15 May 1852; White, *White's Nottingham Directory and Borough Register*, pp. 33–5; R. Mellors, *The Gardens, Parks and Walks of Nottingham and District* (Nottingham, 1926), pp. 46–53; R. Illiffe and W. Baguley, *Victorian Nottingham*, pp. 31–49.

16. D. Gray and V. W. Walker (eds), *Records of the Borough of Nottingham, vol. IX, 1836–1900* (Nottingham: T. Forman, 1956).

17. *Nottinghamshire Review*, 14 May 1852.

18. F. Hanham, *A Manual for the Park; or, A Botanical Arrangement and Description of the Trees and Shrubs in the Royal Victoria Park, Bath* (London: Longman, Green, & Longmans, 1857). We are grateful to Peter Borsay and David Hughes for discussing the Royal Victoria Park and referring us to Hanham's book.

19. Hanham, *Manual*, pp. Iii–iv.

20. Ibid., pp. iv–v, xxxvii–xxxviii.

21. Ibid., dedication.

22. Ibid., pp. viii–ix, vi, 251.

23. W. Barron, Plan of the Worcester Pleasure Grounds (1864), Worcester Record Office, 372/1; see also in the record office, 5003, 8018/6ii, 8018/30 and 899:2; *Worcester Journal*, 2 May 1857; Covins, *The Arboretum Story*, p. 22.

24. *Lincolnshire Chronicle*, 23 August 1872, 30 August 1872; W. K. H. White, *History, Gazetteer and Directory of Lincolnshire* (Sheffield: White & Co., 1872 and 1882), pp. 87, 510.

25. *Lincolnshire Chronicle*, 14 August 1895.

26. F. C. Laird, *A Topographical and Historical Description of the County of Worcester* (London: Sherwood, Neely & Jones, George Cowie & Co., 1820).

27. S. H. Cowell, *The Ipswich Hand Book* (Ipswich, 1849), pp. 57–74, 71–2; Gowing, *Descriptive Handbook*, p. 84; R. Malster, *A History of Ipswich* (Chichester: Phillimore, 2000).

28. R. A. Church, *Economic and Social Change in a Midland Town: Victorian Nottingham 1815–1900* (London: Frank Cass, 1966), pp. 183–7; J. V. Beckett and K. Brand, 'Enclosure, Improvement and the Rise of "New Nottingham" 1845–67', *Transactions of the Thoroton Society*, 98 (1995), pp. 92–111.

29. *Nottinghamshire Review*, 14 May 1852; Gray and Walker, 'Report of the Enclosure Committee', 20 February 1845, *Records of the Borough*, pp. 42–4; Gray and Walker, 'Report of the Enclosure Commissioners', 3 February 1846, *Records of the Borough*, 9, p. 50; Gray and Walker, 'Report of the General Enclosure Committee', 21 October 1847, *Records of the Borough*, 9, p. 55; W. H. Wylie, *Old and New Nottingham* (London: Longman, Brown, Green & Longmans, 1853), p. 33.

30. G. A. Walker, *Gatherings from Graveyards, particularly those of London* (London: Longman & Co., 1839); G. A. Walker, *The Graveyards of London* (London: Longman, 1841); G. Collison, *Cemetery Interment: Containing a Concise History of the Modes of Interment* (London: Longman, Orme, Brown, Green & Longmans, 1840), pp. 150–3; E. Chadwick, *A Supplementary Report on the Results of a Special Inquiry into the Practice of Interment in Towns* (London: HMSO, 1843); R. A. Lewis, *Edwin Chadwick and the Public Health Movement 1832–1854* (London: Chivers, 1968); Briggs, *Victorian Cities*; Dyos and Wolff (eds), *The Victorian City*, vol. 2, pp. 603–738; Fraser, *Urban Politics in Victorian England*, pp. 91–114, 154–77; Fraser (ed.), *Municipal Reform and the Industrial City*; A. Wohl, *Endangered Lives: Public Health in Victorian Britain* (London: Dent, 1983); W. F. Bynum and R. Porter (eds), *Living and Dying in London*, Medical History supplement 11 (London: Wellcome Institute for the History of Medicine, 1991); S. Inwood, *A History of London* (London: Macmillan, 1998), pp. 411–43; L. Schwarz, 'London 1700–1840' and B. Trinder, 'Industrialising Towns 1700–1840' in Clark (ed.), *The Cambridge Urban History of Britain*, vol. 2, pp. 641–72, 805–30; R. Dennis, 'Modern London', in Dauton (ed.), *The Cambridge Urban History*, vol. 3, pp. 95–132; B. Luckin, 'Pollution in the City', in Dauton (ed.), *The Cambridge Urban History*, vol. 3, pp. 207–28; R. Millward, 'The Political Economy of Urban Utilities', in

Daunton (ed.), *The Cambridge Urban History*, vol. 3, pp. 315–50; M. Dupree, 'The Provision of Social Services', in Daunton (ed.), *The Cambridge Urban History*, vol. 3, pp. 351–94; T. Hunt, *Building Jerusalem*, pp. 259–312; Rugg, J., 'Lawn Cemeteries: The Emergence of a New Landscape of Death', *Urban History*, 33 (2006), pp. 213–33.

31. J. C. Loudon, *On the Laying Out, Planting and Managing of Cemeteries* (London: for the author, 1843); R. A. Etlin, 'Père Lachaise and the Garden Cemetery', *Journal of Garden History*, 4 (1984), pp. 211–22; J. Bigelow, *A History of the Cemetery of Mount Auburn* (Boston, MA; Cambridge, MA: J. Munroe & Co., 1860); J. D. Hunt and D. Schuyler, 'Garden Cemeteries', special issue of *Garden History* (1984); B. Rotundo, 'Mount Auburn: Fortunate Coincidences and an Ideal Solution', *Journal of Garden History*, 4 (1984), pp. 255–67; J. S. Curl, 'The Design of Early British Cemeteries', *Journal of Garden History*, 4 (1984), pp. 223–54, p. 226; K. Worple, *Lost Landscapes: Death and the Architecture of the Cemetery in the West* (London: Reaktion Books, 2003); O. Jones, 'Arnos Vale Cemetery and the Lively Materialities of Trees in Place', in Elliott et al. (eds), 'Cultural and Historical Geographies of the Arboretum', pp. 149–71.

32. J. S. Curl, *The Victorian Celebration of Death*, new edition (Stroud: Alan Sutton, 2000); P. Jalland, *Death in the Victorian Family* (1996; Oxford; New York: Oxford University Press, 1999); Worple, *Lost Landscapes*; C. Arnold, *Necropolis: London and its Dead* (London: Pocket Books, 2007); J. Litten, *The English Way of Death: The Common Funeral Since 1450* (1991; London: Robert Hale, 2002); A. Kellehear, *A Social History of Dying* (Cambridge: Cambridge University Press, 2007); N. Llewellyn, *The Art of Death: Visual Culture in the English Death Ritual, c. 1500–1800* (London: Reaktion, 1991); M. E. Hotz, *Literary Remains: Representations of Death and Burial in Victorian England* (New York: State University of New York, 2008).

33. *Horticulturist and Journal of Rural Art and Rural Taste*, 2 (1847–8), pp. 266–71.

34. J. Paxton, 'Remarks on Arboretums', *Paxton's Magazine of Botany and Register of Flowering Plants*, 8 (1841), pp. 41–2; V. Markham, *Paxton and the Bachelor Duke* (London: Hodder & Stoughton, 1935), pp. 40–4, 335–40; Colquhoun, *A Thing in Disguise*, pp. 60–2.

35. Paxton, 'Remarks on Arboretums'; Markham, *Paxton and the Bachelor Duke*, pp. 40–4, 335–40; Colquhoun, *A Thing in Disguise*, pp. 60–2.

36. C. H. J. Smith, *Landscape Gardening: or, Parks and Pleasure Grounds* (New York: for the author, 1852), p. 303.

37. Ibid., pp. 303–6.

38. Ibid., pp. 321–32.

39. Ibid.

40. Ibid., pp. 333–63.

41. McIntosh, *The Book of the Garden*, vol. 1, pp. 584–7.

42. Ibid., vol. 1, pp. 587–90.

43. Ibid.

44. Ibid., vol. 1, pp. 584–7.

45. Ibid.; Elliott, *Victorian Gardens*, pp. 35–7.

46. J. Anderson (ed.), *The New Practical Gardener and Modern Horticulturist* (London: William Mackenzie, 1872), pp. 756–7.

47. Anderson (ed.), *New Practical Gardener*, pp. 756–7.

48. Smith, *Records*, pp. i–xiii, 258–68, quotations p. viii and p. 270; Elliott, *Victorian Gardens*, pp. 71–5, 138–40; Brockway, *Science and Colonial Expansion*, pp. 1–8; Desmond, *Kew*, pp. 39–40, 143–7; Drayton, *Nature's Government*, pp. 152–92.

49. W. Nesfield, 'Observations on the Proposed Arboretum in Conjunction with the Royal Botanic Gardens at Kew', Royal Botanic Gardens, Kew Archives, London; W. A. Nesfield, 'Sketch Plan of the … Pleasure Ground and … National Arboretum, Kew' (1845), Royal Botanic Gardens, Kew Archives, London; Smith, *Records*, pp. 268–72; Elliott, *Victorian Gardens*, pp.71–5; Desmond, *Kew*, p. 169–89.

50. W. Robinson, *The Wild Garden*, 4th edn (London: John Murray, 1894), p. 211; W. Robinson, *English Flower Garden*, 8th edn (London: John Murray, 1900), pp. 718–20; B. Elliott, 'From Arboretum to the Woodland Garden', in Elliott et al. (eds), 'Cultural and Historical Geographies of the Arboretum', pp. 71–83.

51. G. Jekyll, *Wood and Garden* (London: Longmans, Green & Co., 1898); J. Brown, *Gardens of a Golden Afternoon: Edwin Lutyens and Gertrude Jekyll* (Allen Lane: London, 1982), pp. 50–3; Jekyll also assisted E. T. Cooke in preparing the influential *Trees and Shrubs for English Gardens* (London: Country Life, 1902).

52. J. Secord, 'Monsters at the Crystal Palace', in S. De Chadarevian and N. Hopwood (eds), *Models: The Third Dimension of Science* (New Jersey: Stanford University Press, 2004), pp. 138–69; Alberti, *Nature and Culture*; Elliott, *The Derby Philosophers*.

53. Appendix, Part II to the *Second Report of the Commissioners of Inquiry into the State of Large Towns and Populous Districts*, p. 274; Martin, 'Report on the State of Nottingham, Coventry, Leicester, Derby, Norwich, and Portsmouth'; Chadwick, 'Effects of Public Walks and Gardens', pp. 275–6; *Illustrated London News*, 15 July 1843; Jerdan, 'Characteristic Letters'.

54. Hovey, *Magazine of Horticulture*, pp. 122–8, quoted in Simo, *Loudon and the Landscape*, pp. 202–3; Downing, *Rural Essays*, pp. 515–16; F. L., Olmsted, *The Papers of Frederick Law Olmsted: Vol. 3, Creating Central Park 1857–1861*, ed. C. E. Beveridge and D. Schuyler (Baltimore, MD: Johns Hopkins University Press, 1983), p. 359; W. Rybczynski, *A Clearing in the Distance: Frederick Law Olmsted and America in the Nineteenth Century*, 1st edn (New York: Scribner, 2003), pp. 180–2.

9 The Transformation of Victorian Public Arboretums

1. *Walsall Observer*, 9 May 1874; Foster, 'History of the Walsall Arboretum', typescript, held at the Walsall Library; Foster, 'The Development of Public Parks in the West Midlands' (dissertation thesis, Walsall Library), pp. 34–42; A. French, 'The Arboretum' (1989) typescript history, held at the Walsall Library.

2. *Nottingham Journal*, 14 May 1852); *Illustrated London News*, 15 May 1852); Elliott et al., 'The Nottingham Arboretum'; *Lincolnshire Chronicle*, 23 August 1872, 30 August 1872.

3. *The Mail*, 31 August 1872; *Ward's Historical Guide to Lincoln*; *Lincolnshire Chronicle*, 23 August 1872, 30 August 1872; White, *History, Gazetteer and Directory of Lincolnshire*, p. 87 and p. 510.

4. Nottingham Council General Minutes, 9 April 1857, Nottinghamshire County Record Office, Nottingham; Gray and Walker, *Records of the Borough of Nottingham*, p. 120; Elliott et al., 'Nottingham Arboretum'.

5. Gray and Walker, *Records of the Borough*, see entries of 28 April 1857 and 28 May 1857, p. 120; see also entries of 8 June 1857, 30 June 1857, pp. 121, 124.

6.   Gray and Walker, 'Regulations for the Nottingham Arboretum', 15 March 1852, in *Records of the Borough*, p. 89.
7.   Elliott et al. (eds), 'Nottingham Arboretum'.
8.   Elliott, 'Derby Arboretum'; see Loudon, *The Derby Arboretum*, p. 84, where Strutt emphasizes as his first condition that it was to be for 'all classes of the public', and the annual report of the arboretum committee, 1854.
9.   T. Wyborn, 'Parks for the People: The Development of Public Parks in Victorian Manchester', *Manchester Region History Review*, 15 (1995), pp. 3–14.
10. *Westminster Review*, 35 (January 1841), p. 430.
11. *Derby Mercury*, 23 September 1840.
12. Ibid., 23 September 1840.
13. Ibid., 15 March 1854.
14. Report of the Arboretum committee, 9 November 1854.
15. *Derby Mercury*, 7 May 1884.
16. For an analysis of the increase in the power and authority of town councils often through local acts such as those of Leeds (1842), Manchester (1844), Liverpool (1846) and Birmingham (1851), see D. Fraser, *Power and Authority in the Victorian City* (Oxford: Oxford University Press, 1979); D. Eastwood, *Government and Community in the English Provinces, 1700–1870* (Basingstoke: Macmillan, 1997), ch. 6.
17. *Derby and Chesterfield Reporter*, 6 May 1881, 11 November 1881; *Derby Mercury*, 8 November 1882.
18. Gray and Walker, *Records of the Borough*, see entry of 21 March 1859, p. 135; see also entry of 9 May 1859, pp. 135–6.
19. *Nottingham Journal*, 14 May 1852.
20. Ibid., 14 July 1852, 17 August 1853, 14 August 1856, 29 June 1855, 4 June 1860.
21. Gowing, *Descriptive Handbook*, pp. 81–2.
22. Loudon, *The Derby Arboretum*, pp. 80–1.
23. *Derby Mercury*, 5 February 1845.
24. Ibid.
25. Report of the Arboretum management committee, minutes of the Borough of Derby, 9 November 1854.
26. *Walsall Observer*, 9 May 1874; Foster, 'History of the Walsall Arboretum'; Foster, 'Development of Public Parks in the West Midlands', pp. 34–42 (dissertation thesis, Walsall Library); French, 'The Arboretum'.
27. Foster, 'History of the Walsall Arboretum', appendix 7, pp. 161–5.
28. Fletcher and Brown, *The Royal Botanic Garden Edinburgh*, pp. 179–89, 281.
29. Ibid., pp. 179–91, 195, 281, 292–5; Mathew, *The History of the Royal Botanic Garden Library*, pp. 46–9, 60–1, 64.
30. Fletcher and Brown, *Royal Botanic Garden*, pp. 199–203; *The Scotsman* (6 April 1889) quoted in Fletcher and Brown, *Royal Botanic Garden*, p. 203; Mathew, *The History of the Royal Botanic Garden Library*, pp. 64–5, 110–11; Green, *A History of Botany*, pp. 572–7.
31. Fletcher and Brown, *Royal Botanic Garden*, pp. 203–5, 215, 280–1 and plan II; Green, *A History of Botany*, pp. 575–7.
32. Nicholson (ed.), *The Illustrated Dictionary of Gardening*, vol. 3, pp. 445–6; W. Boehm, 'Julius A. Stockhardt (1809–1886) – Pioneer of Agricultural Experimental Stations', *Agricultural Research*, 39 (1986), pp. 1–7.
33. Nicholson, *Illustrated Dictionary of Gardening*, vol. 4 supplement, pp. 450–7.

34. W. Baker, 'On the Sanitary Condition of the Town of Derby', *Local Reports on the Sanitary Condition of the Labouring Population of England* (London: HMSO 1842), pp. 163–4.

35. Report of the Arboretum management committee, minutes of the Borough of Derby, 9 November 1861.

36. *Derby Mercury*, 8 November, 1882.

37. Report of the Arboretum management committee, 9 November 1843.

38. *Derby Mercury*, 7 April 1857, 16 September, 1857; M. Craven, *Illustrated History of Derby* (Derby: Breedon, 1988), pp. 152, 164.

39. *Modern Mayors of Derby* (Derby: Hobson & Son, 1909); Cash book of the Derby Philosophical Society, 1813–1847, Local Studies Library, Derby; T. Bulmer, *History, Topography, and Directory of Derbyshire* (Preston: Thomas Bulmer & Co., 1895), p. 876.

40. *Derby Mercury*, 16 June 1875.

41. Ibid., 7 August 1878.

42. Ibid., 8 November 1882.

43. Minutes of the Arboretum committee, 17 November 1882.

44. P. Ballard, *An Oasis of Delight: The History of the Birmingham Botanical Gardens*, 2nd edn (1983; Studley Books: Brewin, 2003), pp. 37–58; Elliott, *The Derby Philosophers*, pp. 261–5; Alberti, *Nature and Culture*.

Conclusion

1. E. Waugh, *Unconditional Surrender* (1961; London: Penguin, 1971), p. 91.

2. J. Prest, *The Garden of Eden: The Botanic Garden and the Re-Creation of Paradise* (New Haven, CT: Yale University Press, 1981); Merchant, *Reinventing Eden*.

3. J. Grigor, *The Eastern Arboretum, or Register of Remarkable Trees ... in the County of Norfolk* (London: Longman, Brown, Green & Longmans, 1841); E. J. Ravenscroft, *The Pinetum Britannicum*, 3 vols (Edinburgh: W. Blackwood & Sons, 1863–84); Elwes and Henry, *The Trees of Great Britain*; Bean, *Trees and Shrubs*.

4. Forbes, *Pinetum Woburnense*; A. B. Jackson, *Catalogue of the Trees and Shrubs in the Collection of the late ... Sir George Lindsey Holford* (London and Oxford, 1927); H. Clinton-Baker, *Illustrations of Conifers*.

5. H. Veitch, *Veitch's Manual of the Coniferae*, 2nd edn (London: Pollet, 1900); Sargent, *The Silva of North America* (1891–1902); Sargent, *Manual of the Trees of North America*.

6. Letters of F. L. Olmsted to Sir W. Hooker, 29 November 1859 and to the Board of Commissioners of the Central Park, 28 December 1859, in Beveridge and Schuyler (eds), *The Papers of Frederic Law Olmsted*, pp. 232–5; Elliott, *Victorian Gardens*, pp. 115–17.

7. A. J. Downing, *Treatise on the Theory and Practice of Landscape Gardening*, ed. by H. W. Sargent, 5th edn (New York: A. O. Moore & Co., 1859), pp. 437–54; I. Hay, *Science in the Pleasure Ground: A History of the Arnold Arboretum* (Cambridge, MA: Harvard University Press, 1995), pp. 65–8.

8. Schlereth, 'Early North-American Arboreta'; M. Bourke, 'Trees on Trial: Economic Arboreta in Australia', in Elliott et al. (eds), 'Cultural and Historical Geographies of the Arboretum', pp. 217–26.

9. Ruskin, *Modern Painters*, quoted in Ruskin, *Selections*, pp. 79–81; W. G. Collingwood, *The Life of John Ruskin*, 5th edn (London: Methuen, 1905), pp. 61–5, 180–5; J. Dixon Hunt, *The Wider Sea: A Life of John Ruskin* (London: Dent, 1982), pp. 89–96, 129–38.

WORKS CITED

Manuscript Sources

'Abney Park Cemetery Conservation Management Plan', Abney Park Cemetery Trust (London, 1994).

Birmingham Central Library, Archives, West Midlands

Galton Papers (MS3101).

Rules and regulations and catalogues of the Birmingham Botanical and Horticultural Society established in 1829, 1839, 1893 (6685/32, 33, 36).

Cambridge University Library, Cambridge, Archives of the Cambridge Botanic Garden (GBR/0265/BG).

Derbyshire Record Office, Matlock, Strutt family papers (D1564, D2912, D2943M, D3772, D5303).

Eastnor Castle Library, Herefordshire

Eastnor Estate coppice sale records, Eastnor Estate Muniments (1933).

Garden expenditure books kept by W. Coleman (1860–71).

Fitzwilliam Museum, Cambridge, Strutt papers (MSS 48 – 1947).

Glasnevin Botanical Gardens Library, Dublin, printed and manuscript material on the history of the Botanical Garden.

Hankin, A., 'From Aristocrats to Academics: 200 Years in the History of Whiteknights Park', unpublished manuscript history, Local Studies Collection (QR/NW/PG 18), Reading Library, Reading.

Harris Manchester College, Oxford, Oxfordshire, James Martineau papers.

Ipswich Town Library, Suffolk, Collection of printed material concerning the history of parks in the town.

Lambert, A. B., *A Description of the Genus Pinus*, Arnold Arboretum Library [a copy is available on Google Books].

Lincoln Public Library, Collection of newspaper clippings, photographs and other material concerning Lincoln Arboretum.

Lincolnshire County Archives, Lincoln, Lincolnshire

Lincoln Council minutes, 1867–73 (L1/1/4/3–4).

Corporation Committee minutes, 1863–83 (L1/1/8/2–13).

Lincoln Arboretum Committee, 1883–1902 (L1/1/19/1–3).

Linnean Society Library, London, J. E. Smith Correspondence.

Local Studies Library, Derby

Derby Corporation minute books.

Derby Corporation Committee minute books, esp. the Derby Arboretum Committee.

Cash book of the Derby Philosophical Society, 1813–47 (BA 9229–30).

Strutt Family Correspondence (D125/-).

Derby Temperance Society, Annual Reports, 1852–76 (BA 178).

Photographs, newspaper clippings and other material concerning the Derby Arboretum.

National Arboretum, Westonbirt, Tetbury, Gloucestershire

Collection of printed and manuscript material concerning the history of Westonbirt.

Holford, R. S., manuscript memorandum book (*c.* 1860–91).

—, manuscript notebooks.

Neale, J., 'Westonbirt near Tetbury, Gloucestershire ... proofs selected from Twenty seven years practical experience on plants and flowers'.

—, Diary of Jonah Neale.

Nicholas Pearson Associates, *Westonbirt Arboretum and Gardens: Historic Landscape Survey and Restoration Plan* (2003).

Natural History Museum, London, manuscript drawings and notes for the *Arboretum Britannicum*, Sowerby Collection (L MSS Sowerby coll/C48 box no. 40 1838).

Nottinghamshire County Record Office, Nottingham, Nottinghamshire

Nottingham Council General minutes, 1844–54 (CA/3604–13).

Enclosure Commissioner's minutes, 1845–67 (CA/7709–32).

General Enclosure Committee minutes, 1850–1 (CA/CM/60/1).

Enclosure Committee minute book, 1853–61 (CA/CM/51/1).

Royal Botanic Garden Edinburgh, Library and Archives, Midlothian

Manuscript and printed material concerning Royal Botanic Garden history including copies of the Hope Papers.

Williamson, J., 'A Narrative of Experiments Made on Trees' (MSS. F39).

McCoig, M. (attrib.), 'Observations on the Vernal Area' (Hope Papers, GD 253/144/12/8).

Royal Botanic Gardens, Kew, Library and Archives, London

Printed and manuscript material concerning the history of the Royal Botanicl Gardens including 'Sir W. Hooker Letters' (1845–51) correspondence between Hooker, D. Burton, W. A. Nesfield and others (vols 22, 23, 26).

W. A. Nesfield, 'Sketch Plan of the Pleasure Ground and ... National Arboretum, Kew' (1845).

—, 'Observations on the Proposed Arboretum in Conjunction with the Royal Botanic Gardens at Kew'.

Royal Horticultural Society, Lindley Library, London

    Printed reports of the Garden Committee, numbers 1–5 (1823–7).

    Manuscript minutes of the Garden Committee, vols I–VI (1818–30) (M12/01).

Suffolk County Record Office, Ipswich, Suffolk

    Ipswich Borough Council, Council Minute Books (DA 8/151/2–3).

    Minutes of the Estate Committee, Ipswich (DB11/1–22).

    Public Health Committee, 1845–76 (DB 24–).

Walsall Library, West Midlands

    Foster, 'History of the Walsall Arboretum', typescript.

    Foster, 'The Development of Public Parks in the West Midlands' (dissertation thesis, Walsall Library).

    French, A., 'The Arboretum' (1989), typescript history.

Worcestershire Record Office, Worcester, Worcestershire, printed and manuscript material relating to the Worcester Pleasure Grounds Company including a plan of the Worcester Pleasure Grounds (1864) by W. Barron (372/1, 5003, 8018/6ii, 8018/30 and 899:2).

Periodicals

Athenaeum

Bristol Mercury

Cottage Gardener and Country Gentleman's Companion

Derby and Chesterfield Reporter

Derby Mercury

Edinburgh New Philosophical Journal

Edinburgh Review

Gardener's Chronicle

Gardener's Magazine

Horticulturist and Journal of Rural Art and Rural Taste

Hull Packet

Illustrated London News

Journal of the Royal Horticultural Society

Leicester Journal

Lincolnshire Chronicle

Literary World

Magazine of Horticulture

Magazine of Natural History

Morning Advertiser

Nottingham Journal

Nottinghamshire Review

Penny Magazine

Quarterly Journal of Agriculture

Quarterly Review

Stamford Mercury

The Garden

The Gardener's and Farmer's Journal

The Mail

The Times

Walsall Observer

Westminster Review

Worcester Journal

Primary Material

Ablett, W. H., *English Trees and Tree Planting* (London: Smith, Elder & Co., 1880).

Adam, W., *Gem of the Peak* (London: Longman, 1840).

Aiton, W., *Hortus Kewensis*, 3 vols (London: G. Nicol, 1789).

Alcock, R., 'Narrative of a Journey in the Interior of Japan, Ascent of Fusiyama, and Visit to the Hot Sulphur Baths of Aatami in 1860', read on 13 May 1861, *Journal of the Royal Geographical Society*, 31 (1861), pp. 321–55.

—, *The Capital of the Tycoon*, 2 vols (London: Longman, 1863).

Anderson, J. (ed.), *The New Practical Gardener and Modern Horticulturist* (London: William Mackenzie, 1872).

Andrews, H. C., *The Botanist's Repository for New and Rare Plants*, 10 vols (London: T. Bensley, 1797).

Anon., *English Forests and Forest Trees, Historical, Legendary and Descriptive* (London: Ingram, Cooke, 1853).

Baines, T., 'Eastnor Castle, Ledbury, the Seat of Earl Somers', *The Gardeners' Chronicle*, 19 January 1878, pp. 76–8, 107–9.

—, 'Elvaston Castle', *The Gardener's Chronicle* (December 1876), pp. 807, 838.

Baker, W., 'On the Sanitary Condition of the Town of Derby', *Local Reports on the Sanitary Condition of the Labouring Population of England* (London: HMSO, 1842).

Balfour, J. H., *Guide to the Royal Botanic Garden, Edinburgh* (Edinburgh: Edmonston & Douglas, 1873).

Barker, A., 'Eastnor Castle', *Garden*, 21 April 1888, pp. 357–8.

Barker, T. B., *Abney Park Cemetery: A Complete Descriptive Guide to Every Part of this Beautiful Depository of the Dead* (London: Houlston & Wright, 1869).

Barron, W., *The British Winter Garden* (London: Bradbury & Evans, 1852).

Barron, W., and Son, *Select Catalogue of Coniferae and other Ornamental Plants ... Offered for Sale by William Barron and Son* (London and Derby: Bemrose & Sons, 1875).

Bigelow, J., *A History of the Cemetery of Mount Auburn* (Boston, MA; Cambridge, MA: J. Munroe & Co., 1858).

Black's Picturesque Tourist of Scotland (Edinburgh: Adam & Charles Black, 1881).

Blair, R., *The Grave: A Poem by Robert Blair to which is added Gray's Celebrated Elegy written in a Country Churchyard*, ed. by G. Wright (London, 1786).

Bowman, J. E., 'On the Longevity of the Yew, as Ascertained from Actual Section of its Trunk, and on the Origin of its Frequent Occurrence in Churchyards', *Magazine of Natural History*, 2nd series, 1 (1836), pp. 28–35, 85–90.

Bree, W. T., 'Some Account of an Aged Yew Tree in Buckland Churchyard near Dover', *Magazine of Natural History*, 2nd series, 6 (1841), p. 47–51.

Brown, J., *The Forester*, 4th edn (Edinburgh: Blackwood, 1871).

Brown, R., *Prodromus Florae Novae Hollandiae et Insulae Van Diemen* (London: Johnson & Co., 1810).

Bulmer, T., *History, Topography, and Directory of Derbyshire* (Preston: Thomas Bulmer & Co., 1895).

Candolle, A. P. de, *Théorie élémentaire de la botanique*, 2 vols (Paris: Deterville, 1813).

—, *Physiologie végétale*, 3 vols (Paris: Béchet, 1832).

Carter, J., *A Visit to Sherwood Forest* (London: Longman & Co., 1850).

Catalogue of the Library belonging to the Derby Philosophical Society; Derby Mechanics' Institute: Catalogue of Books (Derby, 1851).

J. J. Cartwright (ed.), *The Travels through England of Dr. Richard Pococke Successively Bishop of Meath and of Ossory, during 1750, 1751, and Later Years*, Camden Society, New Series 44 (London: Camden Society, 1889).

Catesby, M., *Hortus Britanno-Americanus* (London: J. Ryall, 1763).

Chadwick, E., 'Effects of Public Walks and Gardens on the Health and Morals of the Lower Classes of the Population', in the *Report on the Sanitary Condition of the Labouring Population of Great Britain* (London: HMSO, 1842), pp. 275–6.

—, *A Supplementary Report on the Results of a Special Inquiry into the Practice of Interment in Towns* (London: HMSO, 1843).

Chambers, W., 'Biographical Sketches; John Claudius Loudon', *Chambers Edinburgh Journal* (May 1844), I, pp. 284–6.

Collison, G., *Cemetery Interment: Containing a Concise History of the Modes of Interment* (London: Longman, Orme, Brown, Green & Longmans, 1840).

Cowell, S. H., *The Ipswich Hand Book* (Ipswich, 1849).

Cramb, A., 'Travelling Notes on Gardens in the Midland Counties', *Gardener's Chronicle*, April 1870, pp. 39–41, 165–8.

Curtis, S., *A Monograph on the Genus Camellia* (London: J. & A. Arch, 1819).

—, *Beauties of Flora: Being a Selection of Flowers Painted from Nature by Eminent Artists* (Gamston, Nottingham, 1820).

Darlington, W., *Reliquiae Baldwinianae: Selections from the Correspondence of the Late William Baldwin, M.D* (1843; New York: Hafner, 1969).

Darwin, E., *The Botanic Garden: A Poem in Two Parts; The Loves of the Plants and the Economy of Vegetation*, 2 vols (London: J. Johnson, 1791).

—, *Phytologia: or the Philosophy of Agriculture and Gardening* (London: J. Johnson, 1800).

Decandolle, A. P., and K., Sprengel, *Elements of the Philosophy of Plants*, trans. (Edinburgh: William Blackwood, 1821).

Don, G., *A General System of Gardening and Botany*, 4 vols (London: C. J. G. & F. Rivington, 1831–7).

Downing, A. J., *Rural Essays* (New York: Leavitt & Allen, 1856).

—, *Treatise on the Theory and Practice of Landscape Gardening*, ed. by H. W. Sargent, 5th edn (New York: A. O. Moore & Co., 1859).

Dublin Botanic Garden, 'A Catalogue of Plants in the Dublin Society's Botanic Garden at Glasnevin', *Transactions of the Royal Dublin Society*, 4 (1800).

Duhamel du Monceau, H. L., *Traité des arbres et arbustes qui se cultivent en France*, 2 vols (Paris: H. L. Guerin & L. F. Delatour, 1755).

—, *La physique des arbres, où il est traité de l'anatomie des plantes et de l' économie végétale*, 2 vols (Paris: H. L. Guerin & L. F. Delatour, 1758).

Eddison, E., *History of Worksop; with Historical, Descriptive and Discursive Sketches of Sherwood Forest* (London: Longman, 1854).

Evelyn, J., *Silva: or a Discourse of Forest Trees and the Propagation of Timber in His Majesties Dominions* (London: J. Martyn & J. Allestry, 1664).

—, *Silva: or, A Dictionary of Forest Trees*, with notes by A. Hunter, 2 vols (London: Henry Colburn, 1825).

Farey, J., *General View of the Agriculture and Minerals of Derbyshire*, 3 vols (London: Sherwood, Neely & Jones, 1811–17).

Forbes, J., *Hortus Woburnensis* (London: Ridgway, 1833).

—, *Pinetum Woburnense* (London: J. Moyes, 1839).

French, J. B., *Walks in Abney Park Illustrated with Life Photographs of Ministers and other Public Men ...* (London: James Clarke & Co., 1883).

Gilpin, W., *Remarks on Forest Scenery, and other Woodland Views* (London: R. Blamire, 1791).

—, *Remarks upon Forest Scenery*, 2 vols (Edinburgh: Fraser, 1834).

Glendinning, R., 'Elvaston Castle: The Seat of the Earl of Harrington', *Gardener's Chronicle*, 49 (1849), pp. 773, 789, 805, 820; (1850), pp. 50, 4, 21, 36, 53, 69, 84, 100.

Glover, S., *History and Gazetteer and Directory of the County of Derby*, 2 vols (Derby: H. Mozley, 1833).

Gordon, G., *The Pinetum: Being a Synopsis of All the Coniferous Plants at Present Known*, 2nd edn (London: Henry G. Bohn, 1875).

Gorham, G. C., *Memoirs of John Martyn FRS and Thomas Martyn, BD, FRS, FLS, Professors of Botany in the University of Cambridge* (London: Hatchard & Son, 1830).

Gowing, R., *Descriptive Handbook of Ipswich* (Ipswich, 1864).

Gray, S. F., *Natural Arrangement of British Plants*, 2 vols (London: Baldwin, Cradock & Joy, 1821).

Grew, N., *An Idea of a Phytological History Propounded* (London: Richard Chiswell, 1673).

Grigor, J., *The Eastern Arboretum, or Register of Remarkable Trees ... in the County of Norfolk* (London: Longman, Brown, Green & Longmans, 1841).

Hales, S., *Statical Essays*, 2 edn, 2 vols (London: W. Innys, 1731–3).

Hanham, F., *A Manual for the Park; or, A Botanical Arrangement and Description of the Trees and Shrubs in the Royal Victoria Park, Bath* (London: Longman, Green, & Longmans, 1857).

Henslow, J. S., *Address Delivered in Ipswich Museum on the 9th March 1848* (Ipswich: S. Piper, 1848).

Higginson, E., *Doctrines and Duties of Unitarians* (Lincoln: W. Brooke, 1820).

Hofland, B., and T. C. Hofland, *A Descriptive Account of the Mansion and Gardens of Whiteknights, a Seat of His Grace, the Duke of Marlborough* (London: privately printed, 1819).

Hooker, W. J., *Flora Scotica* (Edinburgh: Archibald Constable & Co., 1821).

Hooker, W. J., and T. Taylor, *Muscologia Britannica* (London: Longman, Hurst, Rees, Orme & Brown, 1818).

Hoopes, J., *Book of Evergreens* (New York: Orange Judd & Co., 1868).

Hovey, C. M., *Magazine of Horticulture*, 11 (1845), pp. 122–8.

Hull Botanic Garden, *The Addresses of the President and Treasurer at the first General Meeting of the Subscribers to the Hull Botanic Garden; and the Report of the Provisional Committee ... with the Laws of the Institution and a List of the Subscribers* (Hull: J. Perkins, 1812).

Ipswich, *The Visitors and General Guide to Ipswich* (Ipswich, 1869).

Irving, W., *Abbotsford and Newstead Abbey* (London: John Murray, 1835).

Jenyns, L., *Memoir of the Revd. John Stevens Henslow* (London: John van Voorst, 1862).

Jerdan, W., 'Characteristic Letters', *The Leisure Hour* (1 February, 1869), p. 140.

Jekyll, G., *Wood and Garden* (London: Longmans, Green & Co., 1898).

Johns, C. A., *The Forest Trees of Britain*, 2 vols (London: Society for Promoting Christian Knowledge, 1858).

Jones, N., *Life and Death: Discourse on Occasion of the Lamented Death of Joseph Strutt* (London: Green, 1844).

Kaempfer, E., *A History of Japan together with a Description of the Kingdom of Siam*, trans. J. G. Scheuchzer (London: privately printed, 1727).

—, *A History of Japan*, 3 vols (Glasgow, 1906).

Kelly's Directory of Derbyshire, Leicestershire and Rutland, and Nottinghamshire (London: Kelly & Co., 1887).

Kemp, E., *Description of the Gardens at Biddulph Grange* (London: Bradbury & Evans, 1862).

Kennion, E., *An Essay on Trees in Landscape* (London: C. J. Kennion, 1816).

Knight, T. A., *A Selection from the Physiological and Horticultural Papers ... by the Late Thomas Andrew Knight* (London: Longman, Orme, Brown, Green & Longmans, 1841).

Laird, F. C., *A Topographical and Historical Description of the County of Nottingham* (London: Sherwood, Neely & Jones, George Cowie & Co., 1810).

—, *A Topographical and Historical Description of the County of Worcester* (London: Sherwood, Neely & Jones, George Cowie & Co., 1820).

Lambert, A. B., *A Description of the Genus Cinchona*, 1st edn (London: Benjamin & John White, 1797).

—, *A Description of the Genus Pinus*, 1st edn (London: White, 1803).

—, *A Description of the Genus Cinchona*, 2nd edn (London, 1821)

—, *An Illustration of the Genus Cinchona* (London: J. Searle, 1821).

—, *A Description of the Genus Pinus, Illustrated with Figures*, 2nd edn, 2 vols (London: John Hale, 1828).

—, *A Description of the Genus Pinus*, new edn, 2 vols (London: Weddell, 1832).

Lindley, J., *Observations on the Structure of Fruits and Seeds*, trans. from L.-C. Richard, *Analyse du Fruit* (London: John Harding, 1819).

—, *Synopsis of the British Flora Arranged According to the Natural Orders* (London: Longman, Rees, Orme, Brown & Green, 1829).

—, *An Introduction to the Natural System of Botany*, 1st edn (London: Longman, Rees, Orme, Brown & Green, 1830).

—, *An Introduction to the Natural System of Botany*, 2nd edn (London: Longman, Rees, Orme, Brown & Green, 1836).

—, 'Review of J. C. Loudon's *Encyclopaedia of Trees and Shrubs*, from the *Gardener's Chronicle*', quoted in the Catalogue of the Encyclopaedias and Dictionaries of Longman, Brown, Green & Longmans (London, 1845), p. xvi.

—, *The Vegetable Kingdom* (London: Bradbury & Evans, 1846).

Linnaeus, C., *A System of Vegetables...Translated from the Thirteenth Edition of the 'Systema Vegetabilium' of ... Linneus* [sic] ... *by a Botanical Society at Lichfield*, 2 vols (Lichfield: Leigh & Sotheby, 1783).

—, *The Families of Plants ... Translated from the Last Edition of the 'Genera Plantarum'...by a Botanical Society at Lichfield*, 2 vols (Lichfield: John Jackson, 1787).

Liverpool Botanic Garden, *A Catalogue of Plants in the Botanic Garden at Liverpool* (Liverpool: James Smith, 1808).

Loudon, J. C., 'Hints Respecting the Manner of Laying Out the Grounds of the Public Squares in London to the Utmost Picturesque Advantage', *Literary Journal*, 21 (1803), pp. 739–42.

—, *Observations on the Formation and Management of Useful and Ornamental Plantations*, 2 vols (Edinburgh: A. Constable & Co., 1804).

—, *A Short Treatise on Several Improvements Recently Made in Hot-Houses* (Edinburgh: for the author, 1805).

—, *An Immediate and Effectual Mode of Raising the Rental of the Landed Property of England ... by a Scotch farmer, now Farming in Middlesex* (London: Longman, Hurst, Rees & Orme, 1808).

—, *The Utility of Agricultural Knowledge to the Sons of the Landed Proprietors of England ... Illustrated by what has taken place in Scotland ... by a Scotch Farmer and Land Agent* (London: J. Harding, 1809).

—, *Hints on the Formation of Gardens and Pleasure Grounds* (London: John Harding, 1812).

—, *Observations on Laying out Farms in the Scotch Style Adapted to England* (London: J. Harding, 1812).

—, *Encyclopaedia of Gardening*, 2nd edn (London; Longman, Hurst, Rees, Orme, Brown & Green, 1824).

—, *Hortus Britannicus*, 1st edn (London: Longman, Rees, Orme, Brown & Green, 1826).

—, *Encyclopaedia of Plants*, 1st edn (London: Longman, Rees, Orme, Brown & Green, 1829).

—, *Encyclopaedia of Gardening*, 5th edn (London: Longman, Rees, Orme, Brown & Green, 1830).

—, *Hortus Britannicus*, 2nd edn (London, 1830).

—, *Illustrations of Landscape Gardening and Garden Architecture* (London, 1830).

—, *Encyclopaedia of Cottage, Farm and Villa Architecture* (London: Longman, Rees, Orme, Brown, Green & Longmans, 1833).

—, 'Remarks on Laying out Public Gardens and Promenades', *Gardener's Magazine*, 11 (1835), pp. 644–59.

—, *Arboretum et Fruticetum Britannicum*, 8 vols, 1st edn (London: for the author, 1838). [Abbreviated throughout to *Arboretum Brittanicum*].

—, *Suburban Gardener and Villa Companion* (London; Edinburgh: Longman, Orme, Brown, Green & Longmans, W. Black, 1838).

—, 'Elvaston Castle: The Seat of the Earl of Harrington', *Gardener's Magazine*, 15 (1839), pp. 458–60.

—, 'Recollections of a Tour', *Gardener's Magazine*, 15 (1839), pp. 432–49.

—, *The Derby Arboretum ... A Catalogue of the Trees and Shrubs* (London: Longman, Orme, Brown, Green and Longmans, 1840).

—, 'Opinions of the Public Press', appended to J. C. Loudon (ed.), *The Landscape Gardening and Landscape Architecture of the Late Humphry Repton Esquire* (London, 1840), pp. 622–3.

—, *Encyclopaedia of Plants*, 2nd edn (London: Longman, Brown, Green, and Longmans, 1841).

— (ed.), *The Landscape Gardening and Landscape Architecture of the Late Humphry Repton Esq.* (London: for the editor, 1841).

—, *An Encyclopaedia of Trees and Shrubs: Being the 'Arboretum et Fruticetum Britannicum' abridged* (London: Longmans, 1842).

—, *On the Layout, Planting and Managing of Cemeteries* (London: for the author, 1843).

—, *Arboretum et Fruticetum Britannicum*, 8 vols, 2nd edn (London: Longman, Brown, Green, & Longmans, 1844).

—, *Encyclopaedia of Agriculture*, 5th edn (London: Longman, Brown, Green, & Longmans, 1844).

—, *Self Instruction for Young Gardeners*, 1st edn (London; Longman, Brown Green, & Longmans, 1844).

—, 'A Short Account of the Life and Writings of John Claudius Loudon', in Loudon, *Self Instruction for Young Gardeners*, 2nd edn (London: Longman, Brown, Green, & Longmans, 1847).

—, *Self Instruction for Young Gardeners*, 2nd edn (London; Longman, Brown, Green, & Longmans, 1847).

—, *Encyclopaedia of Plants*, new edition edited by Jane Loudon (London: Longman et. al., 1855).

—, *Encylopaedia of Trees and Shrubs* (1842; London: Frederick Warne & Co., 1875).

McIntosh, C., *The Book of the Garden*, 2 vols (Edinburgh: William Blackwood & Sons, 1853).

Main, J., *The Forest Planter and Pruner's Assistant* (London: Ridgway, 1839).

Martin, J. R., Report on the State of Nottingham, Coventry, Leicester, Derby, Norwich, and Portsmouth, in the *Second Report of the Commissioners for Inquiry into the State of Large Towns and Populous Districts*, 2 vols (London: HMSO, 1845), vol. 2.

Masters, M. T., 'List of Conifers and Taxads in Cultivation in the Open Air in Great Britain and Ireland', *Journal of the Royal Horticultural Society*, 14 (1892), pp. 1–18.

—, 'Notes on the Genera of Taxaceae and Coniferae', *Journal of the Linnean Society*, 30 (1893), pp. 1–42.

Modern Mayors of Derby (Derby: Hobson & Son, 1909).

Mirbel, C. F. B de, *Traité d'anatomie et de physiologie végétales*, 2 vols (Paris, 1802).

—, *Éléments de physiologie végétale et de botanique*, 3 vols (Paris, 1815).

Monteath, R., *The Forester's Guide and Profitable Planter,* 2nd edn (Edinburgh: Stirling & Kenney, 1824).

Nicholson, G. (ed.), *The Illustrated Dictionary of Gardening: A Practical and Scientific Encyclopaedia of Horticulture,* 4 vols (London: L. Upcott Gill, 1889).

Nisbet, J. (ed.) *The Forester: A Practical Treatise on Planting and Tending of Forest Trees and General Management of Woodland Estates by James Brown,* 6th edn , 2 vols (Edinburgh: W. Blackwood & Sons, 1894).

Ono, R., and Y. Shimada, *Kai,* 2 vols (Tokyo: Yasaka Shobō, Shōwa, [1763] 1977).

Paxton, J., 'Remarks on Arboretums', *Paxton's Magazine of Botany and Register of Flowering Plants,* 8 (1841), pp. 41–2.

Olmsted, F. L., *The Papers of Frederick Law Olmsted: Vol. 3, Creating Central Park 1857–1861,* ed. C. E. Beveridge and D. Schuyler (Baltimore, MD: Johns Hopkins University Press, 1983).

Pontey, W., *The Forest Pruner, or Timber Owner's Assistant,* 4th edn (Leeds: T. Inchbold, 1826).

Pontin, D. D., *Arboretum Suecicum* (Uppsala, 1759).

Price, U., *Essay on the Picturesque* (Hereford: D. Walker, 1794).

—, *Essays on the Picturesque,* 3 vols (London: J. Mawman, 1810).

Public Walks, Report from the Select Committee on Public Walks (1833), *Parliamentary Papers,* 15 (Dublin: Irish University Press), pp. 2–11.

Quincy, de, A. C. Q., *An Essay on the Nature, the End and the Means of Imitation in the Fine Arts,* trans. J. C. Kent (London, 1837).

Ravenscroft, E. J., *The Pinetum Britannicum,* 3 vols (Edinburgh: W. Blackwood & Sons, 1863–84).

Ray, J., *Historia Plantarum,* 2 vols (London: typis Mariae Clark, prostant apud Henricum Faithorne, 1686).

Robinson, W., *God's Acre Beautiful; or the Cemeteries of the Future* (London: Garden Office, 1880).

—, *The Wild Garden,* fourth edition (London: John Murray, 1894).

—, *English Flower Garden,* 8th edn (London: John Murray, 1900).

Rooke, H., *A Sketch of the Ancient and Present State of Sherwood Forest* (Nottingham: S. Tupman, 1799).

Roscoe, H., *Life of William Roscoe,* 2 vols (London: Cadell, 1833).

Roscoe, W., An *Address Delivered before the Proprietors of the Botanic Garden in Liverpool ... to which are added the Laws of the Institution of a List of the Proprietors* (Liverpool: J M'Reery, 1802).

Ruskin, J., *Modern Painters,* 3rd edn, 5 vols (London, 1860).

—, *Selections from the Writings of John Ruskin* (Edinburgh: Nimmo, 1907).

Sachs, J. von, *History of Botany (1530–1860)*, rev. I. B. Balfour, trans. H. E. F. Garnsey (Oxford: Clarendon Press, 1890).

Sargent, C. S. *The Silva of North America*, 14 vols (New York: Houghton Mifflin, 1891–1902).

—, *Manual of the Trees of North America* (New York: Houghton Mifflin, 1905).

Schlich, W., *A Manual of Forestry*, 5 vols (London: Bradbury, Agnew & Co., 1889).

Searle, J., 'Leaves from Sherwood', in R. White, *Worksop, 'The Dukery' and Sherwood Forest* (London: Simpkin, Marshall, 1875).

Second Report of the Commissioners of Inquiry into the State of Large Towns and Populous Districts (London: HMSO, 1845).

Select Committee on Forestry, *Report from the Select Committee on Forestry* (London: HMSO, 1887).

Sherrard, T., 'A Survey of the Botanic Garden at Glasnevin in the County of Dublin', engraved by J. Taylor, *Transactions of the Dublin Society ... for the year 1799* (Dublin, 1800).

Siebold, P. F. de, and J. G. Zuccarini, *Flora Japonica sive Plantae* (1835–1844) (translated into Japanese by M. Sekura, with a commentary by H. Oba), *Nihon no shokubutsu* (Tokyo: Kodansha, 1996).

Sinclair, G., *Useful and Ornamental Planting* (London: Baldwin and Cradock, 1832).

Sissons, F., *Beauties of Sherwood Forest* (Worksop: Sissons & Sons, 1888).

Smith, C. H. J., *Landscape Gardening: or, Parks and Pleasure Grounds* (London: for the author, 1852).

Smith, J. E., *A Grammar of Botany* (London: Longman, Hurst, Rees, Orme & Brown, 1822).

—, *An Introduction to Physiological and Systematical Botany*, 5th edn (London: Longman, Hurst, Rees, Orme, Brown, & Green, 1825).

—, *An Introduction to Physiological and Systematical Botany*, 6th edn (London: Longman, Hurst, Rees, Orme, Brown, & Green, 1827).

—, *English Flora*, 4 vols (London; Longman, Rees, Orme, Brown & Green, 1829).

—, *Records of the Royal Botanic Gardens, Kew* (London: for the author, 1880).

Smith J., E., and J. Sowerby, *English Botany*, 36 vols (London: J. Davis, 1790).

Smith, P. (ed.), *Memoir and Correspondence of the late Sir James Edward Smith*, 2 vols (London: Longman, Rees, Orme, Brown, Green, & Longman, 1832).

Spencer, W., Sixth Duke of Devonshire, *Handbook of Chatsworth and Hardwick* (London: privately printed, 1844).

Standish, J., and C. Noble, *Practical Hints on Planting Ornamental Trees* (London: Bradbury & Evans, 1852).

Steuart, H., *The Planter's Guide* (Edinburgh: J. Murray, 1828).

Strang, J., *Necropolis Glasguensis* (Glasgow: Atkinson & Co., 1831).

Strutt, J., G., *Sylva Britannica or Portraits of Forest Trees Distinguished for their Antiquity, Magnitude or Beauty* (London: J. G. Strutt, 1826).

Sweet, R., *Hortus Britannicus; or, a Catalogue of Plants Cultivated in the Gardens of Great Britain* (London: J. Ridgway, 1826).

—, *Hortus Suburbanus Londinensis; or a Catalogue of Plants Cultivated in the Neighbourhood of London* (London: J. Ridgway, 1818).

Switzer, S., *The Practical Husbandman*, 2 vols (London: S. Switzer, 1733).

Taylor, J. E. (ed.), *Chronicles of an Old English Oak; or, Sketches of English Life and History, as Reported by those who Listened to them* 1860 (London: Groombridge & Sons, 1860).

—, *Descriptive Handbook of Ipswich and Neighbourhood* (Ipswich: William Hunt, 1873).

—, *Pawsey and Hayes' Illustrated Guide to Ipswich and the Neighbourhood* (Ipswich, 1890).

Thunberg, C. P., *Flora Japonica* (Lipsiae: I. G. Mülleriano, 1784).

—, *Travels in Europe, Africa, and Asia made between the years 1770 and 1779* (London: F. & C. Rivington, 1795).

—, *Plantarum Japonicarum Novae Species* (Upsaliae: Excudebant Palmblad, 1824).

Tyrrell, I., 'Somerset , Lady Isabella Caroline [Lady Henry Somerset] (1851–1921)'; *The Bengal Obituary* (London and Calcutta, 1851), pp. 31, 316.

Ure, A. 'Official Report ... upon Bernhardt's Stove-Furnaces', *Architectural Magazine*, 5 (1836), pp. 31–42.

Wade, W., *Plantae Rariores in Hibernia Inventae* (Dublin: Graisbury & Campbell, 1804).

Walker, G. A., *Gatherings from Graveyards, Particularly those of London* (London: Longman & Co., 1839).

—, *The Graveyards of London* (London: Longman, 1841).

Ward's Historical Guide to Lincoln (Lincoln: Lincolnshire Chronicle, 1900).

Watson, P. W., *Dendrologia Britannica*, 2 vols (London: for the author, 1824–5).

Walter Scott Digital Archive, University of Edinburgh: http://www.walterscott.lib.ed.ac.uk.

West, J., *Remarks on the Management or rather the Mismanagement of Woods, Plantations and Hedgerow Timber* (Newark: J. Perfect, 1842).

White, H. K. W., *The Works of Henry Kirk White of Nottingham* (London: Charles Daly, 1835).

—, *White's Nottingham Directory and Borough Register* (Nottingham: White & Co., 1854).

—, *History, Directory and Gazetteer of Suffolk* (Sheffield: White & Co., 1855).

—, *History, Gazetteer and Directory of Lincolnshire* (Sheffield: White & Co., 1872 and 1882).

Willdenow, C. L., *Berlinische Baumzucht*, 2nd edn (Berlin, 1811).

Willott, C., *Belper and Its People* (Belper: T. Edwards, 1894).

Wrigglesworth, E., *Brown's Illustrated Guide to Hull* (Hull: A. Brown, 1891).

Wylie, W. H., *Old and New Nottingham* (London: Longman, Brown, Green & Longmans, 1853), p. 33.

Secondary Material

Aas, G., and A. Riedmiller, *Trees of Britain and Europe*, trans. M. Walters (London: Harper-Collins, 1994).

Adamson, J. W., *English Education, 1789–1902* (Cambridge: Cambridge University Press, 1930).

Alberti, S., *Nature and Culture: Objects, Disciplines and the Manchester Museum* (Manchester; Manchester University Press, 2009).

Allan, M., *Darwin and his Flowers* (London: Allen Lane, 1977).

Allen, D. E., *The Fern Craze: A History of Pteridomania* (London: Hutchinson, 1969).

—, *The Naturalist in Britain: A Social History* (London: Allen Lane, 1976).

—, 'Natural History in Britain in the Eighteenth Century', *Archives of Natural History*, 20 (1993), pp. 333–47.

—, 'George Bentham's *Handbook for the British Flora*: from Controversy to Cult', *Archives of Natural History*, 30 (2003), pp. 224–36.

Allison, K. J. (ed.), *Victoria History of the County of York: East Riding, vol. I: The City of Kingston-Upon-Hull* (Oxford: Oxford University Press, 1969).

Arnold, C., *Necropolis: London and its Dead* (London: Pocket Books, 2007).

Ayers, P., *The Aliveness of Plants: The Darwins at the Dawn of Plant Science* (London: Pickering & Chatto, 2008).

E. Baigent, 'John Gould Veitch (1839-1870)', *ODNB*.

Bailey, P. C., 'A Mingled Mass of Perfectly Legitimate Pleasures: The Victorian Middle Class and the Problem of Leisure', *Victorian Studies*, 21 (1977), pp. 7–28.

Ballard, P. *An Oasis of Delight: The History of the Birmingham Botanical Gardens*, 2nd edn (1983; Studley: Brewin Books, 2003).

Barber, L., *The Heyday of Natural History* (London: Jonathan Cape, 1980).

Barkley, S. Y., *Trees of Westonbirt School* (Reading: Bradley & Son, 1952).

Barnatt, J., and T. Williamson, *Chatsworth: A Landscape History* (Macclesfield: Windgather Press, 2005).

Barron, W., and Son, *Principal Works carried out by William Barron and Son Ltd* (Derby: *c.* 1930).

Bean, W. J., *Trees and Shrubs Hardy in the British Isles*, 7th edn, 3 vols (London: John Murray, 1951).

Beckett, J. V., and K. Brand, 'Enclosure, Improvement and the Rise of 'New Nottingham' 1845–67', *Transactions of the Thoroton Society*, 98 (1995), pp. 92–111.

Beckett, J. V., et. al. (eds), *Centenary History of Nottingham* (Manchester: Manchester University Press, 1997).

Bellon, R., 'Joseph Hooker Takes a Fixed Post: Transmutation and the Present Unsatisfactory State of Systematic Botany, 1844–60', *Journal of the History of Biology*, 39 (2006), pp. 1–39.

Bentham, G., *Handbook of the British Flora*, 7th edn, rev. J. D. Hooker and A. B. Rendle (Ashford: L. Reeve & Co. Ltd, 1924).

Bermingham, A., *Landscape and Ideology: The English Rustic Tradition* (London: Thames & Hudson, 1987).

—, *Learning to Draw: Studies in the Cultural History of a Polite and Useful Art* (New Haven, CT: Yale University Press, 2000).

—, 'System, Order and Abstraction: The Politics of English Landscape Drawing around 1795' in W. T. J. Mitchell (ed.), *Landscape and Power*, 2nd edn (Chicago, IL; Chicago University Press, 2002), pp. 77–101.

Boehm, W., 'Julius A. Stockhardt (1809–1886) – Pioneer of Agricultural Experimental Stations', *Agricultural Research*, 39 (1986), pp. 1–7.

Bonehill, J., and S. Daniels (eds), *Paul Sandby: Picturing Britain* (Nottingham and London: Royal Academy of Arts, 2009).

Boniface, P., *The Travels of John Claudius Loudon and his Wife Jane* (London: Guild Publishing, 1987).

—, *In Search of English Gardens: The Travels of John Claudius Loudon and his Wife Jane* (London: Century, 1987).

Bonneuil, C., 'The Manufacture of Species: Kew Gardens, the Empire and the Standardisation of Taxonomic Practice in late Nineteenth-Century Botany', in M. N. Bourguet, C. L. and O. Sibum (eds), *Instruments, Travel and Science: Itineraries of Precision from the 17th to the 20th Century* (London: Routledge, 2000), pp. 189–215.

Boulger, G. S., rev. Giles Hudson, 'P. S. Watson', *ODNB*.

Bourke, M., 'Trees on Trial: Economic Arboreta in Australia', in Elliott et al. (eds), 'Cultural and Historical Geographies of the Arboretum', pp. 217–26.

Bowe, P., 'Puckler-Muskau's Estate and its Influence on American Landscape Architecture', *Garden History*, 23 (1995), pp. 192–200.

Bowler, P., *History of the Environmental Sciences* (London: Fontana, 1992).

Bown, D., *Westonbirt: The Forestry Commission's Finest Arboretum* (Baltonsborough: Julian Holland, 1990).

Briggs, A., *Victorian Cities* (London: Pelican, 1968).

Britten, J., 'Banister, John (1650–1692)', rev. Marcus B. Simpson jun., *ODNB*.

Brockway, L., *Science and Colonial Expansion: The Role of the British Botanic Gardens* (New Haven, CT: Yale University Press, 1979).

Brown, J., *Gardens of a Golden Afternoon: Edwin Lutyens and Gertrude Jekyll* (Allen Lane: London, 1982).

Browne, J., *The Secular Ark: Studies in the History of Biogeography* (New Haven, CT: Yale University Press, 1985).

Bynum, W. F., and R. Porter (eds), *Living and Dying in London*, Medical History, supplement 11 (London: Wellcome Institute for the History of Medicine, 1991).

Byrom, C., 'The Pleasure Grounds of Edinburgh New Town', *Garden History*, 23 (1995), pp. 69–90.

Calvert, H., *History of Kingston-Upon-Hull* (Chichester: Phillimore, 1978).

Cannadine, D., *The Decline and Fall of the British Aristocracy* (London: Macmillan, 1992).

Campana, R. J., *Arboriculture: History and Development in North America* (East Lansing, MI; Michigan State University Press, 1999).

Carder, J., *The Sheffield Botanical Gardens: A Short History* (Sheffield: Sheffield Council, 1986).

Chadwick, G. F., *The Works of Joseph Paxton* (London: Architectural Press, 1963).

—, *The Park and Town: Public Landscape in the 19th and 20th Centuries* (London: Architectural Press, 1966).

Chambers, D., *The Planters of the English Landscape Garden: Botany, Trees and the Georgics* (New Haven, CT: Yale University Press, 1993).

—, 'Collinson, Peter (1694–1768)', *ODNB*.

Church, R. A., *Economic and Social Change in a Midland Town: Victorian Nottingham 1815–1900* (London: Frank Cass, 1966).

Clark, P. (ed.), *The Cambridge Urban History of Britain*, 2 vols (Cambridge: Cambridge University Press, 2000).

Cleevely, R. J., 'The Sowerbys and their Publications in the light of the Manuscript Material in the British Museum (Natural History)', *Journal of the Society for the Bibliography of Natural History*, 7 (1976), pp. 343–68.

Clinton-Baker, H., *Illustrations of Conifers*, 2 vols (London: Simson, 1909).

Cole, R. V., *British Trees Drawn and Described*, rev. D. Kempe, 2 vols (London: Hutchinson, 1907).

Collingwood, W. G., *The Life of John Ruskin*, 5th edn (London: Methuen, 1905).

Collins, E. J. T., 'Woodlands and Woodland Industries in Great Britain During and After the Charcoal Iron Era', in J. P. Metaille (ed.) *Protoindustries et histoire des forets* ('Toulouse: Groupement de recherche ISARD, Université de Toulouse-Le Mirai, 1992), pp. 109–20.

Colquhoun, K., *A Thing in Disguise: The Visionary Life of Joseph Paxton* (London: Fourth Estate, 2003).

Conway, H., *People's Parks: The Design and Development of Victorian Parks in Britain* (Cambridge: Cambridge University Press, 1991).

Cooke, E. T., *Trees and Shrubs for English Gardens* (London: Country Life, 1902).

Cooper, C. S., and W. P. Westell, *Trees and Shrubs of the British Isles*, 2 vols (London: J. M. Dent & Sons, 1909).

Covins, F., *The Arboretum Story* (Worcester: Arboretum Residents Association, 1989).

Craven M., *Illustrated History of Derby* (Derby: Breedon, 1988).

Curl, J. S., 'John Claudius Loudon and the Garden Cemetery Movement', *Garden History*, 11 (1983), pp. 133–56.

—, 'The Design of Early British Cemeteries', *Journal of Garden History*, 4 (1984).

—, *The Victorian Celebration of Death*, new edn (Stroud: Alan Sutton, 2000).

Daniels, S., 'The Political Iconography of Woodland', in D. Cosgrove and S. Daniels (eds), *Iconography of Landscape* (Cambridge: Cambridge University Press, 1989), pp. 43–82.

—, *Fields of Vision: Landscape Imagery and National Identity in England and the United States* (Cambridge: Cambridge University Press, 1993).

—, *Humphry Repton: Landscape Gardening and the Geography of Georgian England* (New Haven; Yale University Press, 1999).

Daniels, S., and C. Watkins, 'Picturesque Landscaping and Estate Management: Uvedale Price at Foxley, 1770–1829', *Rural History*, 2:2 (1991), pp. 141–70.

Daunton, M. (ed.), *The Cambridge Urban History*, 3 vols (Cambridge: Cambridge University Press, 2000).

— (ed.), *The Organisation of Knowledge in Victorian Britain* (Oxford: Oxford University Press, 2005).

Delves, A., 'Popular Recreation and Social Conflict in Derby, 1800–1850', in E. Yeo and S. Yeo (eds), *Explorations in the History of Labour and Leisure* (Brighton: Harvester Press, 1981), pp. 89–127.

Dempster, W. J., *Evolutionary Concepts in Nineteenth-Century Britain: Natural Selection and Patrick Matthew* (Edinburgh: Pentland Press, 1996).

Dennis, R., 'Modern London', in Daunton (ed.), *The Cambridge Urban History*, vol. 3, pp. 95–132.

Desmond, A., and J. Moore, *Charles Darwin* (London: Michael Joseph, 1991).

Desmond, R., *Kew: The History of the Royal Botanic Gardens* (London: Harvill Press and Royal Botanic Gardens, Kew, 1995).

Drayton, R., *Nature's Government: Science, Imperial Britain and the 'Improvement' of the World* (New Haven, CT: Yale University Press, 2001).

Dupree, M., 'The Provision of Social Services', in Daunton (ed.), *The Cambridge Urban History*, vol. 3, pp. 351–94.

Dyos, H. J., and M. Wolff (eds), *The Victorian City: Images and Realities*, 2 vols (London: Routledge & Kegan Paul, 1973).

Eastwood, D., *Government and Community in the English Provinces, 1700–1870* (Basingstoke: Macmillan, 1997).

Edlin, H., *Woodland Crafts in Britain* (London: Batsford, 1949).

—, *Trees, Woods and Man*, 3rd edn (London: Collins, 1970).

Elliott, B., *Victorian Gardens* (London: Batsford, 1986).

—, 'From Arboretum to the Woodland Garden', in Elliott et al. (eds), 'Cultural and Historical Geographies of the Arboretum', pp. 71–83.

—, *The History of the Royal Horticultural Society, 1804–2004* (Chichester: Phillimore, 2004).

Elliott, P. A., *The Derby Philosophers: Science and Culture in British Urban Society* (Manchester: Manchester University Press, 2009).

—, *Enlightenment, Modernity and Science: Geographies of Scientific Culture and Improvement in Georgian England* (London: I. B. Tauris, 2010).

Elliott, P., C. Watkins and S. Daniels, 'William Barron (1805–1891) and Nineteenth-Century British Arboriculture: Evergreens in Victorian Industrialising Society', in Elliott, et al. (eds), 'Cultural and Historical Geographies of the Arboretum, *Garden History* (2007), pp. 129–48.

—, 'Nottingham Arboretum: Natural History, Leisure and Public Culture in a Victorian Regional Centre', *Urban History*, 35 (2008), pp. 48–71.

Elsner, J., and R. Cardinal, *Cultures of Collecting* (Cambridge, MA: Harvard University Press).

Elwes H. J., and A. Henry, *The Trees of Great Britain and Ireland*, 7 vols (Edinburgh: privately printed, 1906–13).

Endersby, J., 'Classifying Sciences: Systematics and Status in mid-Victorian Natural History', in Daunton (ed.), *The Organisation of Knowledge in Victorian Britain* (2005), pp. 61–86.

—, *Imperial Nature: Joseph Hooker and the Practices of Victorian Science* (Chicago, IL: Chicago University Press, 2008).

English Heritage, UK database of Historic Parks and Gardens.

Etlin, R. A., 'Père Lachaise and the Garden Cemetery', *Journal of Garden History*, 4 (1984), pp. 211–22.

Finer, S. E., *The Life and Times of Sir Edwin Chadwick* (London: Methuen, 1952).

Fitton R. S., and A. P. Wadsworth, *The Strutts and the Arkwrights, 1758–1830* (Manchester: Manchester University Press, 1958).

FitzRandolph, H., and M. Hay, *The Rural Industries of England and Wales* (Oxford: Oxford University Press, 1926).

Fletcher H. R., and W. H. Brown, *The Royal Botanic Garden Edinburgh, 1670–1970* (Edinburgh: HMSO, 1970).

Forbes, A. C., *English Estate Forestry* (London: Edward Arnold, 1904).

—, *The Development of British Forestry* (London: Edward Arnold, 1910).

Fothergill, P., *Eastnor Castle Estate: A List of Trees in the Castle Grounds* (Eastnor Castle, Ledbury, 1982).

Fowler, J., *Landscapes and Lives: The Scottish Forest through the Ages* (Edinburgh; Canongate, 2002).

Fraser, D., *Power and Authority in the Victorian City* (Oxford; Oxford University Press, 1979).

— (ed.), *Municipal Reform and the Industrial City* (Leicester: Leicester University Press, 1982).

—, *Urban Politics in Victorian England: The Structure of Politics in Victorian Cities* (Leicester: Leicester University Press, 1983).

Fricker, L., 'John Claudius Loudon, the Plane Truth?', in P. Willis (ed.), *Furor Hortensis: Essays on the History of the English Landscape in Memory of H. F. Clark* (Edinburgh: Elysium Press, 1974), pp. 76–88.

Gage, A. T., and W. T. Stearn, *A Bicentenary History of the Linnean Society of London* (London: Academic Press, 1988).

Girouard, M., *The Return to Camelot: Chivalry and the English Gentleman* (New Haven, CT: Yale University Press, 1981).

Golby, J. M., and A. W. Purdue, *The Civilisation of the Crowd: Popular Culture in England, 1750–1900* (London: Batsford Academic & Educational, 1984).

Golinski, J., *Making Natural Knowledge* (Cambridge: Cambridge University Press, 1998).

Gray, D., and V. W. Walker (eds), *Records of the Borough of Nottingham, vol. IX, 1836–1900* (Nottingham; T. Forman, 1956).

Green, J. R., *A History of Botany in the United Kingdom* (London: Dent, 1914).

Grove, R. H., *Green Imperialism: Colonial Expansion, Tropical Island Edens and the Origins of Environmentalism, 1600–1860* (Cambridge: Cambridge University Press, 1995).

Harman, R., and J. Minnis, *Sheffield: Pevsner City Guides* (New Haven, CT: Yale University Press, 2004).

Harrison, B. H., 'Religion and Recreation in Nineteenth Century England', *Past and Present*, 38 (1967), pp. 98–125.

Hartley, B., 'Sites of Knowledge and Instruction: Arboretums and the *Arboretum et Fruticetum Britannicum*', in P. Elliott, C. Watkins, S. Daniels (eds.), *Cultural and Historical Geographies of the Arboretum* special issue of *Garden History* (2007), pp. 28–52.

Hay, I., *Science in the Pleasure Ground: A History of the Arnold Arboretum* (Cambridge, MA: Harvard University Press, 1995).

Hayden, P., *Biddulph Grange, Staffordshire: A Victorian Garden Rediscovered* (London: George Philip with National Trust, 1989).

Hibbert, C., *Queen Victoria in Her Letters and Journals: A Selection* (London: John Murray, 1985).

Hillier, J., and A. Coombes, *The Hillier Manual of Trees and Shrubs* (Newton Abbott: David & Charles, 2002).

Hooper-Greenhill, E., *Museums and the Shaping of Knowledge* (London; New York: Routledge, 1992).

Hotz, M. E., *Literary Remains: Representations of Death and Burial in Victorian England* (New York: State University of New York Press, 2008).

Houghton, W., *The Victorian Frame of Mind, 1830–1870* (New Haven, CT: Yale University Press, 1964).

Hudson, D., and K. W. Luckhurst, *The Royal Society of Arts, 1754–1954* (London: John Murray, 1954).

Hunt, J. D. *The Wider Sea: A Life of John Ruskin* (London: Dent, 1982).

Hunt, J. D., and D. Schuyler (eds), 'Garden Cemeteries', special issue of *Garden History* (1984).

Hunt, T., *Building Jerusalem: The Rise and Fall of the Victorian City* (London: Weidenfeld & Nicolson, 2004).

Hunter, R. A., *Sheffield Botanical Gardens; People, Plants and Pavilions* (Sheffield: Friends of the Botanical Gardens, 2007).

Illiffe, R., and W. Baguley, *Victorian Nottingham: A Story in Pictures, Vol. 10* (Nottingham: Nottingham Historical Film Unit, 1973).

Inkster, I., 'Scientific Culture and Scientific Education in Liverpool prior to 1812 – A Case Study in the Social History of Education', in I. Inkster, *Scientific Culture and the Urbanisation in Industrialising Britain* (Aldershot: Ashgate, 1997), pp. 28–47.

Inkster, I., and J. Morell (eds), *Metropolis and Provinces: Science in British Culture, 1780–1850* (London: Hutchinson, 1983).

Inwood, S., *A History of London* (London: Macmillan, 1998).

MacDonald, J. B., 'The Sowerby Collection in the British Museum (Natural History): A Brief Description of its Holdings and a History of its Acquisition from 1821–1971', *Journal of the Society for the Bibliography of Natural History*, 6:6 (1975), pp. 380–401.

Meynell, G., 'The Royal Botanic Society's Garden, Regent's Park', *London Journal*, 6 (1980), pp. 135–46.

Jackson, A. B., *Catalogue of the Trees and Shrubs in the Collection of the Late ... Sir George Lindsey Holford* (London and Oxford, 1927).

Jackson, A., 'Imagining Japan: The Victorian Perception of Japanese Culture', *Journal of Design History*, 5 (1996), pp. 245–256.

Jackson, G., *Hull in the Eighteenth Century: A Study in Economic and Social History* (Hull: Hull Academic Press, 1972).

Jacques, D., *Georgian Gardens: The Reign of Nature* (London: Batsford, 1983).

Jalland, P., *Death in the Victorian Family* (1996; Oxford; New York: Oxford University Press, 1999).

James, N. D. G., *A History of English Forestry* (Oxford: Oxford University Press, 1981).

—, 'A History of Forestry and Monographic Forestry Literature in Germany, France and the United Kingdom' in P. McDonald and J. Lassoie (eds) *The Literature of Forestry and Agroforestry* (Ithaca, NY; London: Cornell University Press, 1996), pp. 15–44.

Jankovic, V., 'The Place of Nature and the Nature of Place: The Chorographic Challenge to the History of British Provincial Science', *History of Science*, 38 (2000), pp. 79–113.

Jardine, N., 'Inner History; or, How to End Enlightenment', in J. Clark, J. Golinski and S. Schaffer (eds), *The Sciences in Enlightened Europe* (Chicago, IL: Chicago University Press, 1999), 477–94.

Jardine, N., J. A. Secord and E. C. Spary (eds), *Cultures of Natural History* (Cambridge: Cambridge University Press, 1996).

Jarvis, P. J., 'Plant Introductions to England and their Role in Horticultural and Sylvi-Cultural Innovation, 1500–1900' in H. S. A. Fox and R. A. Butler (eds), *Change in the Countryside: Essays on Rural England 1500–1900*, special publication, 10 (London: Institute of British Geographers, 1979), pp. 145–64.

Johnson, N. C., 'Cultivating Science and Planting Beauty: The Spaces of Display in Cambridge's Botanical Garden', *Interdisciplinary Science Reviews*, 31 (2006), pp. 42–57.

Jones, O., 'Arnos Vale Cemetery and the Lively Materialities of Trees in Place', in Elliott et al. (eds), 'Cultural and Historical Geographies of the Arboretum', pp. 149–71.

Jones O., and P. Cloke, *Tree Cultures: The Place of Trees and Trees in their Place* (Oxford: Berg, 2002).

Jordan, H., *The Register of Parks and Gardens: Cemeteries* (London: English Heritage, 2003).

Joyce, P., *A Guide to Abney Park* (London: Abney Park Cemetery Trust, 1994).

Kato, N., *MakinoHyohonkan shozou no Siebold Collection* (Siebold Collection at Makino Herbarium) (Kyoto, 2003).

Kato, N., H. Kato, A. Kihara and M. Wakabayashi (eds), *MakinoHyohonkan shozou no Siebold Collection*, CD ROM Database (Tokyo, 2005).

Kellehear, A., *A Social History of Dying* (Cambridge: Cambridge University Press, 2007).

Kent, A., *Veitch's Manual of Coniferae*, 2nd edn (London: J. Veitch & Sons, 1900).

Kitteringham, G., 'Science in Provincial Society: The Case of Liverpool in the Early Nineteenth Century', *Annals of Science*, 39 (1982), pp. 329–448.

Knight, J. (rev. K. D. Reynolds), 'Maria Foote (Stanhope) (1797–1867)', *ODNB*.

Lasdun, S., *English Park* (London: Andre Deutsch, 1991).

Latour, B., *Science in Action: How to Follow Scientists and Engineers through Society* (Cambridge, MA: Harvard University Press, 1987).

Latour, B., and S. Woolgar, *Laboratory Life: The Construction of Scientific Facts*, 2nd edn (Princeton, NJ: Princeton University Press, 1986).

Le Rougetel, H., 'Miller, Philip (1691–1771)', *ODNB*.

Leakey, R., and R. Lewontin, *The Sixth Extinction: Biodiversity and its Survival* (London: Phoenix, 1998).

Lees-Milne, J., *The Bachelor Duke: 6th Duke of Devonshire 1790–1858* (London: John Murray, 1998).

Lever, C., *They Dined on Eland: The Story of the Acclimatisation Societies* (London: Quiller Press, 1992).

Lewis, R. A., *Edwin Chadwick and the Public Health Movement 1832–1854* (London: Chivers, 1968).

Lightman, B. (ed.) *Victorian Science in Context* (Chicago, IL: Chicago University Press, 1997).

Litten, J., *The English Way of Death: The Common Funeral Since 1450* (1991; London: Robert Hale, 2002).

Little, E. L., Jr., 'Lambert's "Description of the *Genus Pinus*", 1832 edition', *Madrono*, 10 (1949), pp. 33–47.

Livingstone, D., *Putting Science in its Place* (Chicago, IL: Chicago University Press, 2003).

Llewellyn, N., *The Art of Death: Visual Culture in the English Death Ritual, c.1500–1800* (London: Reaktion, 1991).

Lowerson, J., *Sport and the English Middle Classes, 1870–1914* (Manchester: Manchester University Press, 1993).

Luckin, B., 'Pollution in the City', in Daunton (ed.), *The Cambridge Urban History*, vol. 3, pp. 207–28.

Mabberley, D. J., *Jupiter Botanics: Robert Brown of the British Museum* (London: Lubrecht & Cramer, 1985).

MacDonald, J., R. F. Wood, M. V. Edwards and J. R. Aldhous, 'Exotic Forest Trees in Great Britain: Paper Prepared for the Seventh British Commonwealth Forestry Conference, Australia and New Zealand', *Forestry Commission Bulletin*, 30 (London: HMSO, 1957).

MacDougall, E. B. (ed.), *John Claudius Loudon and the Early Nineteenth Century in Great Britain* (Washington, DC: Dumbarton Oaks, 1980).

Major, J. K., *To Live in the New World: Andrew Jackson Downing and American landscape Gardening* (Cambridge, MA: MIT Press, 1997).

Malcolmson, R. W., *Popular Recreations in English Society, 1700–1850* (Cambridge: Cambridge University Press, 1973).

Malster, R., *A History of Ipswich* (Chichester: Phillimore, 2000).

Markham, R. A. D., *A Rhino in the High Street: Ipswich Museum – the Early Years* (Ipswich: Ipswich Borough Council 1990).

Markham, V., *Paxton and the Bachelor Duke* (London: Hodder & Stoughton, 1935).

Mastoris, S., and S. Groves (eds), *Sherwood Forest in 1609: A Crown Survey by Richard Bankes*, Thoroton Society Record Series, 40 (1997).

Mathew, M. V., *The History of the Royal Botanic Garden Library Edinburgh* (Edinburgh: HMSO, 1987).

Mayr, E., *The Growth of Biological Thought* (Cambridge, MA: Harvard University Press, 1982).

McCracken, D. P., *Gardens of Empire: Botanical Institutions of the Victorian British Empire* (Leicester: Leicester University Press, 1997).

McCracken, E., and E. C. Nelson, *The Brightest Jewel: History of the National Botanic Gardens, Glasnevin, Dublin* (Kilkenny: Boethius Press, 1987).

McOuat, G., 'Cataloguing Power: Delineating Competent Naturalists and the Meaning of Species at the British Museum', *British Journal for the History of Science*, 34 (2001), pp. 1–28.

Mellors, R., *The Gardens, Parks and Walks of Nottingham and District* (Nottingham, 1926).

Merchant, C., *Reinventing Eden: The Fate of Nature in Western Culture* (London: Routledge, 2004).

Merrill, L. L., *The Romance of Victorian Natural History* (Oxford: Oxford University Press, 1989).

Miles, C., A., *Christmas in Ritual and Tradition, Christian and Pagan* (London: T. Fisher Unwin, 1912),

Miller, H. S., 'The Herbarium of Aylmer Bourke Lambert', *Taxon*, 19 (1970), pp. 489–553.

Millward, R., 'The Political Economy of Urban Utilities' in Daunton (ed.), *The Cambridge Urban History*, vol. 3, pp. 315–50.

Morrell, J. B., and A. Thackray, *Gentlemen of Science: Early Years of the British Association for the Advancement of Science* (Oxford: Clarendon Press, 1981).

Morris, C., 'The Diffusion of Useful Knowledge: John Claudius Loudon and his Influence in the Australian Colonies', *Garden History*, 32 (2004), pp. 101–23.

Mullins, G., 'Eastnor Castle: Gardener and Gardens', *Gardener's Chronicle* (29 August 1903), p. 155.

Murdoch, A., 'Campbell, Archibald, third duke of Argyll (1682–1761)', *ODNB*.

Johnson, N., 'Names, Labels and Planting Regimes: Regulating Trees at Glasnevin Botanic Gardens, Dublin, 1795–1850', in Elliott et al. (eds), 'Cultural and Historical Geographies of the Arboretum', pp. 53–70.

Naylor, P. (ed.), *Illustrated History of Belper* (Belper: M. G. Morris, 2000).

Naylor, S., 'Provincial Authorities and Botanical Provinces: Elizabeth Warren's Hortus Siccus of the Indigenous Plants of Cornwall', in Elliott et al. (eds), 'Cultural and Historical Geographies of the Arboretum', pp. 84–95.

Neeson, J. M., *Commoners: Common Right, Enclosure and Social in England, 1700–1820* (Cambridge: Cambridge University Press, 1993).

Nelson, E. C., 'So Many Really Fine Plants: An Epitome of Japanese Plants in Western European Gardens', *Curtis's Botanical Magazine*, 16 (1999), pp. 52–68.

Nisbet, J., *The Forester*, 2 vols (Edinburgh: William Blackwood & Sons, 1905).

O'Kane, F., *Landscape Design in Eighteenth-Century Ireland* (Cork: Cork University Press, 2004).

—, 'Educating a Sapling Nation: The Irish Nationalist Arboretum', in P. Elliott, C. Watkins and S. Daniels (eds), 'Cultural and Historical Geographies of the Arboretum', special issue, *Garden History* (2007), pp. 185–95.

Parrish, S. C., *American Curiosity: Cultures of Natural History in the Colonial British Atlantic World* (Chapel Hill, NC: University of North Carolina Press, 2006).

J. Pearson, *Stags and Serpents: The Story of the House of Cavendish* (London: Macmillan, 1983).

Piebenga, S., 'William Sawry Gilpin (1762–1843): Picturesque Improver', *Garden History*, 22 (1994), pp. 175–96.

Piebenga, S., and S. Toomer, 'Westonbirt Arboretum: from Private, Nineteenth-Century Estate Collection to National Arboretum' in Elliott et al. (eds), 'Cultural and Historical Geographies of the Arboretum', special issue, *Garden History* (2004), pp. 113–28.

Prest, J., *The Garden of Eden: The Botanic Garden and the Re-Creation of Paradise* (New Haven, CT: Yale University Press, 1981).

Preston, R., *North American Trees*, 3rd edn (Cambridge, MA: MIT Press, 1976).

Rackham, O., *Trees and Woodland in the British Landscape* (London: Phoenix Press, 2001).

Raven, C. E., *John Ray, Naturalist: His Life and Works* (Cambridge: Cambridge University Press, 1950).

Renkema, H. W., and J. Ardagh, 'Aylmer Bourke Lambert and his 'Description of the *Genus Pinus*', *Linnean Society Journal* – Botany, 48 (1930), pp. 439–66.

Reynolds, K. D., 'Charles Stanhope, fourth Earl of Harrington (1780–1851)', *ODNB*.

Ritvo, H., 'Zoological Nomenclature and the Empire of Victorian Science' in Lightman (ed.), *Victorian Science in Context*, pp. 334–53.

Roberts, W., 'The Centenary of Loudon's Arboretum', *Journal of the Royal Horticultural Society*, 61 (1936), pp. 277–84.

Rodger, R., *Housing in Urban Britain, 1780–1915* (Basingstoke: Macmillan, 1989).

Rose, J., *The Intellectual Life of the British Working Classes* (New Haven, CT: Yale University Press, 2002).

Rotundo, B., 'Mount Auburn: Fortunate Coincidences and an Ideal Solution', *Journal of Garden History*, 4 (1984), pp. 255–67.

Rubinstein, W. D., *Men of Property: The Very Wealthy in Britain since the Industrial Revolution* (London: Social Affairs Unit, 1981).

Rugg, J., 'Lawn Cemeteries: The Emergence of a New Landscape of Death', *Urban History*, 33 (2006), pp. 213–33.

Russell, C., *Science and Social Change, 1700–1900* (London: Macmillan, 1983).

Rybczynski, W., *A Clearing in the Distance: Frederick Law Olmsted and America in the Nineteenth Century*, 1st edn (New York: Scribner, 2003).

Schama, S., *Landscape and Memory* (London: Harper, 2004).

Schlereth, T. J., 'Early North-American Arboreta', in Elliott et al., 'Cultural and Historical Geographies of the Arboretum', special issue, *Garden History* (2007), pp. 196–216.

Schuyler, D., *Apostle of Taste: Andrew Jackson Downing, 1815–1852* (Baltimore, MD: Johns Hopkins University Press, 1999).

Schwarz, L., 'London 1700–1840', in Clark (ed.), *The Cambridge Urban History of Britain*, vol. 2, pp. 641–72.

Sealy, S. G., 'Cuckoos and their Fosterers: Uncovering Details of Edward Blyth's Field Experiments', *Archives of Natural History*, 36 (2009), pp. 129–35.

Seaward, M. R. D., 'William Borrer (1781–1862), father of British Lichenology', *Bryologist*, 105 (2002), pp. 70–7.

Sebag-Montefiore, C., 'Holford, Robert Stayner (1808–1892)', *ODNB*.

Secord, J., 'Monsters at the Crystal Palace', in S. De Chadarevian and N. Hopwood (eds), *Models: The Third Dimension of Science* (New Jersey: Stanford University Press, 2004), pp. 138–69.

Seymour, S., 'Landed Estates, the "Spirit of Planting" and Woodland Management in Later Georgian Britain: A Case Study from the Dukeries, Nottinghamshire' in C. Watkins (ed.) *European Woods and Forests*, pp. 115–34.

Shteir, A. B., *Cultivating Women, Cultivating Science: Flora's Daughters and Botany in England, 1760–1860* (Baltimore, MD: Johns Hopkins University Press, 1996).

—, 'Elegant Recreations? Configuring Science Writing for Women', in Lightman (ed.), *Victorian Science in Context*, pp. 236–43.

Simo, M. L., *Loudon and the Landscape, From Country Seat to Metropolis, 1783–1843* (New Haven, CT: Yale University Press, 1986).

Simpson, J., *The New Forestry* (Sheffield: Pawson & Brailsford, 1903).

Smith, E., *A History of Whiteknights* (Reading: University of Reading, 1957).

Soames, M., *The Profligate Duke* (London: Collins, 1987).

Solman, D., *Loddiges of Hackney* (London: Hackney Society, 1995).

Stearn, W. T., 'Adrian Hardy Haworth, 1768–1833', in A. H. Haworth, *Complete Works on Succulent Plants*, 5 vols (London: Gregg Press, 1965), vol. 1, pp. 9–57.

— (ed.), *John Lindley, 1799–1865: Gardener-Botanist and Pioneer Orchidologist* (London: Woodbridge Antique Collectors Club, 1999).

Stephens, H. M., 'Cocks, Arthur Herbert (1819–1881)', rev. Katherine Prior, *ODNB*.

Sterry, P., *Collins Complete British Trees* (London: Collins, 2007).

Stevens, P. F., *The Development of Biological Systematics: Antoine-Laurent de Jussieu, Nature and the Natural System* (New York: Columbia University Press, 1994).

—, 'J. D. Hooker, George Bentham, Asa Gray and Ferdinand Mueller on Species Limits in Theory and Practice: A mid-Nineteenth Century Debate and its Repercussions', *Historical Records of Australian Science*, 11 (1997), pp. 345–70.

—, 'On Amateurs and Professionals in British Botany in 1858 – J. D. Hooker on Bentham, Brown and Lindley', *Harvard Papers in Botany*, 2 (1997), pp. 125–32.

Stobart, J., 'Culture Versus Commerce: Societies and Spaces for Elites in Eighteenth-Century Liverpool ', *Journal of Historical Geography*, 28 (2002), pp. 471–85.

Symes, M., 'Westonbirt Gardens: A Victorian Elysium', *Garden History*, 18 (1990), pp. 155–73.

—, 'A. B. Lambert and the conifers at Painshill', *Garden History*, 16 (1998), pp. 24–40.

Symes, M., and J. H. Harvey, 'Lord Petre's Legacy: The Nurseries at Thorndon', *Garden History*, 24 (1996), pp. 272–82.

Symes, M., A. Hodges and J. Harvey, 'The Plantings at Whitton', *Garden History*, 24:14 (1986), pp. 138–72.

Tachibana, S., and C. Watkins, 'Botanical Transculturation: Japanese and British knowledge and Understanding of *Aucuba Japonica* and *Larix Leptolepis* 1700–1920', *Environment and History*, 16 (2010), pp. 43–71.

Tachibana, S., S. Daniels and C. Watkins, 'Japanese Gardens in Edwardian Britain: Landscape and Transculturation', *Journal of Historical Geography*, 30 (2004), pp. 364–94.

Taylor, 'Urban Public Parks, 1840–1900: Design and Meaning', *Garden History*, 23 (1995), pp. 201–21.

Thacker, C., *The Genius of Gardening: The History of Gardens in Britain* (London: Weidenfeld & Nicolson, 1994).

Thomas, K., *Man and the Natural World: Changing Attitudes in England 1500–1800* (London: Allen Lane, 1984).

Thompson, B. V. J., *Whiteknights: A History of the University Site* (Reading: University of Reading, 1986).

Tolia-Kelly, D. P., 'Organic Cosmopolitanism: Challenging Cultures of the Non-Native at the Burnley Millenium Arboretum', in P. Elliott, C. Watkins and S. Daniels (eds), 'Cultural and Historical Geographies of the Arboretum', *Garden History*, special issue (2007), pp. 172–84.

Totman, C., *The Green Archipelago: Forestry in Preindustrial Japan* (Berkeley, CA: University of California, 1989).

Townley, H. *English Woodlands and their Story* (London: Methuen, 1910).

Trinder, B., 'Industrialising Towns 1700–1840' in Clark (ed.), *The Cambridge Urban History of Britain*, vol. 2, pp. 805–30.

Tsouvalis J., and C. Watkins, 'Imagining and Creating Forests in Britain, 1890–1939' in M. Agnoletti and S. Anderson (eds), *Forest History: International Studies on Socio-Economic and Forest Ecosystem Change* (Wallingford: Cabi International, 2000), pp. 371–86.

Veitch, H., *Veitch's Manual of the Coniferae*, 2nd edn (London: Pollet, 1900).

Walker, M., *Sir James Edward Smith* (London: Linnean Society of London, 1988).

Walters, S. M., *The Shaping of Cambridge Botany* (Cambridge: Cambridge University Press, 1981).

Walters, S. M., and E. A. Stow, *Darwin's Mentor: John Stevens Henslow, 1796–1861* (Cambridge: Cambridge University Press, 2001).

Watkins, C. (ed.), *European Woods and Forests: Studies in Cultural History* (Wallingford: Cabi International, 1998).

Watts, R., *Women in Science: A Social and Cultural History* (London: Routledge, 2007).

Waugh, E., *Unconditional Surrender* (London: Chapman & Hall, 1961).

Whitehead, D., 'Veterans in the Arboretum: Planting Exotics at Holme Lacy, Herefordshire, in the Late Nineteenth Century', in Elliott et al. (eds), 'Cultural and Historical Geographies of the Arboretum', pp. 96–112.

Williams, D., *The Glasgow Guide* (Edinburg: Canongate Books, 1999).

Wilson, A., 'The Cultural Identity of Liverpool, 1790–1850: The Early Learned Societies', *Transactions of the Historic Society of Lancashire and Cheshire*, 147 (1997), pp. 58–73.

Winsor, M., 'Non-essentialist Methods in pre-Darwinian Taxonomy', *Biology and Philosophy*, 18 (2003), pp. 387–400.

Wohl, A., *Endangered Lives: Public Health in Victorian Britain* (London: Dent, 1983).

Wolf, A., *A History of Science, Technology and Philosophy*, 2 vols (New York: Harper Torch Books, 1963).

Woodward, M., *The Trees of Westonbirt* (London: Westminster Press, 1933).

Worple, K., *Last Landscapes: The Architecture of the Cemetery in the West* (London: Reaktion Books, 2003).

Wyborn, T., 'Parks for the People: The Development of Public Parks in Victorian Manchester', *Manchester Region History Review*, 15 (1995), pp. 3–14.

Yanni, C., *Nature's Museums: Victorian Science and the Architecture of Display* (New York: Princeton Architectural Press, 2005).

Young, C., R., *The Royal Forests of Medieval England* (Philadelphia, PA: University of Pennsylvania Press, 1979).

Young, R. M., *Darwin's Metaphor: Nature's Place in Victorian Culture* (Cambridge: Cambridge University Press, 1985).

INDEX